International Handbook of Mathematics Teacher Education: Volume 4

AF695402

International Handbook of Mathematics Teacher Education (2nd Edition)

Series Editor:

Olive Chapman
University of Calgary
Calgary, Alberta
Canada

This second edition of the *International Handbook of Mathematics Teacher Education* builds on and extends the first edition (2008) in addressing the knowledge, teaching and learning of mathematics teachers at all levels of teaching mathematics and of mathematics teacher educators, and the approaches/activities and programmes through which their learning can be supported. It consists of four volumes based on the same themes as the first edition.

Volume 1: Knowledge, Beliefs, and Identity in Mathematics Teaching and Teaching Development
Despina Potari, *National and Kapodistrian University of Athens, Athens, Greece* and Olive Chapman, *University of Calgary, Calgary, Canada* (eds.)
paperback: 978-90-04-41886-8, hardback: 978-90-04-41885-1,
ebook: 978-90-04-41887-5

Volume 2: Tools and Processes in Mathematics Teacher Education
Salvador Llinares, *University of Alicante, Alicante, Spain* and Olive Chapman, *University of Calgary, Calgary, Canada* (eds.)
paperback: 978-90-04-41897-4, hardback: 978-90-04-41895-0,
ebook: 978-90-04-41896-7

Volume 3: Participants in Mathematics Teacher Education
Gwendolyn M. Lloyd, *Pennsylvania State University, Pennsylvania, USA* and Olive Chapman, *University of Calgary, Calgary, Canada* (eds.)
paperback: 978-90-04-41922-3, hardback: 978-90-04-41921-6,
ebook: 978-90-04-41923-0

Volume 4: The Mathematics Teacher Educator as a Developing Professional
Kim Beswick, *University of New South Wales, Sydney, Australia* and Olive Chapman, *University of Calgary, Calgary, Canada* (eds.)
paperback: 978-90-04-42420-3, hardback: 978-90-04-42419-7,
ebook: 978-90-04-42421-0

International Handbook of Mathematics Teacher Education: Volume 4

The Mathematics Teacher Educator as a Developing Professional

(Second Edition)

Edited by

Kim Beswick and Olive Chapman

BRILL
SENSE

LEIDEN | BOSTON

Cover illustration: Photograph by Kim Beswick

The Library of Congress Cataloging-in-Publication Data is available online at http://catalog.loc.gov
LC record available at http://lccn.loc.gov/

ISBN 978-90-04-42420-3 (paperback)
ISBN 978-90-04-42419-7 (hardback)
ISBN 978-90-04-42421-0 (e-book)

Copyright 2020 by Koninklijke Brill NV, Leiden, The Netherlands.
Koninklijke Brill NV incorporates the imprints Brill, Brill Hes & De Graaf, Brill Nijhoff, Brill Rodopi, Brill Sense, Hotei Publishing, mentis Verlag, Verlag Ferdinand Schöningh and Wilhelm Fink Verlag.
All rights reserved. No part of this publication may be reproduced, translated, stored in a retrieval system, or transmitted in any form or by any means, electronic, mechanical, photocopying, recording or otherwise, without prior written permission from the publisher.
Authorization to photocopy items for internal or personal use is granted by Koninklijke Brill NV provided that the appropriate fees are paid directly to The Copyright Clearance Center, 222 Rosewood Drive, Suite 910, Danvers, MA 01923, USA. Fees are subject to change.

This book is printed on acid-free paper and produced in a sustainable manner.

CONTENTS

PREFACE

It is an honor to follow Terry Wood, series editor of the first edition of the four volume *International Handbook of Mathematics Teacher Education* (2008), as series editor of this second edition of the Handbook. As Terry indicated, she, Barbara Jaworski, Sandy Dawson and Thomas Cooney played key roles in opening up the field of mathematics teacher education "to establish mathematics teacher education as an important and legitimate area of research and scholarship" (Wood, 2008, p. vii). The field has grown significantly since the late 1980s "when Barbara Jaworski initiated and maintained the first Working Group on mathematics teacher education at PME [Psychology of Mathematics Education conference]" (p. vii) and over the last 10 years following the first edition of the Handbook. So, the editorial team, I and the four volume editors (Kim Beswick, Salvador Llinares, Gwendolyn Lloyd, and Despina Potari), of this second edition is honored to present it to the mathematics education community and to the field of teacher education in general.

This second edition builds on and extends the topics/ideas in the first edition while maintaining the themes for each of the volumes. Collectively, the authors looked back beyond and within the last 10 years to establish the state-of-the-art and continuing and new trends in mathematics teacher and mathematics teacher educator education, and looked forward regarding possible avenues for teachers, teacher educators, researchers, and policy makers to consider to enhance and/or further investigate mathematics teacher and teacher educator learning and practice, in particular. The volume editors provide introductions to each volume that highlight the subthemes used to group related chapters, which offer meaningful lenses to see important connections within and across chapters. Readers can also use these subthemes to make connections across the four volumes, which, although presented separately, include topics that have relevance across them since they are all situated in the common focus regarding mathematics teachers.

I extend special thanks to the volume editors for their leadership and support in preparing this handbook. I feel very fortunate to have had the opportunity to work with them on this project. Also, on behalf of myself and the volume editors, sincere thanks to all of the authors for their invaluable contributions and support in working with us to produce a high-quality handbook to inform and move the field of mathematics teacher education forward.

Volume 4, *The Mathematics Teacher Educator as a Developing Professional*, edited by Kim Beswick, focuses on the professionalization of mathematics teacher educators, which, since the first Handbook, continues to grow as an important area for investigation and development. It addresses teacher educators' knowledge, learning and practice with teachers/instructors of mathematics. Thus, as the fourth volume in the series, it appropriately attends to those who hold central roles in mathematics teacher education to provide an excellent culmination to the handbook.

REFERENCE

Wood, T. (Series Ed.), Jaworski, B., Krainer, K., Sullivan, P., & Tirosh, D. (Vol. Eds.). (2008). *International handbook of mathematics teacher education.* Rotterdam, The Netherlands: Sense Publishers.

Olive Chapman
Calgary, AB
Canada

FIGURES AND TABLES

FIGURES

TABLES

KIM BESWICK

MATHEMATICS TEACHER EDUCATORS AS DEVELOPING PROFESSIONALS

An Introduction

This chapter provides an introduction to Volume 4. It begins with an outline of developments in research on mathematics teacher educators over the period since the first edition of the handbook. The bulk of the chapter is structured around the four parts of the volume: Theories and conceptualisations of mathematics teacher educators and their characteristics; Mathematics teacher educators learning in transitions and through collaborations; Mathematics teacher educators learning from practice; and Researching mathematics teacher educators. These parts include a brief account of each chapter with attention drawn to some of the connections among them. The chapter concludes with a summary of the various ways in which there has been progress in our understanding of mathematics teacher educators along with a reminder of the political importance of our work.

***

Since the first edition of the *International Handbook of Mathematics Teacher Education* there has a been a burgeoning of research interest in those whose responsibilities include the education of prospective and practising teachers of mathematics. There have, for example, been discussion groups and working sessions at an International Congress on Mathematics Education (Beswick, Chapman, Goos, & Zaslavsky, 2015), annual conferences of the International Group for the Psychology of Mathematics Education (e.g., Beswick & Chapman, 2013; Beswick, Goos, & Chapman, 2014), and a special issue of the *Journal of Mathematics Teacher Education* (Beswick & Goos, 2018). All of these have focussed on the knowledge of mathematics teacher educators, the nature of that knowledge, and its acquisition and development. In addition, a special issue of the journal *Mathematics Teacher Education and Development* focussed on mathematics teacher educators' inquiries into their practices and their impacts in prospective teachers (Muir, Bragg, & Livy, 2018). The chapters in this book develop the themes concerning mathematics teacher educators' knowledge and the ways in which their work and their reflection on it contribute to their development.

© KONINKLIJKE BRILL NV, LEIDEN, 2020 | DOI: 10.1163/9789004424210_001

One seemingly superficial difference between this volume and the corresponding volume in the first edition of the handbook is the much greater number of authors involved. Whereas there were 23 contributors to the first edition volume, there are 54 authors involved in this volume, and this in spite of the number of chapters being fewer (14 compared with 17). Even if we discount 22 of the 23 authors of the Schoenfeld et al. chapter there are 32 contributors this time. The extent of collaboration evident in the authorship of chapters in this volume is consistent with a broadening of the conversation about mathematics teacher educators that spans many countries, and a movement away from individual mathematics teacher educators reflecting alone on their learning to undertaking this reflective work as a collaborative enterprise. It suggests that the work of mathematics teacher educators may be entering the mainstream of mathematics education research. The reflections and theoretical contributions of individuals are, of course, still vital and these also feature in the current volume.

Development of the field is reflected in many chapters included in this volume. Two of these, by Thornton, Beaumont, Lewis, and Penfold (Chapter 2), and Abboud, Robert, and Rogalski (Chapter 6) arise from experiences of teaching or participating in education programs for prospective mathematics teacher educators. Such programs are recognised elsewhere in the volume as needed (Appova, Chapter 8) but rare (Wu, Yao, & Cai, Chapter 9). Thornton et al. and Abboud et al. focus on the learning of the authors from their experiences in these contexts and, although Abboud et al. provide a quite detailed description of their program, this is not the main focus of these chapters. The ways in which the design of programs for mathematics teacher educators relates to conceptualisations of mathematics teacher educator knowledge and varies with context largely remains to be explored, although these chapters and that of Appova make a contribution to the discussion of what a curriculum for mathematics teacher educators might comprise, what appropriate aims for such programs might be, and the extent to which standardisation across programs might be desirable.

Goodchild (Chapter 12) extends the range of mathematics teachers with whom mathematics teacher educators work in his reflections on his learning from working with teachers of university mathematics who in most cases have no connection with teacher education. Extension in a different way is offered by Wu et al. (Chapter 9) who argue that the teaching practice of mathematics teacher educators involves considerably more than their work directly with prospective or practising mathematics teachers.

In her introduction to the corresponding volume of the first edition of the handbook, Jaworski (2008) pointed to a shift in the theoretical orientation of mathematics education research from a largely individual psychological focus towards accounting for the social contexts in which teaching, and learning occur. She noted the increased use of sociocultural theories and greater attention to political and policy issues. These trends have continued and are in evidence in the chapters of this volume. Two of the chapters, those by Goodchild (Chapter 12), and Goos

(Chapter 3), provide updates on these authors' contributions to the first edition. The progression in their research reflects development in the field.

This volume is organised in four parts:

- Theories and conceptualisations of mathematics teacher educators and their characteristics
- Mathematics teacher educators learning in transitions and through collaborations
- Mathematics teacher educators learning from practice
- Researching mathematics teacher educators

These are discussed in turn in the following sections.

THEORIES AND CONCEPTUALISATIONS OF MATHEMATICS TEACHER EDUCATORS AND THEIR CHARACTERISTICS

The chapters in this part illustrate the growing use of sociocultural theories and community of practice ideas in mathematics education research (Jaworski, 2008), with sociocultural theories, as well as psychological views, in evidence. Merrilyn Goos, for example, provides an account of how zone theory (Valsiner, 1997) can be applied not only to the development of mathematics teachers (Goos, 2008) but also to the development of mathematics teacher educators, and how mathematics teacher educators with differing backgrounds (e.g., mathematicians and mathematics education researchers) can be viewed as constituting communities of practice (Wenger, 1998). She argues that encounters at the boundaries of these communities, especially if sustained over time, offer potential for learning by mathematics teacher educators on both sides of the boundary. Roza Leikin, on the other hand, builds her model of mathematics teacher educators' knowledge and skills on conceptions of teacher knowledge founded in constructivist views of learning that, although acknowledging the role of the social context in learning, portray knowledge as an attribute of the individual.

In her chapter, Leikin describes several conceptualisations of mathematics teacher educator knowledge, including that of Goos (2009) using zone theory, that have in common a hierarchical relationship between mathematics teacher educators' knowledge and the knowledge of school mathematics teachers. She makes the important point that the precise nature of the relationship between layers of these hierarchies depends upon the specific goals and activities of the mathematics teacher educators concerned, as well as the characteristics of the mathematics teachers with whom they are working. Mathematics teacher educators may be university-based mathematicians, mathematics education researchers, or expert school mathematics teachers. Despite their shared objective of developing mathematics teachers' knowledge and skills for teaching mathematics in schools, the specific goals that mathematics teacher educators from these groups have for their interactions with mathematics teachers (prospective or practising) and the activities in which they engage, vary to the extent that there may be minimal overlap between the knowledge an affective characteristics of different communities of mathematics teacher educators. Rather, each community has distinctive knowledge, attitudes, and beliefs. Leikin illustrates this by considering the starkly different views of representatives of three communities of mathematics teacher educators of the educative value of a specific mathematical task. In doing this, Leikin extends the notion of mathematical horizon (Ball & Bass, 2009) in two ways. First, consistent with the ways in which many researchers have conceptualised mathematics teacher educators' knowledge as hierarchically related to the knowledge of teachers of school mathematics, she points to the need for mathematics teacher educators to be aware of the mathematical horizon for the mathematics teachers with whom they work. Second, she highlights the psychological horizon at both levels. That is, both mathematics teachers and mathematics teacher educators need to take account not only of the mathematical knowledge of those they teach, but also of their relevant psychological attributes that, together with their mathematical knowledge, define their mathematical potential (in the case of school students) or professional potential (in the case of mathematics teachers).

Steve Thornton, Nicola Beaumont, Matt Lewis and Colin Penfold consider how the transition from mathematics teacher to mathematics teacher educator occurs. They draw upon a range of reflective self-studies of this experience including Krainer's (2008) account in the previous edition of this volume. In addition, Murray and Male's (2005) ideas of moving from expert to novice, and of being a novice but assumed to be expert resonated with both the experiences reported in the literature and with these authors' experiences as participants in, or teachers of, a course aimed at facilitating that transition. They conclude that much of what occurs in becoming a mathematics teacher educator represents a shift in professional identity that requires becoming comfortable with ambiguity and uncertainty.

MATHEMATICS TEACHER EDUCATORS LEARNING IN TRANSITIONS AND THROUGH COLLABORATIONS

The four chapters that comprise this part together highlight the importance to their learning of collaboration among mathematics teacher educators, and the particular opportunity for learning afforded by the transition from mathematics teacher to mathematics teacher educator.

For Olive Chapman, Signe Kastberg, Elizabeth Suazo-Flores, Dana Cox, and Jennifer Ward, learning arises from their collaboration to produce their chapter. In it they consider three self-based inquiry methodologies – narrative inquiry (Clandinin & Connelly, 2000), self-study (Feldman, 2003), and autoethnography (Ellis & Bochner, 2000) – and reflect on how their use of these approaches has afforded insights into their own practices and into issues of mathematics teacher education more broadly. They claim that, despite the potential of these methodologies, reports of their use remain under-represented in the literature due to a lack of recognition of the validity of self-based methodologies. Key to re-dressing this perception is attention to the theoretical underpinnings of each of the methodologies (Chapman et al., Chapter 7, this volume).

Laurinda Brown and Alf Coles approach their learning as mathematics teacher educators in terms of enactivism, according to which knowing is doing, in that action in familiar situations does not require thought. For them, learning occurs when this immediate knowledge of what to do breaks down requiring new behaviours to be enacted. Analogously to the hierarchical relationships that Leikin observes in a range of conceptualisations of mathematics teacher education knowledge that have been built on conceptualisations of mathematics teacher knowledge, Brown and Coles describe how their enactivist perspective on learning can be applied to students learning mathematics, mathematics teachers learning to teach, and mathematics teacher educators learning to teach mathematics teachers.

Maha Abboud, Aline Robert, and Janine Rogalski distinguish between mathematics teacher educators whose role is to develop the practice of mathematics teachers, and mathematics education researchers whom they term 'didacticians.' The program for prospective mathematics teacher educators that they describe is pragmatic in its focus and not aimed at producing researchers. Although they characterise mathematics teachers, mathematics teacher educators, and didacticians as constituting three levels of analysis they explicitly reject a hierarchical or nested relationship among them. Rather, they contend, the practices of mathematics teachers are the common focus of each of these groups.

The transitions from prospective to practising mathematics teacher is of central interest to all mathematics teacher educators including Judy Anderson and Deborah Tully. Unique about the activities that they describe is deliberate attention to connections and collaboration among prospective and practising mathematics teachers, between mathematics teacher educators with differing backgrounds (university mathematicians and mathematics education researchers), and between mathematics teacher educators and both prospective and practising mathematics teachers. Indeed, they consider mathematicians, mathematics education researchers and practising teachers all to be mathematics teacher educators because they all support the development of prospective mathematics teachers. In this context practising teachers acting as mathematics teacher educators can be viewed as boundary spanners between the communities of prospective teachers and university-based mathematics teacher educators (Anderson & Tully). The initiatives Anderson and Tully describe were designed to create 'third spaces' (Gutierrez, 2008) in which a single community of practice could be built. In that community prospective teachers could develop their identities as mathematics teachers, practising teachers could develop as mathematics teacher educators, and mathematics teacher educators could learn about their practice.

MATHEMATICS TEACHER EDUCATORS LEARNING FROM PRACTICE

Several chapters in this part focus on teaching university level mathematics to prospective teachers, as in the cases of James Mendoza Álvarez, Kathryn Rhoads, and Theresa Jorgensen and Aina Appova and also practising teachers in the case of Alvarez et al. Simon Goodchild also reports on his learning as a mathematics teacher educator in a higher education context. Unlike others in this volume, he is concerned with developing teachers of university mathematics rather than of school mathematics. The students with whom these mathematics teachers work are not necessarily prospective teachers. Indeed, in most cases the students are prospective engineers or economists and the like (Goodchild). In contrast with this, Appova considered the goals and practises of mathematics teacher educators engaged in developing prospective teachers' pedagogical content knowledge in the context of mathematics content courses. Her model of pedagogical content knowledge for both mathematics teachers and mathematics teacher educators draws upon well-known conceptualisations of pedagogical content knowledge for teachers (e.g., Ball, Thames, & Phelps, 2008; Grossman, 1990; Schoenfeld, 2010; Shulman, 1987) and represents them as hierarchically related. Appova argues that including Schoenfeld's idea of orientations (encompassing beliefs and values) towards teaching mathematics in her model of pedagogical content knowledge is consistent with longstanding thinking in science education (e.g., Magnussen, Krajcik, & Borko, 1999) and more recent conceptualisations of pedagogical content knowledge for both mathematics teachers and mathematics teacher educators (e.g., Chick & Beswick, 2018).

Alvarez et al. conducted design-based research into their teaching of a Function and Modelling course. They described how they used three tasks with both prospective and practising teachers and how the tasks evolved as they learned from their practice in implementing them. They found the themes that Zaslavsky (2008) identified among the goals of mathematics teacher educators for teacher learning useful in analysing their findings, noting that each of the themes was relevant in relation to at least one and sometimes all three of the tasks and for both groups of teachers. Like Appova, Alvarez et al. situate their work in a hierarchical view of mathematics teacher educator knowledge in relation to mathematics teacher knowledge. Choice and use of tasks are important to both groups even though the goal of task use shifts from learning mathematics in the case of mathematics teachers to learning about teaching mathematics in the case of mathematics teacher educators. In both cases tasks can provide an important way of developing insight into learners' (school students or teachers) thinking and learning needs and hence can be important drivers of teacher or teacher educator development.

Goodchild's chapter builds on his contribution to the original edition of this volume (Goodchild, 2008) by offering a reflection on the next stage of his journey as a mathematics teacher educator. In addition to extending he scope of mathematics teacher education and development to working with teachers of mathematics in higher education, Goodchild also offers a nuanced categorisation of mathematics teachers in this context. These teachers include mathematics education researchers and research mathematicians, as well as others employed to teach mathematics based on qualifications in another mathematically rich discipline such as physics, or who specialise in teaching mathematics in programs that rely on mathematics (e.g., engineering). As both Goos and Leikin identify in relation to mathematics teacher educators, the different communities of mathematics teachers in higher education that Goodchild describes are likely to have differing cultures. That is, they vary in the goals of their mathematics teaching, and have differing beliefs about both mathematics and its teaching. These differences present challenges for mathematics teacher educators in higher education. Goodchild also notes the challenge presented by university cultures in which research may be regarded as more important than teaching, but points to promising signs that this may be changing.

Research about mathematics teacher educators has considered how the knowledge they require differs from that needed by mathematics teachers – that is, what are the knowledge implications of the change of content being taught from school mathematics to how to teach school mathematics? Goodchild opens a new field of inquiry by essentially raising the question of how the knowledge needed by mathematics teacher educators changes as the level of mathematics that the teachers being worked with are teaching changes from school mathematics to university mathematics. Of course, this question could be asked in relation to mathematics teacher educators teaching prospective or practising teachers of primary school mathematics compared with secondary mathematics teachers, but it appears not to have been. Goodchild also notes concomitant differences in the context of teaching and the characteristics of the teachers in schools and higher education, but these factors differ arguably as much from primary to secondary school mathematics teaching. It could be that the relative familiarity to mathematics teacher educators of school contexts as sites of mathematics teaching has led to the differences being taken for granted along with the adaptations that mathematics teacher educators make as a result of them. Whatever the reason, Goodchild's chapter represents a valuable stimulus to further research in higher education and for mathematics education researchers to problematise the familiar in school mathematics teacher education and development.

Alan Schoenfeld and his co-authors present the Teaching for Robust Understanding (TRU) framework and collectively outline the many ways in which it and its associated tools have been used to support school mathematics teachers, and how, as a consequence, the mathematics teacher educators have, themselves learned and developed. In this work Schoenfeld et al. are attempting to address issues of scale around mathematics teacher educators' work. As they say, there are many more mathematics teachers than can be supported by mathematics teacher educators. Like several other authors in this volume (e.g., Anderson & Tully; Goos), Schoenfeld et al. see promise in teacher learning communities, with these authors particularly emphasising the need to support these groups through the provision of robust and adaptable tools.

Yingkang Wu, Yiling Yao, and Jinfa Cai take a broad view of the teaching practice of mathematics teacher educators that encompasses their work with prospective and practising mathematics teachers, and also their (usually prior) teaching of school mathematics, their research related to teaching, and their own professional development. Wu et al. consider mathematics teacher educators' knowledge, competencies, and beliefs to be in a cyclical relationship with the entirety of mathematics teacher educators' teaching practice. Like Goodchild, Wu et al. observe that mathematics teacher educators vary greatly in terms of the foci of their work, and the contexts in which they work. Their identification of three types of mathematics teacher educators in the Chinese context suggest the potential for further exploration of national and cultural differences in the ways in which mathematics teacher educators are characterised, and the scope of their work is defined. There is potential to for comparative studies to highlight taken for granted aspects of these and other aspects of the field.

These potentials are illustrated by Rosemary Callingham, Yershat Sapazhanov, and Alibek Orynbassar in their comparison of the pathways by which five Australian and three Khazakhstan mathematics teacher educators arrived at their current roles. They use Bronfenbrenner's (1977) ecological systems theory to highlight the influences on the trajectories of the participants in their study of personal experiences of learning mathematics in school and subsequently teaching mathematics in schools, institutional requirements such as for university entry, differing educational systems structures, and cultural and societal values in relation to mathematics and mathematics teaching. They introduce the interesting concept of a mathematical tourist whose disposition is to explore and appreciate the discipline, and they wonder how the importance of such an inclination compares with that of traditionally measured mathematical ability.

RESEARCHING MATHEMATICS TEACHER EDUCATORS

The single chapter in this part, by Margaret Marshman, presents an urgent call for research on mathematics teacher educators that can provide an evidence base from which to critique the escalating accountability and regulatory demands on teacher educators, including mathematics teacher educators, that characterise the educational environments in many countries. These pressures add to the demands presented by the rise of out-of-field teaching in many of the same countries, the need to focus on research that impacts mathematicians in particular (Goodchild), as well as the inherent complexity of the central tasks of mathematics teacher educators. Marshman calls particularly for research on ways in which diverse communities of mathematics teacher educators such as those identified by Goodchild, Goos, and Wu et al. can work together to develop a unified voice that can challenge policy directions, as well as sending coherent and consistent messages to teachers. She also points to a need for mathematics teacher educators to consider how they can contribute to developing mathematics teachers who are able to both influence and respond effectively to regulatory interventions. The mathematics education community as a whole needs to become better able to communicate with and influence policy makers (Marshman).

A challenge in answering Marshman's call is the perception that exists according to Chapman et al. even among those who make decisions about what gets published such as some journal editors, that small studies and particularly self-based studies are of little value. It would seem prudent both to work towards changing that perception by following Chapman et al.'s advice in relation to attending to the theoretical underpinnings of the methodologies employed, while also working to broaden the range of approaches to researching mathematics teacher educators. Although, as Wu et al. noted the predominance of small, exploratory, qualitative studies of mathematics teacher educators is understandable for a field that remains relatively early in its development, there are signs among the contributions to this volume of a wider range of possibilities. Self-based methodologies remain a valuable means by which mathematics teacher educators can research themselves and arrive at self-understanding that improves practice, while also pointing to broader lessons for mathematics teacher education (Chapman et al.), but combining experiences gleaned from such studies to identify broader lessons as Chapman et al. have done adds to their value. Abboud et al. illustrated how an 'outsider' can be useful in researching mathematics teacher educators' learning including by adding a degree of objectivity. The study that Alvarez et al. reported was partially funded and this allowed them to have a graduate student video-record classes while at other times a member of the author team was available to observe. These measures also add to (perceived) objectivity. There is a need to continue to develop the methodological breadth of research in this field.

The prevalence of studies in this volume that arose from collaborations, or that investigated collaborations among differing categories of mathematics teacher educators and between mathematics teacher educators and teachers bode well for the prospect of stronger, more unified responses in the future.

CONCLUSION

The chapters that comprise this volume demonstrate the important development that has occurred in our understandings of mathematics teacher educators, their roles and the contexts of their work. The trends that Jaworski (2008) identified in mathematics education research have contributed to this development. We have more nuanced understanding of who mathematics teacher educators are and can be; and there is beginning to be attention to specific aspects of mathematics teacher educators' work, such as their use of tasks (Alvarez). It seems likely that, just as conceptualisations of mathematics teacher educator knowledge have mirrored those of mathematics teacher knowledge, the trajectory of research interests in relation to mathematics teacher educators will follow that of research concerning mathematics teachers. The beginnings of international comparative studies are also represented here (Callingham et al.). These studies and others that focus on familiar ideas in unfamiliar contexts, such as Goodchild's account of working with teacher of university level mathematics, can make the taken for granted visible. This volume also sees the emergence of research into programs for educating mathematics teacher educators. Despite the progress that has been made our understanding of mathematics teacher educators is still in its infancy with each aspect of progress presenting potential for further expansion of the field.

Marshman presented compelling reasons for continuing to build our research base in relation to mathematics teacher education and to forge links across the boundaries between the diverse groups of mathematics teacher educators. As accountability and regulatory regimes become more pervasive and stringent it is vital that mathematics teacher educators can explain and defend, and also promote their research-based practice. Vital also is the continued development of the range of research approaches applied to the field.

REFERENCES

Ball, D. L., & Bass, H. (2009). *With an eye on the mathematical horizon: Knowing mathematics for teaching to learners' mathematical futures.* Paper presented at the 43rd Jahrestagung der Gesellschaft für Didaktik der Mathematik, Oldenburg, Germany. Retrieved August 15, 2018, from https://eldorado.tudortmund.de/bitstream/2003/31305/1/003.pdf

Ball, D. L., Thames, M. H., & Phelps, G. (2008). Content knowledge for teaching: What makes it special? *Journal of Teacher Education, 59*(5), 389–407.

Beswick, K., & Chapman, O. (2013). Mathematics teacher educators' knowledge. Discussion group 3. In A. M. Lindmeier & A. Heinze (Eds.), *Proceedings of the 37th Conference of the International Group for the Psychology of Mathematics Education* (Vol. 1, p. 215). Kiel, Germany: PME.

Beswick, K., Chapman, O., Goos, M., & Zaslavsky, O. (2015). Mathematics teacher educators' knowledge for teaching. In S. J. Cho (Ed.), *The proceedings of the 12th International Congress on Mathematical Education: Intellectual and attitudinal challenges* (pp. 629–632). Springer Science+Business Media.

Beswick, K., Goos, M., & Chapman, O. (2014). Mathematics teacher educators' knowledge (Working session 4). In S. Oesterle, C. Nichols, P. Liljedahl, & D. Allan (Eds.), *Proceedings of the joint meeting of PME 38 and PME-NA 36* (Vol. 1, p. 254). Vancouver, Canada: PME.

Bronfenbrenner, U. (1977). Toward an experimental ecology of human development. *American Psychologist, 32*(7), 513–531.

Chick, H., & Beswick, K. (2018). Teaching teachers to teach Boris: A framework for mathematics teacher educators pedagogical content knowledge. *Journal of Mathematics Teacher Education, 21*(5), 475–499.

Clandinin, J., & Connelly, M. (2000). *Narrative inquiry: Experience and story in qualitative research.* San Francisco, CA: Jossey-Bass.

Ellis, C., & Bochner, A. P. (2000). Autoethnography, personal narrative, reflexivity: Researcher as subject. In N. K. Denzin & Y. S. Lincoln (Eds.), *Handbook of qualitative research* (pp. 733–768). Thousand Oaks, CA: Sage.

Feldman, A. (2003). Validity and quality in self-study. *Educational Researcher, 32*(3), 26–28.

Goodchild, S. (2008). A quest for 'good' research. In B. Jaworski & T. Wood (Eds.), *The international handbook of mathematics teacher education Volume 4: The mathematics teacher educator as a developing professional* (pp. 201–220). Rotterdam, The Netherlands: Sense Publishers.

Goos, M. (2008). Sociocultural perspectives on learning to teach mathematics. In B. Jaworski & T. Wood (Eds.), *International handbook of mathematics teacher education* (Vol. 4, pp. 75–91). Rotterdam, The Netherlands: Sense Publishers.

Goos, M. (2009). Investigating the professional learning and development of mathematics teacher educators: A theoretical discussion and research agenda. In R. Hunter, B. Bicknell, & T. Burgess (Eds.), *Crossing divides: Proceedings of the 32nd Annual Conference of the Mathematics Education Research Group of Australasia* (Vol. 1). Palmerston North, NZ: MERGA.

Grossman, P. (1987). A tale of two teachers: The role of subject matter orientation in teaching. *Knowledge Growth in a Profession Publication Series.* Stanford, CA: Stanford University, School of Education.

Gutierrez, K. D. (2008). Developing sociocultural literacy in the third space. *Reading Research Quarterly, 43*(2), 148–164.

Jaworski, B. (2008). Development of the mathematics teacher educator and its relation to teaching development. In B. Jaworski & T. Wood (Eds.), *The international handbook of mathematics teacher education Volume 4: The mathematics teacher educator as a developing professional* (Vol. 4, pp. 335–361). Rotterdam, The Netherlands: Sense Publishers.

Krainer, K. (2008). Reflecting the development of a mathematics teacher educator and his discipline. In B. Jaworski & T. Wood (Eds.), *International handbook of mathematics teacher education: The mathematics teacher educator as a developing professional* (Vol. 4, pp. 177–199). Rotterdam, The Netherlands: Sense Publishers.

Magnusson, S., Krajcik, J., & Borko, H. (1999). Nature, sources, and development of pedagogical content knowledge for science teaching. In J. Gess-Newsome & N. G. Lederman (Eds.), *Examining pedagogical content knowledge* (pp. 95–132). Dordrecht: Kluwer Academic Publishers.

Muir, T., Bragg, L. A., & Livy, S. (2018). Engagement and impact: A focus on mathematics teacher educators' studies into practice. *Mathematics Teacher Education and Development, 20*(3), 1–3.

Murray, J., & Male, T. (2005). Becoming a teacher educator: Evidence from the field. *Teaching and Teacher Education, 21*, 125–142. https://doi.org/10.1016/j.tate.2004.12.006

Schoenfeld, A. H. (2010). *How we think: A theory of goal-oriented decision making and its educational applications*. New York, NY: Routledge.

Shulman, L. S. (1987). Knowledge and teaching: Foundations of the new reform. *Harvard Educational Review, 57*(1), 1–22.

Valsiner, J. (1997). *Culture and the development of children's action: A theory of human development* (2nd ed.). New York, NY: John Wiley & Sons.

Wenger, E. (1998). *Communities of practice: Learning, meaning and identity*. Cambridge: Cambridge University Press.

Zaslavsky, O. (2008). Meeting the challenges of mathematics teacher education through design and use of tasks that facilitate teacher learning. In B. Jaworski & T. Wood (Eds.), *The international handbook of mathematics teacher education: The mathematics teacher educator as a developing professional* (Vol. 4, pp. 93–114). Rotterdam, The Netherlands: Sense Publishers.

Kim Beswick
School of Education
University of New South Wales, Sydney

PART 1

THEORIES AND CONCEPTUALISATIONS OF MATHEMATICS TEACHER EDUCATORS AND THEIR CHARACTERISTICS

ROZA LEIKIN

1. HOW FAR IS THE HORIZON?

Teacher Educators' Knowledge and Skills for Teaching High School Mathematics Teachers

The overarching goal of mathematics teacher educators is to facilitate the education and life-long learning and professional development of mathematics teachers. Mathematics teacher educators are usually experts in mathematics teaching. However, this expertise is not sufficient – in their profession mathematics teacher educators require special (added) knowledge, skills, attitudes and beliefs. In this chapter, I analyze the state of the art in research in mathematics teacher educators' proficiency. I pay special attention to the connections between theory and practice of mathematics teacher educators. The chapter includes a review of models of mathematics teacher educators' knowledge and expertise and explores how these models differ from the models of mathematics teachers' knowledge and expertise. This analysis is accompanied by examples from mathematics teacher educators' practices of different kinds.

INTRODUCTION

Mathematics teacher educators are responsible for the sustainability of high quality mathematics teaching. They prepare prospective teachers for their future work with students, support practising teachers in advancing and sustaining excellence in mathematics teaching, and supervise changes in mathematical content and teaching approaches during educational reforms. The mathematics teacher educator's work is extremely complex due to the multiple foci of attention and multiple communities of practice involved in mathematics teacher education. The complexity of mathematics teacher educators' mission has increased in the modern era due to the technological, scientific and societal developments that happen nowadays at an alarming rate. These developments suggest that the goal of educational systems should be not only to develop students' knowledge and skills in a particular content area, but also to advance students' ability to learn, and to promote their self-regulation and communicative skills, their motivation and self-esteem, as well as their flexibility in both cognitive and social fields. Teacher educators in general and mathematics teacher educators in particular

© KONINKLIJKE BRILL NV, LEIDEN, 2020 | DOI: 10.1163/9789004424210_002

have to support teachers' adaptation to the changing world in their work with new generations of students (Pellegrino & Hilton, 2012).

The question of mathematics teacher educators' knowledge and proficiency has been the focus of mathematics education researchers for more than two decades (beginning with the Psychology of Mathematics Education [PME] conference, 1999). The questions under consideration in these studies included (but not been limited to) the following that are relevant to the discussion in this chapter:

- Who are the professionals best suited to teach mathematics teachers?
- What types of activities can allow them to attain their complex goals?
- What knowledge and skills that are essential for proficient work with mathematics teachers?

In the broad landscape of mathematics teacher educators' practice, this chapter focuses on the education of secondary school mathematics teachers and mathematics teacher educators, whose goal is to facilitate initial preparation and promote the life-long learning and professional development of mathematics teachers.

To address the three questions above, I analyze existing research on mathematics teacher educators' knowledge and proficiency through the lens of the hierarchy between mathematics teacher educators' and secondary school mathematics teachers' knowledge and practices as they appear in mathematics the education literature. Then I consider the complexity and multiplicity of goals and activities of mathematics teacher educators as associated with the variety of communities of practice of mathematics teachers and mathematics teacher educators. I describe a mathematical activity that I employed many times with practising and prospective mathematics teachers. Using this task as an example, I present stories of three mathematics teacher educators from different communities of practice: one mathematics education researcher, one professional mathematician and one expert mathematics teacher. These stories reflect some of the core differences between the three mathematics teacher educators' communities.

To address the third question in this chapter, I suggest that the concept of '*horizon knowledge*' (cf. Ball & Bass, 2009) be extended to include the areas of *advanced mathematical knowledge* (as suggested earlier in Zazkis & Mamolo, 2011; Zazkis & Leikin, 2010) and, as suggested in this chapter, *psychological knowledge* associated with teaching and learning of mathematics. I argue that mathematics teachers' horizon knowledge of mathematics and of psychology are integral components of mathematics teacher educators' knowledge and skills and explain this argument using the model of mathematics teacher educators' knowledge and skills in terms of teachers' mathematical potential and challenging content for mathematics teachers.

BACKGROUND

Hierarchy between Mathematics Teacher Educators' and Mathematics Teachers' Knowledge and Skills

Over the past two decades, studies of mathematics teacher educators' competencies, knowledge and skills have indicated a hierarchical relationship between mathematics teacher educators' and mathematics teachers' competencies (e.g., Chick & Beswick, 2018; Goos 2009; Jaworski, 2008; Perks & Prestage, 2008; Zaslavsky & Leikin, 2004). Such a hierarchy presumes that to teach mathematics in school, mathematics teachers' competencies should include deep, broad and robust mathematical knowledge of school mathematics and far beyond (what you teach), accompanied with didactical proficiency (how you teach) and broad and deep psychological knowledge of students and learning processes (who you teach and how they learn). In turn, mathematics teacher educators are expected to be as competent as mathematics teachers in all these areas, and in addition, to understand the structure and the complexity of mathematics teachers' knowledge. Mathematics teacher educators must have knowledge and skills far beyond mathematics teachers' knowledge and skills, allowing them to manage teachers' learning and professional development in an engaging manner. Figure 1.1 illustrates three models that exemplify the hierarchy between the elements that are essential for mathematics teacher educators' and secondary school mathematics teachers' (MTs' in Figure 1.1) proficiency.

The first model, "the extended Teacher-Educators' Triad," is based on Jaworski's (1992) Teaching Triad, which comprises three components of proficient mathematics teachers' practice. These are: mathematical challenge, which involves stimulating mathematical thinking, inquiry and learning, management of students' learning associated with the creation of a learning environment and norms, and sensitivity to students that allows matching between mathematical challenge, management of learning and students' mathematical competencies. The extended model embraces factors which are critical for mathematics teacher educators' proficiency: management of teachers' learning and sensitivity to teachers and challenging content for mathematics teachers that includes the mathematics teachers' teaching triad (as suggested in Zaslavsky & Leikin, 2004) and mathematical challenge for teachers (Leikin, Zazkis, & Meller, 2018). The addition of mathematical challenge to the previously suggested extended model was based on the finding that mathematicians consider mathematical challenge situated in advanced mathematical knowledge (Zazkis & Leikin, 2010) to be central to their mission of teacher education.

The second model presented in Figure 1.1 is the extended Steinbring's (1998) model of the teaching process. Steinbring argued that this process is not linear but cyclic. The cycle starts with mathematics teachers' knowledge, which leads to

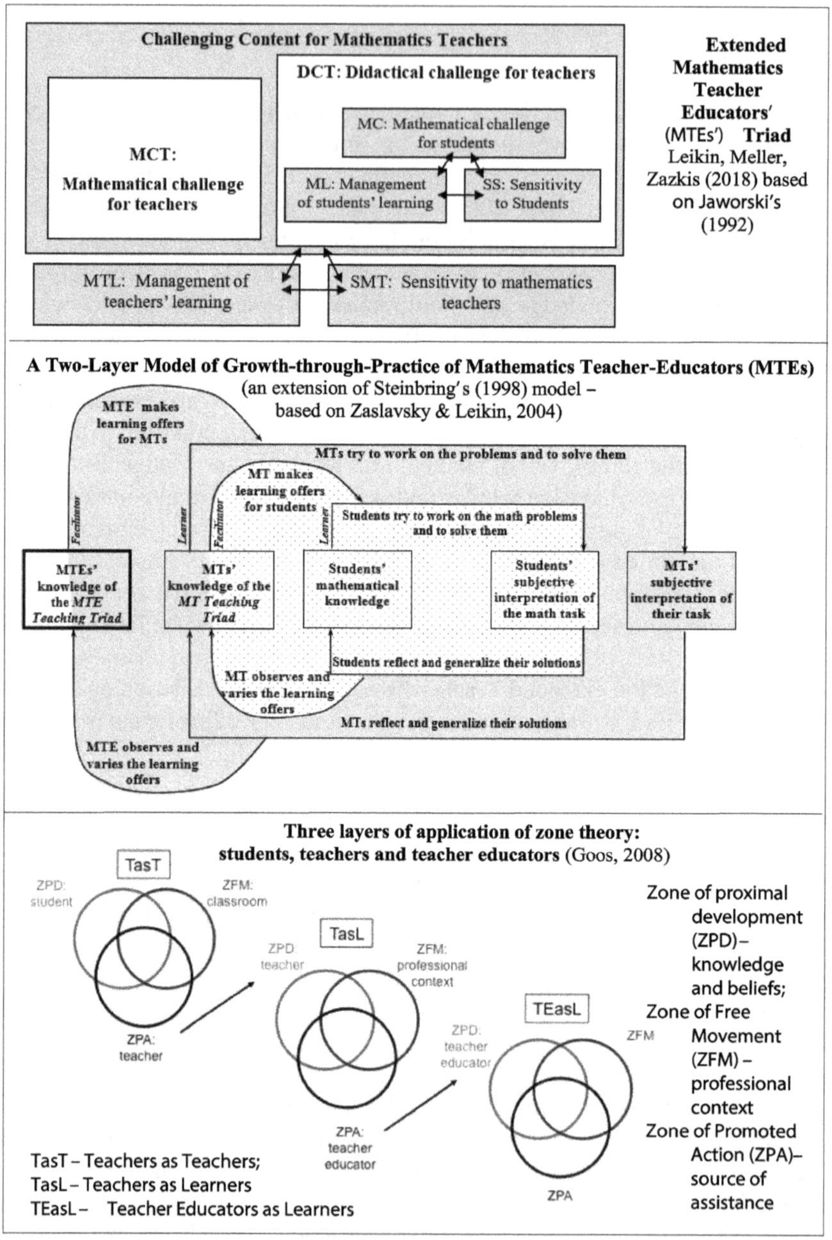

Figure 1.1. Examples of hierarchical models of mathematics teacher educators' (MTEs') knowledge and practice

"learning offers" that teachers suggest to their students (similar to teachers' role of devolution of a good problem to students, as described in Brousseau 1997). Based on their knowledge, students interpret the learning offers and work on the tasks that the mathematics teachers devolve to them. This process leads to the development of students' knowledge, while simultaneously, mathematics teachers critically analyze students' work, resulting in the development of mathematics teachers' knowledge and skills as well. The extension suggested by Zaslavsky and Leikin (2004) considers Steinbring's cycle as an internal cycle in the context of teachers' professional development that encompasses teachers' learning-through-teaching (Leikin & Zazkis, 2010) while mathematics teacher educators suggest "learning offers" to mathematics teachers in order to encourage teachers' learning. Mathematics teachers, in turn, interpret these offers and work on them aiming at the professional development while their own practice (the internal cycle of the extended model) is in the background of these mathematics teachers' learning experiences. The extended model describes and explains development through practice of mathematics teacher educators who are expert teachers themselves.

The third model presented in Figure 1.1 is suggested by Goos (2009) and is rooted in Vygotsky's (1978) theory of zone of proximal development and extended by Valsiner (1997). It covers the continuing development of knowledge and beliefs of the participants (referred as zone of proximal development), professional context (referred to as zone of free movement) and sources of assistance to learners (referred to as zone of promoted action). Goos (2009) claims that zone theory brings teaching, learning and context into the same discussion and can be applied in three connected layers. The first layer considers teacher-as-teacher (TasT in Figure 1.1) while zone of free movement/zone of promoted actions structure student learning. The second layer focuses on teacher-as-learner (TasL in Figure 1.1) with zone of free movement/ zone of promoted actions that structure mathematics teachers' professional learning; (iii) a teacher-educator-as-learner (TEasL in Figure 1.1) that presumes continuing development of mathematics teacher educators' knowledge and beliefs within their zone proximal development, the way in which mathematics teacher educators' professional contexts constrain their actions (zone of free movement), and the opportunities to learn that are opened to mathematics teacher educators (zone of promoted actions). The obvious hierarchy between mathematics teachers' and mathematics teacher educators' learning processes in this model is expressed in the connection of teacher-as-learner and teacher-educator-as-teacher. This hierarchical model suggests explanations for the transformation of mathematics teachers' practices of teaching (students), learning to improve teaching (from practice and from teacher educators), becoming a mathematics teacher educator and learning as a mathematics teacher educator to improve teacher education.

The three models described above put different theoretical lenses on mathematics teachers' and mathematics teacher educators' proficiency through emphasis on the centrality of mathematics teachers' and mathematics teacher educators' knowledge in promoting teachers' learning through making learning offers for mathematics teachers that integrate mathematical and didactical components. The hierarchy between mathematics teachers' and mathematics teacher educators' knowledge presumes that the structure of mathematics teacher educators' knowledge and skills is similar to that of mathematics teacher educators and integrates mathematics teachers' knowledge and skills. However, mathematics teacher educators' knowledge has added to mathematics teachers' knowledge and the hierarchy differs for mathematics teacher educators from different communities of practice.

Focusing on mathematics teacher educators who are mathematics education researchers, Jaworski (2008) argues that while "qualities required of teacher educators are in many respects the same as those required of mathematics teachers" (p. 1) there is additional knowledge required from mathematics teacher educators. This includes knowledge of theory of learning and teaching, knowledge of the educational system and skills associated with ways in which this additional knowledge can be implemented in the educational setting for mathematics teachers. In addition, as mentioned before, following analysis of mathematician's conceptions about what should be taught to high school mathematics teachers, Leikin et al. (2018) added 'mathematical challenge for teachers' to the challenging content for teachers as was suggested by Zaslavsky and Leikin (2004) in the earlier model that reflected knowledge and skills of mathematics teacher educators who are expert mathematics teachers themselves. That is to say, mathematics teacher educators' conceptions, knowledge and skills are a function of the community of practice of mathematics teacher educators to which they belong and of the mathematics teachers with whom they work.

DIVERGENCE OF COMMUNITIES OF PRACTICE OF MATHEMATICS TEACHER EDUCATORS AS ASSOCIATED WITH THEIR GOALS AND ACTIVITIES

In line with Leontiev's (1983) activity theory, mathematics teacher educators' practice is determined by the specific goals and conditions of particular educational programs for mathematics teachers, which include corresponding activities, actions and tasks. Jaworski (2008) stressed that while mathematics teacher educators' goal is to provide mathematics teachers with relevant experiences that foster construction of their knowledge, the specific goals of mathematics teacher educators can differ with respect to their work in different settings. Mathematics teacher educators work with mathematics teachers from different communities of practice. The distinctions between the communities are associated, for example, with mathematics teachers' educational history (e.g., prospective, practising, or second-career mathematics teachers), the grade level at which teachers work (primary, mid and high school), the level of mathematics that mathematics teachers teach (e.g., special high-level mathematics classes, classes for students with difficulties or regular level classes) or teachers' proficiency (novice teachers versus expert teachers).

During different stages of their education and working careers, mathematics teachers meet mathematics teacher educators from different communities: mathematicians who teach mathematics in universities, mathematics educational researchers, instructional designers and expert mathematics teachers. Mathematics teacher educators' activities can have different goals depending on their professional community. For example, their activities can be directed at the development of mathematics teachers' mathematical knowledge and skills associated with problem-solving proficiency, task-generation ability, understanding of different types of mathematical tasks and questions, and deep understanding of specific topics from the mathematical curriculum and beyond. Mathematics teacher educators can also aim for advancement of teachers' didactical knowledge and skills associated with development of students' mathematical knowledge, including classroom settings, norms and scaffolding practices. Alternatively, and no less important, mathematics teacher educators' goals can include promoting teachers' 21st century skills, including critical thinking, flexibility and collaboration (Pellegrino & Hilton, 2012).

Examples of differences in the goals of activities of mathematics teacher educators associated with differences in communities of practice of mathematics teachers and of mathematics teacher educators can be seen in several studies. In Appova and Taylor's (2019) study, mathematics teacher educators are research mathematicians with the primary goal of helping mathematics teachers to enhance their pedagogical content knowledge. This study demonstrates that mathematicians assume the role of mathematics teacher educators when teaching mathematics courses for teachers and that the goals and practices of mathematics teacher educators who are expert mathematics teachers differ significantly from the practices of mathematics teacher educators who are mathematics faculty. Following this analysis Appova and Taylor deduce that expertise in mathematics teachers' education presumes both expertise in teaching prospective mathematics teachers and in teaching K-12 mathematics. Chen, Lin, and Yang (2018) examined a teacher education program in which mathematics teacher educators were educational researchers. The main goal of this program was development of mathematics teachers' understanding of the meaning of conjecturing and their ability to draw distinctions between different types of conjecturing. Mathematics teachers' ability to design conjecturing activities was considered an essential component of mathematics teachers' knowledge and skills and consequently the mathematics teacher educators in this study were experts in the examined conjecturing teaching approach. In Chick and Beswick (2018), mathematics teacher educators, both researchers and expert teachers, consider developing mathematics teachers' pedagogical content knowledge to be a central goal of mathematics teachers' education. Similarly to the models analyzed in the previous section, Chick and Beswick stressed the hierarchy of pedagogical content knowledge of mathematics teacher and mathematics teacher educator and considered mathematics teacher educators' pedagogical content knowledge a meta-knowledge of subject matter and pedagogical content knowledge. According to this study mathematics teacher educators need to know what teachers need to know, and also need to know how this knowledge can be developed as well as the content to be taught to mathematics teachers. Even (1999) lead a program with the overarching goal of developing a community of mathematics teacher educators capable of supporting national educational reform. Expert mathematics teachers were recruited to develop their knowledge and skills as mathematics teacher educators with a special emphasis on understanding the nature of the mathematics teaching and learning as suggested by the reform. Even stressed the centrality of mathematics teacher educators' leadership skills and of creating a supportive professional reference group that allows the mathematics teacher educators to lead the reform.

By focusing on different communities of mathematics teacher educators, it becomes obvious that the hierarchy between knowledge and skills of mathematics teachers and mathematics teacher educators is a function of the communities of mathematics teachers and mathematics teacher educators under consideration. Practices of mathematics teacher educators from different communities can be tangent to each other, that is having only few common actions, while each mathematics teacher educators' community can have special (relative to others) knowledge, skills, attitudes and beliefs. Because of the links between mathematics teacher educators' professional communities and their goals and activities, the hierarchy described above is not always preserved. Oftentimes the importance or even relevance of knowledge of educational theories (specific to the community of educational researchers) or knowledge of university mathematics (specific to research mathematicians) is not obvious for mathematics teachers (Zazkis & Leikin, 2010) and mathematics teachers feel confused by the connections between what they learn and what they do.

In what follows, I introduce three mathematics teacher educators from different communities of practice – mathematics education researchers, mathematicians and expert high school mathematics teachers – and, using their views on one particular mathematical discovery by a prospective mathematics teacher, illustrate the differences outlined above.

Meeting Mathew, Merav and Eti – Representatives of Three Different Mathematics Teacher Educator Communities of Practice

Mathew, Merav and Eti (pseudonyms) are mathematics teacher educators who belong to the three different communities of mathematics teacher educators. All of them work with high school mathematics teachers. Merav, Mathew and Eti oftentimes work together on teacher education projects and task design projects. The stories are written using field notes taken during conversations with mathematics teacher educators and when interesting events happened during the courses and workshops for teachers.

Merav is a mathematics education researcher with a PhD in mathematics education who works with both prospective and practising mathematics teachers.

Merav specializes in researching teachers' knowledge and professional development and is interested in the advancement of inquiry-based teaching in Grades K-12. She believes that computer software, for example Geo-Gebra, can promote teaching and learning processes in any classroom. She also believes that high school mathematics teachers should know mathematics beyond what is in the school curriculum and thinks that advanced mathematical knowledge (cf. Zazkis & Leikin, 2010; Zazkis & Mamolo, 2011) is a critical component of high school mathematics teachers' proficiency. Merav directs her activities towards changing mathematics teachers' instructional approaches and attaining teachers' understanding of their students based on analysis of research papers with an eye to reflective analysis of their own practice. When teachers perform such an analysis, connections to mathematical content can be less explicit and the connections between research and practice may not be clear enough for mathematics teachers (Lin, Yang, Hsu, & Chen, 2018).

Mathew is a mathematician who teaches mathematical courses in the university mathematics department. Prospective mathematics teachers study mathematics in his courses together with other participants. Mathew directs his activity at the education of a new generation of mathematicians and at developing robust and deep mathematical knowledge in individuals whose profession requires a strong mathematical background. Usually he does not design his lectures specifically for mathematics teachers. Mathew, like Merav, believes that mathematics teachers must know a lot more than what they teach to their students, and that they can use mathematics that they study in university to rouse the interest of stronger students and to be able to operate according to the school children's questions. In this, his views align with those of mathematicians in other studies of mathematics teacher educators (e.g., Leikin et al., 2018; Appova & Taylor, 2019). Mathew thinks that studying in a mathematics department gives future teachers a rather large body of knowledge. According to him, university mathematics courses allow teachers to attract high school students to all kinds of interesting things in advanced mathematics. They mainly provide high school mathematics teachers with tools for the development of rigorous thinking and allow deeper and more interesting teaching of mathematics in school. Mathew thinks that even though his courses are not designed for teachers, they can see a kind of didactical modelling of "good teaching" in his course that includes elements such as learning from examples, proving, requiring the use of rigorous language and presenting beautiful proofs.

Eti is an expert mathematics teacher who teaches high school mathematics and works with other high school mathematics teachers as an mathematics teacher educator. Eti completed an MA in mathematics education and has more than 10 years of experience in conducting courses and workshops for mathematics teachers with different levels of expertise and developing technology-rich activities for mathematics teachers. When working with other mathematics teachers, Eti aims to share with them her successful mathematics teachers' experiences, and to work with them to analyse situations that they consider to be problematic in their classes.

Eti attempts to develop mathematics teachers' reflective skills, advance teachers' knowledge of mathematical concepts that they teach, and help them to get access to ready-to-use instructional materials that she finds useful in her teaching practice. In contrast with Mathew, Eti pays less attention to mathematics teachers' mathematical knowledge beyond the curriculum, and in contrast to Merav, the use of research in mathematics education is rarely integrated in her courses.

Story 1: Merav Introduces Activity to Prospective Mathematics Teachers and Is Excited about Yoni's Discovery

This incident took place at one of the workshops with prospective high school mathematics teachers at a teaching certificate program. The group included 32 participants. Merav asked participants to work on Task 1 (Figure 1.2). The task was borrowed from Leikin (2014). The choice of this task was based on the observation that problem solving through investigations is an effective activity directed at the advancement of mathematical knowledge and creativity. This kind of task has also been shown to advance sensitivity to the excitement and the difficulties that school students can experience while learning mathematics.

Several of the prospective mathematics teachers successfully performed two proofs (Proof 1.1 and Proof 1.2) and presented them during the subsequent group discussion. They found these proofs "comprehensible, but not easy to perform." They admitted that in the two proofs similarity of triangles was used to find ratio of segments rather naturally. However, they had difficulty with the auxiliary constructions – especially in Proof 1.2. Some of them argued they "would never think to use that kind of a construction."

The discussion also included analysis of the cognitive processes involved in production of the proofs using Duval's (1998) theory that connects difficulty in geometry with "shifts" between different figures, especially when they play different roles in proofs. For example, segment *ED* in Proof 1.2 (Figures 1.2a and 1.3a) was attended as a midline in the triangle *ABC*, as a side in triangle *MED* and as a part of the side *EK* in triangle *HEK*. This kind of discussion was directed at drawing connections between the prospective mathematics teachers' study of theory and their own practice – to develop awareness of the importance of mathematics education research.

Following the proving activity, the prospective teachers turned to investigation of the given figure in DGE. Of the more than 40 discovered properties published in Leikin (2014), they collectively found around 15 properties, mostly related to ratios of segments and areas of different figures. Then, with Merav's guidance, they formulated problems that required proving properties P1–P6 depicted in Figure 1.2. Discussion of the posed problems focused on the levels of complexity of the posed problems, with "complexity" based mainly on conceptual density and the length of proofs.

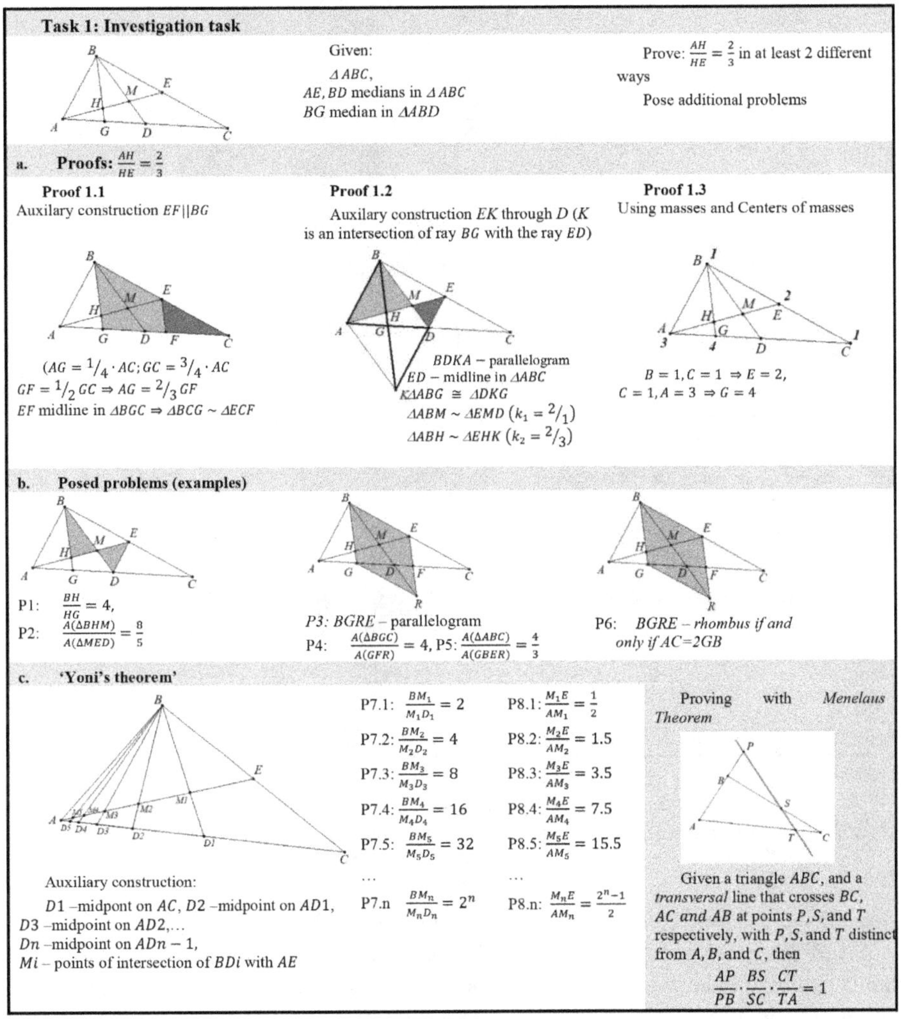

Figure 1.2. Investigation task. (a) Multiple solutions. (b) Problems posed by teachers. (c) Yoni's theorem

At the end of the session, Yoni (one of the prospective teachers – pseudonym) said that he "found something interesting but does not know how to prove it." He described his discovery as follows:

> While in the given problem *BD* is the median of the given triangle ABC and BG is the median of the "half triangle" ABD, I continued constructing medians in the following "half triangles." I measured the ratios and found that they represent a geometric series.

a. Stages in Proof 1.2

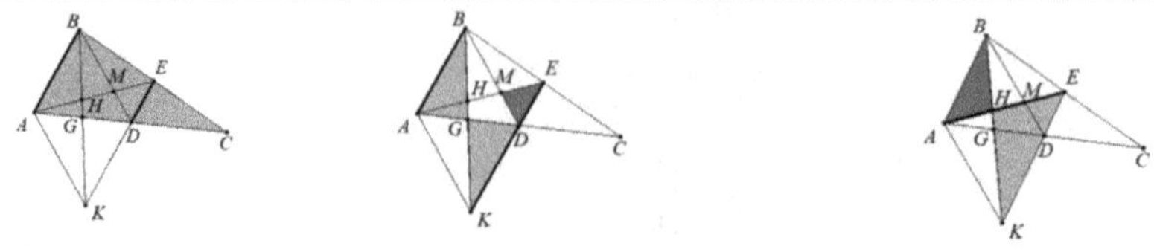

DE – midline in ΔABC	$ABDK$ – parallelogram	$\Delta AHB \sim \Delta EHK$
$DE = \frac{1}{2}AB$	$DE = \frac{1}{2}DK \Rightarrow DE = \frac{1}{3}EK \Rightarrow AB = \frac{2}{3}EK$	$\frac{AH}{HE} = \frac{2}{3}$

b. Choosing *transversal* line and a triangle to use Menelaus Theorem when proving 'Yoni's theorem'

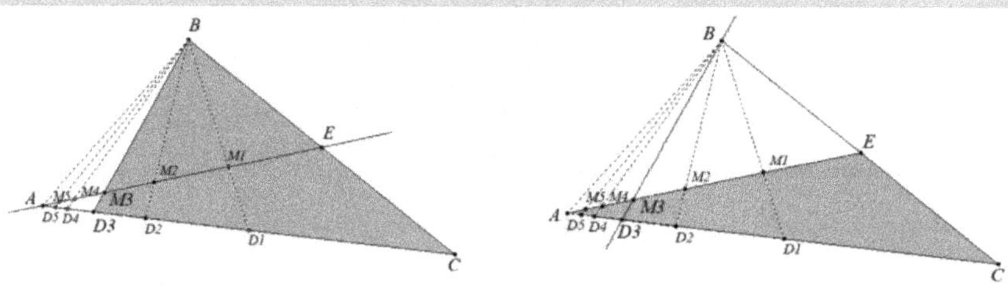

Proofs using Menelaus Theorem

AE is a the transversal line for ΔD_iBC ($i = 3$)	BD_i is a the transversal line for AEC ($i = 3$)
$\frac{BM_i}{M_iD_i} \cdot \frac{D_iA}{AC} \cdot \frac{CE}{EB} = 1$	$\frac{AM_i}{M_iE} \cdot \frac{EB}{BC} \cdot \frac{CD_i}{D_iA} = 1$
$\frac{D_iA}{AC} = \frac{1}{2^i}; \frac{CE}{EB} = 1 \Rightarrow \frac{BM_i}{M_iD_i} = 2^i$	$\frac{CD_i}{D_iA} = \frac{1}{2^n - 1}; \frac{EB}{BC} = \frac{1}{2} \Rightarrow \frac{AM_i}{M_iE} = \frac{2^n - 1}{2}$

Figure 1.3. Shifts in attention. (a) "different roles of the segment ED" when proving Problem 1. (b) Focusing on different types of triangles when proving Yoni's theorem using Menelaus theorem

(See Figure 1.2: properties P7.1–P7.n with notations changed – D1, D2, …, Dn.)

All of the participants and Merav were excited by this finding. Merav reported that she "never encountered such a property in mathematics textbooks." Since Yoni studied for an M.Sc. in mathematics, Merav felt disappointed that he did not find a proof to this property.

There was no time remaining in the workshop and the prospective teachers were asked to prove this property as a bonus task at home. The proofs were presented at the discussion forum on the course website. The property was called 'Yoni's theorem' until the end of the course.

Story 2: Mathew Finds Yoni's Discovery Trivial!

Merav was thrilled by Yoni's theorem and told Mathew how the session developed. Mathew liked Task 1 very much and said it is very rich. Nevertheless, he found

Yoni's theorem to be trivial since it followed immediately from Menelaus' Theorem (Figure 1.2). He also noticed that the original problem can be "easily solved" using masses and centers of mass. Merav was disappointed that she did not think of the mass proof and the use of Menelaus' theorem herself and decided to share this experience with the prospective teachers in the course.

Story 3: Eti Develops Workshop for Mathematics Teachers Using Yoni's Discovery but Thinks It Is Not Good for All the Teachers

After her conversation with Mathew, Merav suggested that Eti use Task 1 in the courses for experienced practising mathematics teachers that Eti managed at the time. Eti prepared a PowerPoint presentation that introduced Task 1, proofs 1.1 and 1.2 and designed applets in Dynamic Geometry Environment "to perform collective investigation." She used this presentation in a "thinking together" manner. She argued that this was an "exemplary" task for teachers before they were involved in investigation of other problems by themselves.

Eti thought that Problem 1 as a proof problem might be used by the teachers in their classes "since the proof problem originally is taken from the school textbook." She also decided to demonstrate to teachers, "shifts" in the role of the segment *ED* as a midline in the triangle *ABC*, as side in the triangle *MED* and as a part of the side *EK* in triangle *HEK* (Figure 1.3a). However, she decided that reading Duval (1998; as suggested by Merav) could be too complicated for teachers in her courses. She liked the mass proof suggested by Mathew a lot and prepared a slide with a reminder of mass center.

Furthermore, Eti was skeptical about whether "the teachers will find time for the investigation task." At her workshop with teachers she used 'collective investigation' and performed dragging and measuring in her applet according to teachers' suggestions. She directed them toward the properties that are "more interesting" (Figure 1.2, P1–P6). She argued that "this is an excellent activity for teachers, but not for students" since teachers have to focus on the proofs and proving that eventually students need to be able to complete for the matriculation examination. Thus, investigations can be done by more advanced students at home or in classroom when they have spare time.

While preparing her presentation, Eti noticed that while Problem 1 required proving the ratio $\frac{\mathrm{AH}}{HE}=\frac{2}{3}$ of the segments on the median AE, Yoni examined ratios of segments on the "medians that intersect AH." Thus, she examined ratios of segments created by Yoni's medians and found that these ratios are terms in the series $\frac{2^n-1}{2}$ (Figure 1.2 – properties P8.1–P8.n). Eti was very happy about this discovery and included it in her presentation.

However, she decided to present 'Yoni's theorem' and her discovery without proof. She included a slide with Menelaus's theorem and together with Merav they created a slide that demonstrated how Menelaus's theorem can be used to prove

'Yoni's theorem' and her own discovery (Figure 1.3b). Merav suggested to her that using Menelaus theorem in proofs of the series P7.i and P8.i (Figures 1.2c and 1.3b) is very effective for the discussion of "shifts of attention" when proving geometry problems. However, Eti decided to allow teachers to understand the proof based on the slide and said that "probably this is good for teachers but not for all teachers and definitely for advanced students only."

Three Mathematics Teacher Educators' Communities Have Different Views on the Same Mathematical Activity

The three stories illustrate huge differences in the views on the educative power of Task 1 held by Merav, Mathew and Eti. These differences can be considered typical for the members of three mathematics teacher educators' communities of practice: mathematicians, mathematics education researchers and expert mathematics teachers. For Merav, mathematical challenge, mathematical inquiry and psychological aspects associated with mathematical thinking and learning are central in her work with mathematics teachers. She believed that mathematics teachers' personal experience would lead them to be willing to change teaching approaches based on the personal enjoyment and understanding of the importance of this experience from the psychological point of view. Story 1 demonstrates that for her, mathematical activities served as the basis for the discussion of the associated cognitive processes as, for example, Duval's (1998) theory of the role of visualization in proofs and proving in geometry. Obviously, Merav was sorry that Yoni did not prove his property but agreed with him that the proof was not trivial. Mathew, like Merav, found the discovery interesting, and especially exciting because the teachers themselves discovered the property. However, in contrast to Merav, he believed mathematics teachers should know Menelaus' theorem and be able to prove the discovered property immediately. Moreover, Mathew suggested a proof of Problem 1 that used mass (Proof 1.3 – Figure 1.2), which required rich knowledge "beyond the curriculum." For him, university mathematics is a critical basis for good teaching of mathematics in school. Eti also found the initial task that Merav devolved to her mathematics teachers challenging and worth using in her work as a mathematics teacher educator. She also decided to adopt Yoni's Theorem for her workshop. However, she made an adaptation of Problem 1 and its development. In contrast to Merav, she did not believe that a task of this type could be used by the teachers in their classes and thought "it is good for teachers but not for students." She also did not believe that discussion of psychological aspects of learning of mathematics would be interesting for mathematics teachers in her course. These differences in Merav's, Mathew's and Eti's conceptions are definitely related to the teaching and learning experiences that characterize the three communities of practice to which they belong. I argue that these distinctions between Merav's, Mathew's and Eti's conceptions can be explained in terms of mathematics teacher educators' knowledge at the 'mathematical horizon' and 'psychological horizon.'

MATHEMATICS TEACHER EDUCATORS' KNOWLEDGE AT MATHEMATICAL AND PSYCHOLOGICAL HORIZONS

The concept of horizon knowledge of mathematics is borrowed from Ball and Bass (2009) who "define *horizon knowledge [of mathematics]* as an awareness … of the large mathematical landscape in which the present experience and instruction is situated" (p. 6). Knowledge at the mathematical horizon for high school mathematics teachers is associated with extension of the concept of 'horizon knowledge' (Ball & Bass, 2009) to teachers' advanced mathematical knowledge (Zazkis & Leikin, 2010; Zazkis & Mamolo, 2011), that is knowledge beyond the high school mathematical curriculum that belongs primarily to university mathematics. 'Yoni's theorem' presented above, the mass proof to Problem 1, and Menelaus theorem belong to mathematics teachers' knowledge at the mathematical horizon. For Mathew, these concepts are accessible, and his mathematical horizon lies far beyond, while for Merav and Eti these concepts are situated at their mathematical horizon. However, it can be speculated that these concepts are more accessible for Merav than for Eti.

As with the concept of horizon knowledge of mathematics, I define *horizon knowledge of psychology* as an awareness of the large psychological landscape in which the present experience and instruction is situated. Knowledge at the psychological horizon for teaching mathematics includes a broad range of theories and concepts, including the concepts of zone of proximal development (Vygotsky, 1934/1982), mathematical challenge (Leikin, 2014), and student's mathematical potential. For example, a mathematics teacher's knowledge of the role of domain-general cognitive skills such as working memory, attention, inhibition, mental flexibility for mathematical processing can be seen as the mathematics teachers' knowledge at a 'psychological horizon.' The stories of Merav and Eti, described earlier in this chapter, demonstrate that while Duval's (1998) theory is used by Merav in her workshop with prospective mathematics teachers and is important for her, for Eti this theory is at the very far horizon and is not accessible at the moment. Figure 1.4 depicts distinctions between the conceptions held by Mathew (representative of mathematics teacher educators – mathematicians), Merav (representative of mathematics teacher educators – mathematics education researchers) and Eti (representative of mathematics teacher educators – expert mathematics teachers) in terms of knowledge at mathematical and psychological horizons.

I suggest that 'horizon knowledge of mathematics' and 'horizon knowledge of psychology' for teaching are integral parts of mathematics teacher educators' knowledge and skills that determine the quality of their work with mathematics teachers. Moreover, it is desirable that mathematics teachers' horizon knowledge be in the zone of accessibility for mathematics teacher educators.

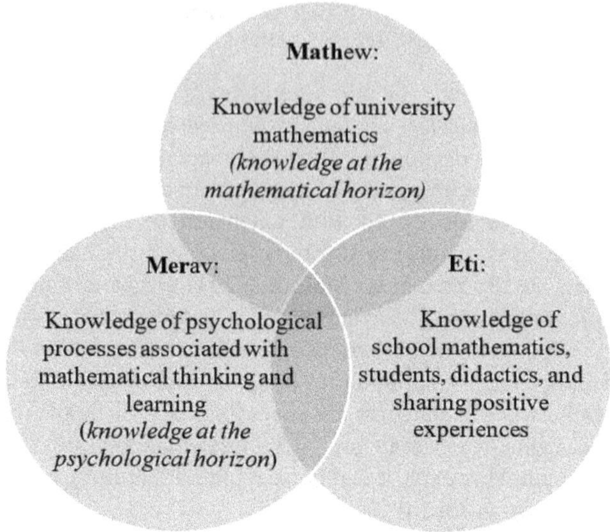

Figure 1.4. Distinctive characteristics of the three mathematics teacher educators' communities of practice

MATHEMATICS TEACHER EDUCATORS' KNOWLEDGE AND SKILLS IN TERMS OF MATHEMATICS TEACHERS' PROFESSIONAL POTENTIAL AND OF MATHEMATICAL CHALLENGE

The vision of mathematics teachers' horizon knowledge of mathematics and psychology as integral components of mathematics teacher educators' knowledge is linked to underscoring the structure and nature of mathematics teacher educators' knowledge and skills in terms of mathematics teachers' professional potential, students' mathematical potential and mathematical challenge. The construct of students' mathematical potential is a crucial element of mathematics teachers' knowledge and comprises the multiplicity of factors that mathematics teachers have to take into account in their instructional practices. Students' mathematical potential integrates students' knowledge and skills, which include domain-specific (e.g., mathematical problem-solving skills) and domain general (e.g., working memory, mental flexibility) affective characteristics associated with mathematics, students' personalities and the learning opportunities provided to them. When mathematics teachers design learning opportunities for students they must take into account students' mathematical potential, and must be aware that students' learning is not only a function of their mathematical knowledge and skills but depends on domain general cognitive traits, motivation, beliefs, and learning histories as well. Thus, mathematics teachers' knowledge at the psychological horizon is essential for their proficient work. Moreover, learning opportunities that mathematics teachers design for their students have to be challenging to students.

Mathematical challenge is a mathematical difficulty that a person is able and willing to overcome (Leikin, 2014, 2018). It is a function of three major factors in addition to mathematical potential (Figure 1.5): the first factor is associated with conceptual and structural characteristics of mathematical tasks, and the two other factors that contribute to variations in mathematical challenge are socio-mathematical norms and didactical setting. Mathematics teachers' horizon knowledge of mathematics is critical for teachers' proficiency in realizing and advancing students' mathematical potential.

While knowledge at mathematical and psychological horizons is central to mathematics teachers' proficiency, mathematics teacher educators are responsible for advancing mathematics teachers' professional potential. Mathematics teachers' professional potential is a complex function of their knowledge, skills and creativity in mathematics, didactics and psychology of mathematics teaching and learning, mathematics teachers' beliefs, motivation and personality as well as their own leaning and teaching experiences. Mathematics teacher educators' knowledge has to include all of these elements. Furthermore, activities that mathematics teacher

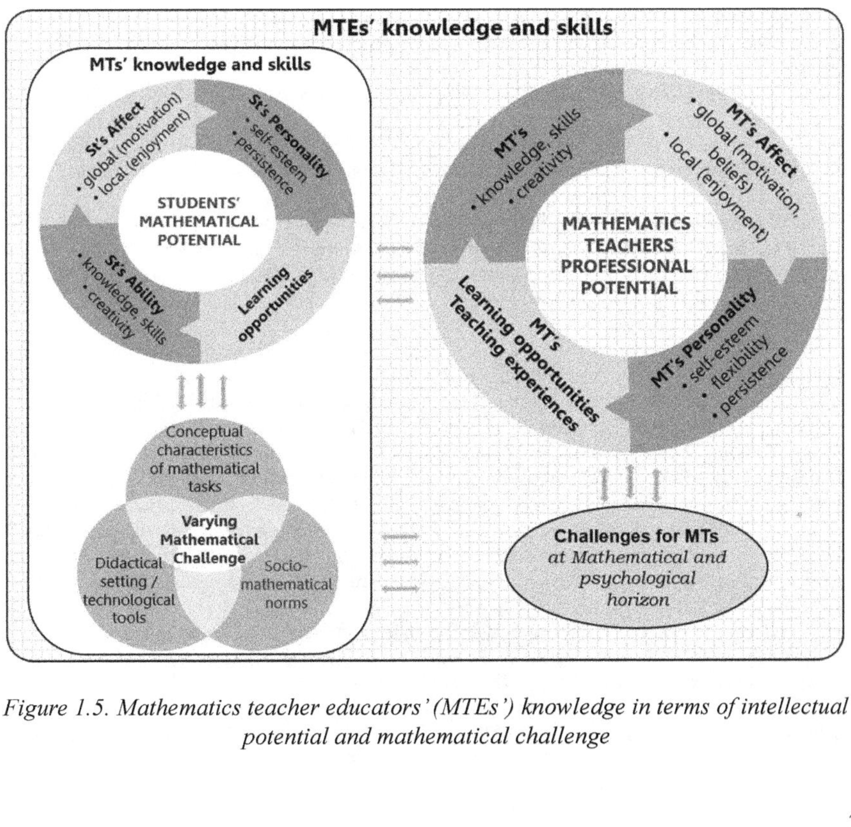

Figure 1.5. Mathematics teacher educators' (MTEs') knowledge in terms of intellectual potential and mathematical challenge

educators design for mathematics teachers have to be challenging for them, lead to analysis of all the components of students' mathematical potential and develop components of teachers' professional potential. Thus, knowledge at mathematical and psychological horizons are integral for mathematics teacher educators' knowledge and skills (Figure 1.5).

Note though, that – taking into account the distinctions between Merav, Mathew and Eti – we should be aware of the differences between mathematics teacher educators from different communities of practice and the complementary nature of the contribution of mathematics education researchers, mathematicians and expert teachers to the education of mathematics teachers. Thus, in many cases collaboration between mathematics teacher educators in the educational programs for mathematics teachers seems to be critical in order to develop all the components of teachers' professional potential.

REFERENCES

Appova, A., & Taylor, C. E. (2019). Expert mathematics teacher educators' purposes and practices for providing prospective teachers with opportunities to develop pedagogical content knowledge in content courses. *Journal of Mathematics Teacher Education, 22*(2), 179–204.

Ball, D. L., & Bass, H. (2009). *With an eye on the mathematical horizon: Knowing mathematics for teaching to learners' mathematical futures*. Paper presented at the 43rd Jahrestagung der Gesellschaft für Didaktik der Mathematik, Oldenburg, Germany. Retrieved from https://eldorado.tudortmund.de/bitstream/2003/31305/1/003.pdf

Brousseau, G. (1997). *Theory of didactical situations in mathematics*. Dordrecht: Kluwer Academic Publishers.

Chen, J. C., Lin, F. L., & Yang, K. L. (2018). A novice mathematics teacher educator–researcher's evolution of tools designed for in-service mathematics teachers' professional development. *Journal of Mathematics Teacher Education, 21*(5), 517–539.

Chick, H., & Beswick, K. (2018). Teaching teachers to teach Boris: A framework for mathematics teacher educator pedagogical content knowledge. *Journal of Mathematics Teacher Education, 21*(5), 475–499.

Duval, R. (1998). Geometry from a cognitive point of view. In C. Mammana & V. Villani (Eds.), *Perspectives on the teaching of geometry for the 21st century: An ICMI study* (pp. 37–52). Dordrecht: Kluwer Academic Publishers.

Even, R. (1999). The development of teacher leaders and inservice teacher educators. *Journal of Mathematics Teacher Education, 2*(1), 3–24.

Goos, M. (2009). Investigating the professional learning and development of mathematics teacher educators: A theoretical discussion and research agenda. In R. Hunter, B. Bicknell, & T. Burgess (Eds.), *Crossing divides: Proceedings of the 32nd Annual Conference of the Mathematics Education Research Group of Australasia* (Vol. 1, pp. 209–216). Palmerston North, NZ: MERGA.

Jaworski, B. (1992). Mathematics teaching: What is it? *For the Learning of Mathematics, 12*(1), 8–14.

Jaworski, B. (2008). Development of the mathematics teacher educator and its relation to teaching development. In B. Jaworski & T. Wood (Eds.), *The international handbook of mathematics teacher education: The mathematics teacher educator as a developing professional* (Vol. 4, pp. 335–361). Rotterdam, The Netherlands: Sense Publishers.

Leikin, R. (2014) Challenging mathematics with multiple solution tasks and mathematical investigations in geometry. In Y. Li, E. A. Silver, & S. Li (Eds.), *Transforming mathematics instruction: Multiple approaches and practices* (pp. 59–80). Dordrecht: Springer.

Leikin, R. (2018). Part IV: Commentary – Characteristics of mathematical challenge in problem-based approach to teaching mathematics. In A. Kanjander, J. Holm, & E. J. Chernoff (Eds.), *Teaching and learning secondary school mathematics: Canadian perspectives in an international context* (pp. 413–418). Cham: Springer.

Leikin, R. (in press). Characterization of mathematics teacher educators' knowledge in terms of teachers' professional potential and challenging content for mathematics teachers. In M. Goos & K. Beswick (Eds.), *The learning and development of mathematics teacher educators – International perspectives and challenges*. New York, NY: Springer.

Leikin, R., Zazkis, R., & Meller, M. (2018). Research mathematicians as teacher educators: Focusing on mathematics for secondary mathematics teachers. *Journal of Mathematics Teacher Education, 21*(5), 451–473.

Leontiev, L. (1983). *Analysis of activity: Psychology* (Vol. 14). Moscow: Moscow State University – Vestnik MGU.

Lin, F. L., Yang, K. L., Hsu, H. Y., & Chen, J. C. (2018). Mathematics teacher educator-researchers' perspectives on the use of theory in facilitating teacher growth. *Educational Studies in Mathematics, 98*(2), 197–214.

Pellegrino, J. W., & Hilton, M. L. (2012). *Educating for life and work: Developing transferable knowledge and skills in the 21st century*. Washington, DC: The National Academies Press. doi:10.17226/13398

Perks, P., & Prestage, S. (2008). Tools for learning about teaching and learning. In B. Jaworski & T. Wood (Eds.), *The international handbook of mathematics teacher education: The mathematics teacher educator as a developing professional* (Vol. 4, pp. 265–280). Rotterdam, The Netherlands: Sense Publishers.

Steinbring, H. (1998). Elements of epistemological knowledge for mathematics teachers. *Journal of Mathematics Teacher Education, 1*(2), 157–189.

Valsiner, J. (1997). *Culture and the development of children's action: A theory of human development.* (2nd ed.). New York, NY: John Wiley & Sons.

Vygotsky, L. S. (1934/1982). *Thought and language*. Moscow: Pedagogika.

Vygotsky, L. S. (1978). *Mind in society: The development of higher psychological processes*. Cambridge, MA: Harvard University Press.

Zaslavsky, O., & Leikin, R. (2004). Professional development of mathematics teacher educators: Growth through practice. *Journal of Mathematics Teacher Education, 7*(1), 5–32.

Zazkis, R., & Leikin, R. (2010). Advanced mathematical knowledge in teaching practice: Perceptions of secondary mathematics teachers. *Mathematical Thinking and Learning, 12*(4), 263–281.

Zazkis, R., & Mamolo, A. (2011). Reconceptualizing knowledge at the mathematical horizon. *For the Learning of Mathematics, 31*(2), 8–13.

Roza Leikin
Department of Mathematics Education
University of Haifa

STEVE THORNTON, NICOLA BEAUMONT, MATT LEWIS AND COLIN PENFOLD

2. DEVELOPING AS A MATHEMATICS TEACHER EDUCATOR

Learning from the Oxford MSc Experience

Transitioning from teaching to teacher education is a challenging process, involving both the development of knowledge required for teacher education and learning from practice. This chapter draws from the experiences of teachers transitioning to teacher educators during the Masters of Science (Teacher Education) program at the University of Oxford. Data are drawn from the participants' own reflections and include a focus on both theoretical and practical understandings. In the course we drew on Rowland's Knowledge Quartet as a framework for thinking about how we develop as teacher educators as well as how colleagues with whom the participants worked develop as teachers. The chapter describes how the participants were able to use the Knowledge Quartet to highlight issues important in their context, and hence to research their practice with a focus on what mattered most to the prospective and practising teachers with whom they worked.

INTRODUCTION

Transitioning from teaching to teacher education is a challenging process, involving both the development of knowledge through *learning about* (theory), *learning how* (practice) and *reflecting* (seeking to understand and improve one's practice). This chapter draws from the experiences of three people who were making, or had begun to make, the transition from teachers to teacher educators during and after the Master of Science (Teacher Education) at the University of Oxford. Data are drawn from the participants' own reflections, collected through an unstructured, free-flowing discussion some two years after completing the course. The chapter highlights the importance of three key ideas encountered during the MSc course that have become salient in the participants' varying contexts: obtaining buy-in and understanding the reasons for resistance, rigorously challenging our own and prevailing beliefs, and celebrating ambiguity.

© KONINKLIJKE BRILL NV, LEIDEN, 2020 | DOI: 10.1163/9789004424210_003

THE UNIVERSITY OF OXFORD MASTER OF SCIENCE IN TEACHER EDUCATION

In 2014 the University of Oxford instituted a Master of Science in Teacher Education (MSc in Teacher Education) for current and future mathematics and science teacher educators (Childs, Hillier, Thornton, & Watson, 2014). The course was developed to capture a growing interest internationally in equipping new teacher educators to better prepare prospective or practising teachers of mathematics and science. The initial cohort included teacher educators from developing countries, curriculum leaders, school-based personnel with responsibility for a cohort of school-based prospective teachers, or others interested in mathematics or science teacher education. The geographic spread of the cohort required that the course be delivered at distance via an online learning platform, which was enhanced by annual week-long face-to-face residentials.

The aims of the course were:

- To develop familiarity with research and professional debates associated with teacher education in mathematics and science;
- To learn about pedagogy for teacher education in these subjects in a variety of settings;
- To acquire a repertoire of methods for transforming the subject knowledge of teachers and educators for teaching purposes;
- To introduce participants to the quality assurance and research standards and methods that characterise the research fields of subject education; and
- To equip participants to continue professional and academic dialogue with others in the field (Childs et al., 2014, p. 50).

The course was structured as two units for each of two years, followed by a dissertation dealing with an issue in the participants' own context. The first year focussed on the nature of mathematics and science, what it means to teach the subject, and what knowledge a teacher needs. In the second year the focus shifted to how teachers learn and the knowledge needed by a teacher educator. Together the four units shaped each participant's individual research proposal and study, culminating in a dissertation.

Unit 1 asked what is science/mathematics and what it means to learn/do science/ mathematics? Personal views were challenged and looked at through the eyes of a teacher educator. In considering these areas participants began to consider the knowledge base that they, as teacher educators, needed to acquire in order to develop practising and prospective teachers' understanding of the nature of their subject and of learning in science and mathematics. A particular focus was developing an "academic mindset," that is, reading and reviewing research and professional literature and considering its application in their context

Unit 2 asked what does it mean to teach the subject and what knowledge does a subject teacher need? Using a framework of the Knowledge Quartet (Rowland &

Zazkis, 2013) participants looked at teachers' foundation, transformation, connection and contingency knowledge in action. A particular focus was conceptual analysis of specific ideas in science and mathematics and observation of teachers to look at the issues and challenges of teaching of these concepts in context.

Unit 3 moved the focus to how teachers learn and the design of an effective teacher education course. Participants began by considering how, as teacher educators, they could contribute to developing the beliefs and knowledge base discussed in Units 1 and 2. By reviewing research on teacher learning and models of teacher learning, and through consideration of teacher learning in their own context, participants were able to identify key areas of strength and key areas for development, and begin to design relevant teacher education experiences.

In Unit 4 participants considered their own knowledge as developing teacher educators. Professional and academic literature helped participants develop their understanding of the role of teacher educators, their knowledge base and the design of teaching in teacher education courses in science and mathematics.

Each of the participants brought these ideas together in a research study and dissertation. The studies were contextual in that they addressed the needs identified by each participant in their own situation, and hence dealt with issues such as the need to develop teachers' content knowledge, the need to develop pedagogical strategies that challenged the way teachers themselves had learned, and the challenges faced in promoting change at scale. What became clear is that regardless of context, each participant faced a similar set of issues in promoting cultural change in teachers' perceptions of mathematics and teaching, and that teacher educators therefore need to have knowledge that crosses disciplinary boundaries, including mathematics, sociology, philosophy, cognition and adult learning.

As Childs et al. (2014) say, little is known about the process through which a teacher becomes a teacher educator, nor about the knowledge base needed, how that might be developed, or how they might go about researching their own practice. The course design was, therefore, responsive and flexible, contingent on the reflections and experiences of the participants. It is not the purpose of this chapter to describe or critique the MSc in Teacher Education, nor to recommend aspects related to course design for new teacher educators. Rather, the central issue addressed in this chapter is the challenge of transitioning from a teacher to a teacher educator, and what experiences, either in a formal course such as the MSc in Teacher Education or subsequently in an educational context, are seminal in that transition.

LITERATURE REVIEW

The challenges associated with becoming a teacher educator have received increasing attention in recent years in both the mathematics and science education literature (e.g., Jaworski & Wood, 2008; Loughran, 2013). Much of this work has focussed on developing an identity as a teacher educator, often through self-study (Berry, 2007; Davey, 2013). Indeed, the journal *Studying Teacher Education* is sub-titled *A journal of self-study of teacher education practices*. Other studies (Bullough, 2005; Murray & Male, 2005; Trent, 2013) have focussed on changing and developing identities among small groups of teacher educators as they have made the transition from teacher to teacher educator. Others have taken a sociocultural approach, locating the work of becoming a teacher educator in a community of practice (Goos, 2009; Williams, Ritter, & Bullock, 2012).

What is significant in this work is that becoming a teacher educator, whether in mathematics, science or some other discipline, requires much more than being an expert teacher, or possessing deep subject matter or even pedagogical knowledge – it appears to entail a fundamental shift in thinking that makes the tacit explicit, that views knowledge as uncertain or problematic rather than as certain or predictable, and that celebrates the tensions and synergies between theory and practice (Loughran, 2013). Such a shift necessarily presents conflicts, boundaries and obstacles in both the personal and professional domains; in short, the journey from teacher to teacher educator represents, for many, a "rocky road" (Wood & Borg, 2010, p. 17).

Two key issues emerge in the literature, that Murray and Male (2005) term *expert become novice*, and *novice assumed to be expert*. Negotiating the complexities of these issues is central to developing a *pedagogy of teacher education* (Loughran, 2013), in which the teacher educator simultaneously looks inward to examine their own teaching expertise with a scholarly lens, and outward to examine research and scholarship in the field of teacher education through the lens of their own expertise.

The issue of expert become novice is particularly salient in those situations where the expert teacher, often by virtue of their expertise, moves into a new context as a teacher educator. It is often naively assumed that expertise in teaching school students automatically translates into expertise in teaching teachers (Williams et al., 2012). Yet, the reality is often very different (Ritter, 2007). Expert teachers moving into teacher education may feel alienated or de-skilled, particularly if they have moved from a position of leadership and authority in the school setting. As one participant in Murray and Male's (2005) study reported,

> there were strong feelings for me of masquerading and being about to be found out as an impostor. I felt de-skilled – it was as if all my years of teaching experience had fallen away and I was left feeling inadequate and exposed in this strange world. (p. 130)

The issues of novice assumed to be expert arises particularly where it is assumed that, by virtue of being in the academy, the new teacher educator automatically possesses a range of research and teaching skills that will enable them to become active researchers within a very short period of time. The research and publish imperatives of many university environments places new teacher educators into precarious positions for which their expertise in a previous occupation does not serve them well. Again, as one participant in Murray and Male's (2005) study reported, "I've got lots of street credibility with the students and some staff because of my practical knowledge of schools, but I don't have any kind of research record, so to others I'm just a waste of space" (p. 133).

Negotiating the transition from expert teacher to teacher educator takes time. It requires identifying the inter-relationships between what is taught and how it is taught, developing an overt knowledge not only of the subject matter but also of the pedagogical principles underpinning teaching practice. It requires a shift from *first-order practitioners*, school teachers whose role is to optimise learning for their students, to *second-order practitioners*, teacher educators whose role is to empower others to optimise learning for their students (Murray & Male, 2005). In short, it requires the development of a pedagogy of teacher education (Loughran, 2013), as opposed to a pedagogy of mathematics, or English, or science.

In conceptualising a pedagogy of teacher education, Loughran (2013) does not call for a set of rules or procedures. Rather, he argues that through study of the wealth of literature about good teaching and learning, foundations for practice in teacher education might be developed that are responsive to the issues, needs and concerns of participants. Such a pedagogy makes the unseen seen and brings into question the taken-for-granted. This, after all, is at the heart of studying teaching and learning. Simultaneously shining a light on what we do as teachers through scholarly study and questioning what might be commonly accepted in the light of our own wisdom and practice enables the teacher educator to combine "the interdependent worlds of *teaching about teaching* and *learning about teaching*" (Loughran, 2013, pp. 173, 174). Necessarily, the pedagogy of teacher education described by Loughran is complex and changeable, marked by questioning and uncertainty.

The development of a pedagogy of teacher education is vividly captured by Berry (2007) in her self-study of becoming a teacher educator over time. She describes the shifts in her perceptions of knowledge and expertise in her "sacred story" of becoming a teacher educator, questioning the taken-for-granted, and making the invisible explicit. Berry explains that where knowledge was previously seen as certain or predictable, the transition to teacher educator required a view of knowledge about teaching as something that is uncertain, messy or problematic. Rather than being delivered as tips and tricks of teaching located outside the knower, she came to see knowledge about teaching as something developed through personal reflection and theorising experience, constructed inside the person. Teaching expertise defined by smoothness of delivery was replaced by expertise that enabled the complexities and uncertainties of teaching to be revealed to others. What mattered most for Berry was no longer the what and how of teaching, but the why.

The contrasts and disruptions described by Berry (2007) appear to stem in large part from the relatively sudden disruption to established ways of thinking and being caused by a new role in a new environment. In similar accounts given by Tzur (2001) and Krainer (2008) the disruptions are less dramatic, arguably because each stage of the process involved boundary crossing between domains. In his self-reflective analysis, Tzur (2001) describes four stages of development distinguished by the domain of the activity: learning mathematics as a student; learning to teach mathematics as a teacher; learning to teach mathematics educators as a teacher educator; and learning to teach mathematics teacher educators as a mentor. For Tzur the jump between domains was less stark and dramatic than those experienced by the new teacher educators described above. At each stage of this development Tzur was actively involved in teaching, studying and researching. As a student, he tutored other students who struggled with the mathematics, as a student of teaching he undertook mini research projects examining how students learned specific topics, and as a teacher educator and mentor of teacher educators he began to integrate teaching and research to better understand teachers' experiences. While Tzur may not have experienced the feelings of dislocation so vividly described by Berry (2007), Davey (2013) or those in the studies by Trent (2013), he does recognise the conceptual leaps that are necessary in moving from one stage of development to another. These conceptual leaps involve making the invisible explicit – in the case of becoming a teacher it is making the misconception and incomplete understanding of students visible, in the case of becoming a teacher educator it is noticing that teachers often fail to see the potential of activities designed to reveal deep understanding.

In his chapter in the previous edition of this Handbook, Tzur (2008) describes his growing awareness that teachers with whom he was working failed to see the epistemological potential of lessons that he, as teacher educator, had co-planned and implemented. He describes the initial disbelief among teachers that the lessons would lead to deep understanding, and the reasons for which the lessons appeared counter-intuitive to teachers. He introduces a new construct, Profound Awareness of the Learning Paradox (PALP), to describe a conception-based approach to teaching that is rooted in epistemologically related pedagogies. Tzur argues that developing and becoming familiar with profound awareness of the learning paradox is essential in becoming a mathematics teacher educator.

Krainer (2008) recounts a similar story of becoming a mathematics teacher educator. Like Tzur, Krainer's journey does not appear to be marked by dramatic feelings of dislocation or inadequacy. He developed an interest in student learning even as a student himself and wrote a Diploma thesis as part of his initial teacher education program. Research was thus integral to becoming a teacher. He joined a group of university staff members and experienced teachers to plan a professional development program, thus beginning what seems an almost natural journey to teacher educator.

Krainer (2008) introduces a seven-layer nested model describing research domains in mathematics education, with each domain encompassing and building on knowledge in previous domains but asking new questions and having new foci. At the centre is mathematics itself, which although remaining the focus of research, does not specifically deal with questions of mathematics education. The first mathematics education research domain, ME1, focuses on mathematical content from an educational point of view. In this domain questions such as which mathematical concepts or ideas should be introduced in school mathematics and what tools or technologies might be used, are raised. ME2 focuses on applications and history of mathematics and how they might contribute to make mathematics more meaningful to learners. ME3 focuses on individual students, their possible misconceptions and the strategies we might use to overcome them. ME4 looks at the context of learning, focusing on interactions between students and teachers and the negotiation of classroom social and sociomathematical norms. These research domains are clearly within the area of interest of every teacher of mathematics. It is when we get to ME5 that the research questions typically become those of teacher educators. ME5 focuses on how teachers learn, their beliefs, knowledge and practices and the impact of teacher education in shaping and enhancing the work of teachers. ME6 focuses on the learning of teacher educators. Interestingly, Krainer claims that the research questions here are the same as for teachers, but that the subject tends to be the self rather than others. ME7 focuses on mathematical abilities of systems and societies as a whole, and the impact of educational policy or curriculum.

Although there are differences in the extent to which becoming a teacher educator represents a radical boundary-crossing into a new way of thinking or being, what is common is that transitioning from teacher to teacher educator requires consideration of a different set of questions. The focus shifts to consideration of how teachers learn, of why some teachers are receptive to change and others less so and of the connection between research and teaching. Each of the studies discussed above has reflected, to a greater or lesser degree, an increased uncertainty about truth, particularly as it applies to teaching and learning. Each has also described a process of deep reflection and interaction. Such a process may not be for everyone – as Tzur (2001) says, "being a good teacher does not necessarily imply being a good teacher educator" (p. 275).

RESEARCH METHODOLOGY AND PARTICIPANTS

The conversation reported below took place in a London bar over dinner. The first author, Thornton (Steve) was the course tutor for the first 18 months of the course and was visiting the United Kingdom from Australia. Each of the three co-authors is an experienced successful teacher of mathematics who also has considerable background in working with teachers. Author 2, Beaumont (Nicola) was based in London as Head of Department in a private school also responsible for School Direct candidates. Nicola's role required a blend of practical and theoretical knowledge, as much of the traditional university-based study undertaken by prospective teachers is located within her school. Author 3, Lewis (Matt) was a mathematics consultant employed by a national government organisation in the United Kingdom. He worked with practising teachers to develop content knowledge and pedagogical content knowledge, particularly among those who may lack confidence. Author 4, Penfold (Colin) had for the past eight years been working in several countries in the Middle East, Africa and Brunei as a consultant on a range of issues including teacher development and mathematics. He had worked to promote systemic change, particularly in broadening the range of pedagogical approaches used by teachers who were accustomed to transmission approaches rather than more flexible, student-centred ones.

A fourth potential participant, also based in London in a School Direct role, was unable to be present at the last minute. The remaining four of the initial cohort of eight participants in the MSc were based outside of the United Kingdom and were also unable to be present. Two of these were based in a Papua New Guinea university and worked in a teacher education course, a third was based at a South African school and had a leadership role in mathematics education, including working with prospective teachers, while the fourth worked in a university in the United States of America, teaching first year undergraduate mathematics with a brief to assist other mathematics lecturers to examine their pedagogy. The diversity of people in the MSc course and the diverse contexts on which they worked, added enormously to the richness of the academic discussion during the course, but what was most striking were the similarities in the issues they faced, rather than the differences.

It is noteworthy that the participants had had very little contact with each other since the conclusion of the course, except at their graduation. The conversation thus served both as a time for research and reflection, and as an opportunity to meet socially. This necessitated a very free-flowing unstructured conversation in which participants could set and respond to the agendas as they arose.

As part of the invitation each participant was sent an email with four general questions:

- What was the biggest challenge you faced in transitioning from a teacher to a teacher educator?
- What parts of the MSc helped as you considered those challenges?
- Were there any theoretical ideas that particularly stood out during the MSc experience?
- What changed as a result of being part of the MSc?

The group then met in the London bar for two to three hours, including a period of 40 minutes during which the conversation was recorded and subsequently transcribed. The first author (Thornton) read through the transcription, tracking the flow of the conversation, and highlighting key themes and generalisations. A first draft of this paper, along with the full transcript, was then sent to each of the participants with an invitation to add, delete or edit.

As is evident in the results, the initial questions served merely as a discussion starter. The conversation rapidly diverged from the issues raised in the questions, following a path that evolved during the conversation. The research methodology is, therefore, emergent and responsive, allowing the issues to arise naturally in an academic, but informal conversation. Although the conversation was necessarily restricted in time, the natural setting ensured that those issues that were raised were, for these participants, those that were most salient in considering the transition from teacher to teacher educator.

Such a methodology fits firmly in the tradition of participant perspective research described, for example, by Kemmis (2012). Kemmis contrasts a *spectator* perspective, which examines a practice from the outside to identify how they act and are situated within a network of practices, with a *participant* perspective, which seeks to give insight into the practice itself. The participant perspective asks questions such as:

- What do I think my pedagogical practice is and should be? and
- How might I see myself developing the excellence of my own practice and at the same contribute to the collective development of the practice?

The participant perspective thus contributes not only to enhancing teacher education and our knowledge about teacher education, but itself transforms the participants and foregrounds the development of their own identities and self-understanding.

The use of free-form emergent group conversation, although in one sense dictated by the setting, is a powerful means of obtaining participant voice. We suggest that traditional structured or semi-structured interviews privilege the voice of the researcher, no matter how much they strive to obtain authentic participant views. They necessarily follow an agenda that is more or less set by the researcher, rather than an emergent and responsive agenda set by the participants. In effect the conversation became a "yarning circle" (Donovan, 2015), a narrative methodology developed from Australian Aboriginal community gatherings in which everyone has the right to speak without interruption. Participants face each other, in a circle or in this case around a table in the bar, tell their stories and respond to those of others. While at face value it may appear to be a focus group, it differs in that each person is in effect simultaneously the participant and the researcher.

RESULTS AND DISCUSSION

We present the data gathered from the conversation as a meander through the salient themes that emerged, rather than as a synthesis or analysis. We do this to capture the richness and nuance of a living, emergent conversation that we hope situates the reader as an active listener, or fly-on the-wall, rather than as a passive consumer of the inevitably detached interpretation of the researcher.

"Getting the Buy-In"

To the initial question of the biggest challenge faced in moving from being a teacher to teacher educator Nicola responded, "Getting the buy-in from teachers." She explained that although she felt she had learned a lot from doing the MSc, she did not feel as if the teachers in her school had been on the journey with her. They had been happy to participate in the research that was part of the dissertation or other assignments that Nicola had completed but had not necessarily seen any reason to change their practice. They had not seen Nicola as a resource that they might use to help them in their development as teachers. This was exacerbated by the constraints of the school environment, which reduced the time that Nicola had to meet with her colleagues to 35 minutes per week. So, although Nicola would plan to model good teaching using some of the ideas developed through the MSc, she found that these ideas did not necessarily resonate with the other teachers in her faculty.

It is significant that Nicola's comment was heartfelt and instantaneous. As a teacher in an independent school there are high expectations regarding curriculum outcomes and student achievement. Her experience of resistance to change is not uncommon; as März and Kelchtermans (2013) point out teachers' interpretation and sense-making about change stem from three factors: their personal beliefs, their normative ideas about good teaching, and their structural reality. Teachers who hold instrumentalist views of mathematics (Beswick, 2005), including, in Nicola's case, teachers whose initial expertise was in engineering, are likely to emphasise performance and skill mastery over understanding and problem solving; successful teachers, such as those whose students have typically achieved at high levels in external examinations are likely to resist calls for more student-centred or investigative pedagogies; while external demands on teachers for compliance are likely to result in lower levels of risk-taking. Faced with these realities it is unsurprising that obtaining buy-in was at the forefront of Nicola's mind.

Nicola's observations mirror those of the Hong Kong language teachers in Trent's (2013) study, who reported that although their colleagues had been supportive and had expressed enthusiasm for new ideas or approaches, they had not followed them through. Like the teachers in Nicola's school, the demands of meeting a teaching schedule or preparing students for examinations were more pressing priorities that impacted on the teachers' willingness to engage with new ideas.

"It's Just a Matter of Scale"

On the other hand, Nicola did obtain interest and buy-in from the prospective teachers with whom she worked as part of the School Direct program, or from those who were on school-based placements. Nicola observed that in working with prospective teachers "you essentially have people who really do want to work with you." Matt agreed that in his context of working across several schools, it was the newly qualified teachers who were most interested in exploring new ideas as they did not feel as if "they knew the answers already." He had not anticipated the resistance from more experienced teachers.

Nicola observed that although in general the prospective teachers had a sound mathematical knowledge as they had been successful at mathematics themselves, they did not necessarily know how to transform that knowledge in a way that would be helpful for children. They also failed to notice when opportunities arose for them to build on student responses in-the-moment to promote deeper learning. Colin observed that in his context of working with teachers from developing countries, many lacked a broad mathematical experience or had little confidence in their own subject matter knowledge. As a result, they, too, failed to notice opportunities to build on students' responses.

In discussing this issue, Nicola, Matt and Colin referred to one of the key readings in the MSc that dealt with the Knowledge Quartet (Rowland & Zazkis, 2013). The Knowledge Quartet identifies four aspects of knowledge that teachers of mathematics require. Foundation Knowledge is the knowledge, beliefs and understanding that are acquired in the academy, consisting primarily of knowledge about mathematics, but also beliefs about mathematics, its value and its purposes. Transformation Knowledge includes knowing which representations might be best to illustrate a particular concept, what materials may be best and how best to explain ideas to students – it is the capacity to make mathematics meaningful to students. Connection Knowledge concerns the coherence of mathematics, and how decisions are made with regard to planning and sequencing across lessons or series of lessons that take into account learning progressions and cognitive demand. These aspects of knowledge feed into the fourth level, Contingency Knowledge, which concerns classroom events that are often impossible to plan for. It is the readiness to respond to children's ideas as they arise, being prepared to deviate from a set agenda where appropriate.

Colin explained that the teachers in his context had had very limited exposure to ways of teaching and learning that differed from very traditional teacher exposition and student practice of mechanical skills. As a result, their opportunities to develop contingency knowledge were very limited. The readings had helped him to structure the way he thought about teacher knowledge and to recognise the importance of a sound Foundation Knowledge, but he was left with the question of how Contingency Knowledge could be developed in a systematic and structured way. "When we talk about contingency, what are the components of contingency? How do we develop it, can it be developed?" Nicola observed that although the prospective teachers were keen to learn, those whose experience of learning mathematics was very traditional often shut down ideas that they did not understand before trying them.

The issues faced by Nicola, Matt and Colin were, therefore, similar. They were all keen to implement and share ideas they had gained through their own experience and through their study in the MSc. They had all recognised the importance of, and also the difficulties of, developing higher levels of knowledge about teaching in the people with whom they worked, and had made sense of some of these difficulties through their reading in the MSc. They all found that new teachers were willing to learn and try new ideas and were disappointed that many more experienced teachers were often resistant. In Colin's context of working in developing countries where almost all mathematics instruction had been very instrumental, the scale of the problem was enormous. Large numbers of teachers had limited knowledge of mathematics, and although there was an appetite to learn, change was very hard. In Nicola's case prospective teachers who were challenged with something new tended to shut down and be sceptical about the potential to improve student learning. For Nicola and Matt, the biggest challenge was that of working with experienced teachers who were resistant to change. As Colin said, "It's just a matter of scale."

"It's Strange, That was before We Started, That's Had a Big Impact"

In thinking about the importance of structuring his thinking through ideas such as the Knowledge Quartet (Rowland & Zazkis, 2013), Matt brought the conversation back to the impact of the MSc in his own development. He explained that although his job required him to work with teachers, prior to undertaking the MSc he had no access to readings, colleagues, tutors or theories that helped to make sense of the learning of teachers. The act of thinking about how teachers learn had been "revelatory" for Matt.

The readings that were expected in the MSc were intentionally chosen to both confirm and confront. As Colin said, "One of the good things about it was that it wasn't the case that everything necessarily chimed with the way I felt." Indeed, Nicola was extremely critical of some of the articles she had read, particularly one that described teacher expertise in terms of a set of behaviours that accumulate over time (Berliner, 2004). She was uncomfortable about the notion of "expert teacher" as something that could be reduced to a set of more or less well described behaviours – "actually, the whole thing didn't feel very, it just didn't feel very nice." However, the fact that she had read the article made her think much more deeply about what does make an expert teacher and how she might work with colleagues to improve their practice.

Matt similarly criticised what he saw as the somewhat over-generalised model of teacher expertise described by Berliner (2004). For him, the most helpful reading was a model of teacher professional growth (Clarke & Hollingsworth, 2002) that the MSc students had been asked to read prior to even commencing the course. The model provides an interconnected and non-linear view of how teachers grow, identifying mechanisms by which change in one domain is associated with change in another. It describes how external sources of stimulus (the external domain) might influence changes in teachers' knowledge and beliefs (the personal domain). In turn a teacher's knowledge and beliefs impact on how they view the external domain. Knowledge and beliefs, together with external stimuli, lead to professional experimentation (the domain of practice), which in turn impacts on teachers' beliefs and practices. Together the personal domain and domain of practice produce salient outcomes for teachers and students (the domain of consequence).

Initially the Clarke and Hollingsworth (2002) model had seemed overly complex. As Matt said, "So here's a diagram, which is meant to explain so much … and it didn't really chime with me at all." Yet, over time, the model helped to explain why the external stimulus alone did not result in teacher change, and why trying something and noticing what results can be such a powerful agent of change.

> It's not better because you did what someone else told you to do, it's better because of what you notice, it's better because of the changes you made in your practice, or that the teacher made in their practice, and that applies to so many of the different contexts in which I work with teachers. (Matt)

Similarly, Nicola observed that the model suggests that there are many avenues and mechanisms that contribute to teacher growth, whereas before the course she had thought "I'll say something, you'll listen." Colin developed his own adaptation of the model for his context to incorporate reflection, enactment and the relationship between theory and practice. Thus, models such as that of Clarke and Hollingsworth (2002) had a profound impact in helping Nicola, Matt and Colin to make sense of their context, even though it was initially complex and was introduced before the course started. As Matt said, "It's strange, that was before we started, that's had a big impact."

"This Is Bona Fide Research"

The reflection on the issues associated with change, with teacher knowledge, and the impact of readings prompted Colin to raise what he termed an "inquiry question." The three participants, indeed the eight participants in the MSc course, had had very limited contact with each other since the conclusion of the course. Due to their geographical separation, contact during the course had also been limited. Yet, the things they were discussing, and the things that made an impact, were essentially the same. "Now, my question is why? Why are those things chiming with us? … Is it because it sort of related to where we were? Is it because somehow it really challenged us cognitively? What was it?" (Colin).

Matt's response was, "I just think, isn't it because this is bona fide research, like research doesn't have to be a randomised control trial with a particular effect size for it to be something valid, effective and makes a difference." Matt argued that research is about ideas, and how things might work. Research is culturally embedded, and although Nicola, Matt and Colin were each working in different contexts, their backgrounds were similar and hence they interpreted things in a similar way.

In the literature on becoming a teacher educator discussed earlier in this chapter, one of the critical issues was that of the novice assumed to be expert. For the new teacher educators writing about their own experiences, or who were the subjects of case studies, becoming part of the research culture represented a major obstacle in assuming the identity of a teacher educator. Yet, the participants in the MSc already saw themselves as researchers and saw research as more than randomised trials.

During the induction week at the beginning of the course, each participant was asked to critique two research articles, with the aim of inducting them into an academic mindset. One of the papers they were asked to critique reported on a randomised trial in which psychology students had been asked to learn a concept in either a concrete or abstract environment (Kaminski, Sloutsky, & Heckler, 2008). The paper reported that those students who had learned in an abstract environment were more able to transfer their knowledge to alternative contexts, whereas those who had learned in a concrete environment were restricted to being able to answer questions relating to that context. The paper was controversial as it recommended that mathematics learning happens most effectively if abstract ideas are introduced first. In their critique of the paper, the MSc students examined issues such as definitions of terms used, the author's use of judgmental language, the use of repetition, self-referencing, the number of references in support of or contrary to a particular position to identify bias, and the context in which the research was conducted.

Although critiquing a paper was, for many, a new experience, it engendered a spirit of inquiry that enabled the participants to look at research in a more critical and nuanced way. As Nicola said "I would have looked at [the research] before, and would have said "Yes, this is gospel, let's jump in. Let's just go and do all of those things. Whereas I think now I'm just a little more cynical." For Colin, the exercise of critiquing research articles had provided a systematic way of reading. Instead of replying to a question that might be posed by one of the teachers with whom he works "from the gut," he would now immediately think that he needed to explore it further.

"There Is No Silver Bullet"

The research discussion raised the question of what constitutes evidence, particularly in a political context that emphasises the importance of "evidence-based decisions." As Nicola observed, schools are happy to "jump on the next bandwagon of the next big thing" that claims to be evidence-based, without necessarily questioning the context in which it is located or their own context. Perhaps unfairly, we termed it the "Hattification of education," a reference to the tendency to simplistically accept the effect sizes reported in Hattie's (2013) meta-analyses.

Nicola again raised the issue of context, arguing that many of the factors discussed in research such as that reported by Hattie (2013) do not relate to her context. She maintained that there are times when the wisdom of experience is more significant and more relevant than large-scale results out of context. This mirrored Matt's dislike of "the evidence narrative" which, he argued, tends to reduce scepticism and, in that sense, diminish rather than enhance the scientific endeavour.

However, Colin challenged us by suggesting that if we automatically dismiss "the Hattie stuff" that is as bad as automatically accepting it. He advocated that we ought to seek other research that confirms or challenges the conclusions in one piece of research and bring together quantitative and qualitative results. Indeed, he noted that Hattie himself had written about how his research had been over-simplified or misused, citing the research on feedback (Hattie & Timperley, 2007), which, although having a large effect size, can be either positive or negative depending on how it is offered. Matt related this to his own experience of examining feedback in the *Increasing Competence and Confidence in Algebra and Multiplicative Structures* project led by Jeremy Hodgen. He maintained that "there is nuance in everything" and that sometimes we need a "little bit more guile" to understand what is actually at the heart of the research.

Thus, there was a profound sense that each piece of research should be examined and evaluated on its merits, taking into account both the context in which it was conducted and the context in which the reader finds herself. As Matt said, "I'm not looking for a silver bullet, I'm not looking for the one intervention that is always going to work … that's what teach like a champion's like."

"I'm More Consistently Aware That I Don't Know What the Answer Is"

The unscripted and emergent discussion ranged freely over issues such as teacher resistance, context, policy, the nature of research, and our own attitude to knowledge. It highlighted the importance of the MSc in initially valuing their expertise as teachers through discussion of the nature of mathematics and how students learn mathematics. It highlighted the importance of readings in providing explanatory structures, and the value of critique in problematising knowledge. Ultimately it had perhaps raised more questions than it had answered. Part way through the discussion Colin's words were "I'm more confused than ever," but he later modified this to say, "I'm more consistently aware that I don't know what the answer is." This was not an admission of inadequacy or insecurity, but rather an acceptance and celebration of an inquiring mindset that relentlessly seeks ways of helping teachers to work more effectively in their context.

CONCLUSIONS AND IMPLICATIONS

The conversation reported above reflects, albeit without intention, many of the issues highlighted in the literature. What is particularly interesting is that this happened in a collegial, informal, unstructured setting rather than as part of a structured reflection with a more formal intent. For Berry (2007), or for the participants in Murray and Male's (2005) study, becoming a teacher educator represented a sudden disruption to established ways of thinking and a shift from knowledge that was certain or predictable to something messy, uncertain or problematic. Or in Colin's words "I'm more consistently aware that I don't know what the answer is."

For Tzur (2001) the transition involved making the invisible explicit in order to help teachers see the epistemological potential of planned teaching and learning episodes. This was not a matter of acting on externally generated research, but of taking on the role of researcher as integral to teaching. Or as Matt said, "there is no silver bullet," and we need to look at the nuances of research in each context.

Krainer's (2008) nested model encompasses the full range of research domains in mathematics education, from mathematics itself through the learning of individual students to the work of teachers, teacher educators and ultimately the role of systems. In Krainer's model the research on how teachers learn, and their beliefs, knowledge and practices (ME5) is replicated in the learning of teacher educators (ME6). What was evident in the conversation among the MSc participants was the extent to which the course and their interactions throughout the course had caused them to question their beliefs, knowledge and practice as well as the extent to which they could influence others. Like Nicola they had become "a little more cynical," but in a way which promoted them to question their practice and prior beliefs about teachers and change.

The issues of transitioning from classroom teacher to teacher educator identified in the literature address aspects of both the situational self, that is the self as generated through interactions with others and the setting, and the substantive self, that is the self formed through a core of "solid and unchanging beliefs about self" (Wood & Borg, 2010, p. 20). The impetus for Nicola, Matt and Colin to undertake the MSc was situational in that each had recently undertaken a new role, which in Colin's case also involved culturally and educationally diverse contexts.

In the conversation reported above the focus quickly moved from situational issues ("getting the buy-in," "it's just a matter of scale") to substantive issues ("that's had a big impact," "this is bona fide research"). One might, therefore, argue that the two selves are aligning, which, in the literature, is held to be the point at which the transition to the new role is complete. However, we would argue that the transition is never complete – that the role of teacher educator is always and inevitably one of becoming, characterised by an ever-growing restlessness and uncertainty. Nicola, Matt and Colin accepted and indeed celebrated that uncertainty ("there is no silver bullet," "I'm more consistently aware that I don't know what the answer is").

Such a view flies in the face of programs such as Teach First or Teach for Australia that minimise exposure to university-based study in favour of extended school placement, or indeed the very notion of being "classroom ready" (Craven et al., 2014). While we accept that there are practical aspects of teaching that can be communicated by the expert teacher and, to a greater or lesser degree, assimilated by the prospective or practising teacher, we maintain that in essence teaching is a complex blend of art, craft and science, infused with an attitude of perpetual inquiry.

Even more so, we argue that becoming a teacher educator is complex and multi-faceted. The naïve assumption that expert teachers make expert teacher educators is not borne out in practice and runs the risk of reducing teacher education to little more than a transmission of tips and tricks that perpetuate a technical view of teaching. As Murray and Male (2005) suggest, too strong a sense of professional identity as a school teacher may even restrict an individual's development as a teacher educator.

As pointed out in the literature research into how teachers transition to teacher educators is limited. In most cases this happens in situ, with little or no formal or even informal support for the new teacher educator. As a result, new teacher educators are left to fend for themselves to come to terms with a culture of academic inquiry that permeates academia in a way that is not nearly as evident in teaching itself.

The extent to which formal courses such as the MSc can facilitate the transition from teacher to teacher educator remains unresearched and unknown. However, the value of the experiences encountered by Nicola, Matt and Colin during the Oxford MSc was that they enabled them to continue to build on and problematise their professional identities as teachers into identities as teacher educators. Rather than focusing on mechanistic or pragmatic aspects of teacher education, one goal of any such course must, therefore, be to promote "an intellectual commitment to uncertainty" (Thornton, 2015, p. 216).

REFERENCES

Berliner, D. C. (2004). Describing the behavior and documenting the accomplishments of expert teachers. *Bulletin of Science, Technology and Society, 24*(3), 200–212. https://doi.org/10.1177/0270467604265535

Berry, A. (2007). *Tensions in teaching about teaching: A self-study of the development of myself as a teacher educator.* Dordrecht: Springer.

Beswick, K. (2005). The beliefs/practice connection in broadly defined contexts. *Mathematics Education Research Journal, 17*(2), 39–68.

Bullough, R. V. J. (2005). Being and becoming a mentor: School-based teacher educators and teacher educator identity. *Teaching and Teacher Education, 21*, 143–155. https://doi.org/10.1016/j.tate.2004.12.002

Childs, A., Hillier, J., Thornton, S., & Watson, A. (2014). Building capacity: developing a course for mathematics and science teacher educators. In K. Maaß, B. Barzel, G. Törner, D. Wernisch, E. Schäfer, & K. Reitz-Koncebovski (Eds.), *Educating the educators: International approaches to scaling-up professional development in mathmatics and science education* (pp. 47–55). Münster: WTM Verlag.

Clarke, D., & Hollingsworth, H. (2002). Elaborating a model of teacher professional growth. *Teaching and Teacher Education, 18*, 947–967.

Craven, G., Beswick, K., Fleming, J., Fletcher, T., Green, M., Jensen, B., … Rickards, F. (2014). *Action now: Classroom ready teachers – Report of the Teacher Education Ministerial Advisory Group (TEMAG).* Canberra.

Davey, R. (2013). *The professional identity of teacher educators: Career on the cusp?* London: Routledge.

Donovan, M. J. (2015). Aboriginal student stories, the missing voice to guide us towards change. *Australian Educational Researcher, 42*(5), 613–625. https://doi.org/10.1007/s13384-015-0182-3

Goos, M. (2009). Investigating the professional learning and development of mathematics teacher educators: A theoretical discussion and research agenda. In R. Hunter, B. Bicknell, & T. Burgess (Eds.), *Crossing divides: Proceedings of the 32nd Annual Conference of the Mathematics Education Research Group of Australasia* (Vol. 1, pp. 209–216). Palmerston North, New Zealand: MERGA.

Hattie, J. (2013). *Visible learning: A synthesis of over 800 meta-analyses relating to achievement.* London: Routledge.

Hattie, J., & Timperley, H. (2007). The power of feedback. *Review of Educational Research, 77*(1), 81–112. https://doi.org/10.3102/003465430298487

Jaworski, B., & Wood, T. (Eds.). (2008). *The international handbook of mathematics teacher education: The mathematics teacher educator as a developing professional* (Vol. 4). Rotterdam, The Netherlands: Sense Publishers.

Kaminski, J. A., Sloutsky, V. M., & Heckler, A. F. (2008). The advantage of abstract examples in learning math. *Science, 320*, 454–455.

Kemmis, S. (2012). Researching educational praxis: Spectator and participant perspectives. *British Educational Research Journal, 38*(6), 885–905.

Krainer, K. (2008). Reflecting the development of a mathematics teacher educator and his discipline. In B. Jaworski & T. Wood (Eds.), *International handbook of mathematics teacher: The mathematics teacher educator as a developing professional* (Vol. 4, pp. 177–200). Rotterdam, The Netherlands: Sense Publishers.

Loughran, J. (2013). *Developing a pedagogy of teacher education: Understanding teaching & learning about teaching.* London: Routledge.

März, V., & Kelchtermans, G. (2013). Sense-making and structure in teachers' reception of educational reform. A case study on statistics in the mathematics curriculum. *Teaching and Teacher Education, 29*, 13–24.

Murray, J., & Male, T. (2005). Becoming a teacher educator: Evidence from the field. *Teaching and Teacher Education, 21*, 125–142. https://doi.org/10.1016/j.tate.2004.12.006

Ritter, J. K. (2007). Forging a pedagogy of teacher education: The challenges of moving from classroom teacher to teacher educator. *Studying Teacher Education, 3*(1), 5–22. https://doi.org/10.1080/17425960701279776

Rowland, T., & Zazkis, R. (2013). Contingency in the mathematics classroom: Opportunities taken and opportunities missed. *Canadian Journal of Science, Mathematics and Technology Education, 13*(2), 137–153. https://doi.org/10.1080/14926156.2013.784825

Thornton, S. (2015). *Slow maths: A metaphor of connectedness for school mathematics* (Unpublished PhD thesis). Australian National University, Canberra.

Trent, J. (2013). Becoming a teacher educator: The multiple boundary-crossing experiences of beginning teacher educators. *Journal of Teacher Education, 64*(3). https://doi.org/10.1177/0022487112471998

Tzur, R. (2001). Becoming a mathematics teacher-educator: Conceptualizing the terrain through self-reflective analysis. *Journal of Mathematics Teacher Education, 4*, 259–283. https://doi.org/10.1023/A:1013314009952

Tzur, R. (2008). Profound awareness of the learning paradox: A journey towards epistemologically regulated pedagogy in mathematics teaching and teacher education. In B. Jaworski & T. Wood (Eds.), *International handbook of mathematics teacher: The mathematics teacher educator as a developing professional* (Vol. 4, pp. 137–156). Rotterdam, The Netherlands: Sense Publishers.

Williams, J., Ritter, J., & Bullock, S. M. (2012). Understanding the complexity of becoming a teacher educator: Experience, belonging, and practice within a professional learning community. *Studying Teacher Education, 8*(3), 245–260. https://doi.org/10.1080/17425964.2012.719130

Wood, D., & Borg, T. (2010). The rocky road: The journey from classroom teacher to teacher educator. *Studying Teacher Education, 6*(1), 17–28. https://doi.org/10.1080/17425961003668914

Steve Thornton
Australian Academy of Science (formerly University of Oxford)

Nicola Beaumont
Trinity School, Croydon

Matt Lewis
National Centre for Excellence in the Teaching of Mathematics

Colin Penfold
Education Development Trust

MERRILYN GOOS

3. THEORETICAL PERSPECTIVES ON LEARNING AND DEVELOPMENT AS A MATHEMATICS TEACHER EDUCATOR

This chapter updates and extend my contribution to the first International Handbook on Mathematics Teacher Education. My first edition chapter discussed sociocultural perspectives on teaching mathematics and proposed a socioculturally oriented theoretical framework – drawing on Valsiner's zone theory – for understanding the work of mathematics teacher educator-researchers. This chapter surveys and critiques current theoretical positions, both cognitive and sociocultural, used to research the learning and development of mathematics teacher educators. It then illustrates two sociocultural frameworks (zone theory; boundary practices) for researching the learning of mathematics teacher educators who were engaged in different initial teacher education projects, and consider what can be learned from these conceptualisations of learning and developing as a mathematics teacher educator.

INTRODUCTION

This chapter takes as its point of departure my contribution to the first edition of the *International Handbook of Mathematics Teacher Education* (Goos, 2008). The first edition chapter considered what can be learned from research that takes a sociocultural perspective on conceptualising "learning to teach." In that chapter I analysed selected sociocultural studies that drew on discourse, situative, and community of practice perspectives, and I then presented an alternative sociocultural approach that adapted Valsiner's (1997) zone theory of child development to investigate the learning and development of prospective and practising mathematics teachers. I concluded by speculating on how zone theory might provide a sociocultural framework for understanding the work of mathematics teacher educators. Thus, although the focus of the first edition chapter was on learning to teach mathematics, it also opened up a new space for thinking about the role of mathematics teacher educators in promoting this learning as well as examining the learning and development of mathematics teacher educators themselves.

The present chapter surveys current theories and conceptualisations of mathematics teacher educators and how they learn. However, its main purpose is to consider how sociocultural perspectives contribute to our understanding of mathematics teacher

© KONINKLIJKE BRILL NV, LEIDEN, 2020 | DOI: 10.1163/9789004424210_004

educator learning. In taking this position I adopt Lerman's (1996) definition of sociocultural approaches to learning and teaching as involving "frameworks which build on the notion that the individual's cognition originates in social interactions (Harré & Gillett, 1994) and, therefore, the role of culture, motives, values, and social and discursive practices are central, not secondary" (p. 4).

The chapter begins with an overview of cognitive and sociocultural frameworks that have so far informed research into mathematics teacher educator development. The remainder of the chapter explores the potential for two complementary sociocultural perspectives – a change perspective and a practice perspective – to stimulate further research. I illustrate the use of each perspective in my own research with mathematics teachers and mathematics teacher educators. The chapter concludes with some observations about questions and future directions arising from this analysis.

CONCEPTUALISING MATHEMATICS TEACHER EDUCATOR LEARNING AND DEVELOPMENT

Cognitive Frameworks: Developing Knowledge through Reflection

Ten years ago, at the time of publication of the first *Handbook*, little attention had been given to research on teacher educators in general and there were few published studies of the development of mathematics teacher educators in particular. Even (2008) suggested that disregard for the education of mathematics teacher educators, by comparison with education of mathematics teachers, mirrored earlier research in mathematics education that was more concerned with students' learning than teachers' learning. However, in the intervening period researchers in mathematics education have begun to investigate questions about the nature and development of mathematics teacher educator expertise. A series of discussion groups convened at the Psychology of Mathematics Education (PME) conferences and the 12th International Congress of Mathematical Education (ICME-12) (Beswick, Chapman, Goos, & Zaslavsky, 2015; Beswick, Goos, & Chapman, 2014; Goos, Chapman, Brown, & Novotna, 2010, 2011, 2012), and a recent special issue of the *Journal of Mathematics Teacher Education* (Volume 21(5), published in 2018), provide evidence of growing interest in this area.

Many investigations into the nature of mathematics teacher educator expertise seek to understand the various types of *knowledge* needed for this role. This approach parallels previous research that has conceptualised mathematical knowledge for teaching as comprising combinations of, and interactions between, content knowledge and pedagogical content knowledge (Ball, Thames, & Phelps, 2008; Baumert et al., 2010; Shulman, 1987). For example, Chick and Beswick (2018) adapted Chick's (2007) framework for describing the pedagogical content knowledge needed by those who teach mathematics to school students in order to create a corresponding framework for the pedagogical content knowledge needed by mathematics teacher educators who are teaching prospective teachers of mathematics. Similarly, Zazkis and Mamolo (2018) were interested in the affordances of advanced mathematical content knowledge held by mathematics teacher educators, which enabled them to work with prospective teachers in flexible and contingent ways and extend their thinking towards the "mathematical horizon" (cf. Ball et al., 2008). These examples illustrate the common assumption underlying much research into mathematics teacher educator knowledge that this is a form of meta-knowledge encompassing at least some of the knowledge types needed by school teachers of mathematics (Beswick et al., 2015).

The concept of meta-knowledge defining mathematics teacher educator expertise naturally leads to the question of how such knowledge is acquired. Just as the knowledge needed by mathematics students, mathematics teachers, and mathematics teacher educators is seen to be organised in overlapping layers, each of which extends and operates on the knowledge held at the previous layer, so too is the process of becoming knowledgeable understood through recursive relationships between these layers of knowledge. For example, Zaslavsky and Leikin (2004) developed a three-layered hierarchical model of learning, where each successive layer contains the knowledge of mathematics learners, mathematics teachers, and mathematics teacher educators respectively. There is also space for a fourth layer representing the knowledge of educators of mathematics teacher educators. Theoretical approaches found in early studies of mathematics teacher educator development were largely based on constructivist views of teaching and learning, in particular, the notion of *reflective practice* as a means of establishing relationships between activity and consequences to explain how human beings advance their thinking. A recursive and reflective flavour permeates these accounts of mathematics teacher educator learning, such as the self-studies of mathematics teacher educator developmental trajectories provided by Krainer (2008) and Tzur (2001), as well as meta-studies where mathematics teacher educators analysed their own learning as part of a larger teacher professional development project (e.g., Diezmann, Fox, de Vries, Siemon, & Norris, 2007; Even, 2008). This approach continues to inform more recent studies, such as that described by Masingila, Olanoff, and Kimani (2018) in which two novice mathematics teacher educators reflected on their practice while being mentored by a more experienced colleague. Mathematics teacher educators are also well positioned to learn from reflection on their research with teachers, even though this learning is often left unacknowledged and unarticulated (Jaworski, 2001).

Reflective practice is claimed to lead to greater awareness of the personal theories motivating one's practice. However, because sociocultural theories of learning take into account the settings in which practice develops, the latter perspective may have more to offer to researchers who wish to study the complexity of social practices and situations that engender learning in teacher educators.

Sociocultural Frameworks: Participating in Practices That Shape Identities

Sociocultural perspectives on learning grew from the work of Vygotsky (1978) in the early 20th century. One of Vygotsky's key claims concerns the social origins of higher mental functions, and in order to explain how a child's interaction with an adult or more capable peer stimulates mental functions that are yet to mature he introduced the concept of the zone of proximal development (ZPD). A further claim inherent in Vygotsky's theoretical approach arises from his advocacy for a genetic or developmental method: in other words, to understand mental phenomena it is necessary to study the process of growth and change rather than the product of development. Research that followed this *change* perspective increasingly focused on the relationship between individuals and their environment, encompassing cultural practices, institutional contexts, and the role of personal histories, beliefs and values in shaping individual identities.

An extension to Vygotsky's original conceptualisation of the zone of proximal development was formulated by Valsiner (1997), who proposed a theory of child development that introduced two additional "zones" – the zone of free movement (ZFM) and the zone of promoted action (ZPA) – to incorporate the social setting and the goals and actions of participants. Valsiner viewed the zone of proximal development as a set of possibilities for development that are in the process of becoming realised as individuals negotiate their relationship with the learning environment and the people in it. The zone of free movement structures an individual's access to different areas of the environment, the availability of different objects within an accessible area, and the ways the individual is permitted or enabled to act with accessible objects in accessible areas. The zone of promoted action comprises activities, objects, or areas in the environment in respect of which the individual's actions are promoted. The zone of free movement and zone of promoted action are dynamic and interrelated, forming a *zone of free movement/zone of promoted action* complex that is constantly being re-organised through interaction between people in the learning environment. It is this *zone of free movement/zone of promoted action* complex that steers development along a set of possible pathways, although individuals can still exercise agency in changing their environment or relationship with people in it in order to achieve their emerging goals. Valsiner's work exemplifies the *change* perspective within sociocultural research, and his theoretical ideas have been taken up by researchers in mathematics education to study the learning of both school students and teachers (e.g., Blanton, Westbrook, & Carter, 2005; Bansilal, 2011; Geiger, Anderson, & Hurrell, 2017; Goos, 2013; Hussain, Monaghan, & Threlfall, 2013).

In addition to influencing the *change* perspective within sociocultural research, Vygotsky was one of several theorists whose work underpinned development of a *practice* perspective, such as the concept of situated learning in a community of practice (Lave & Wenger, 1991; Wenger, 1998). Although this concept arose from studying informal learning in apprenticeship and other out-of-school contexts, community of practice models have been effectively applied in mathematics education research focused on school classrooms and teacher professional learning (e.g., Gómez, 2002; Goos & Bennison, 2008; Graven, 2004). Recent research on the learning and development of mathematics teacher educators has begun to invoke the community of practice concept, for example, in the study of Masingila et al. (2018) in which novice and experienced participants reflected on their practice together as well as individually.

The purpose of this chapter is to consider how the two lines of sociocultural inquiry identified above, the *change* perspective based on Valsiner's (1997) zone theory and the *practice* perspective informed by Wenger's (1998) ideas about communities of practice, can make a theoretical contribution to understanding the nature and development of mathematics teacher educator expertise. In doing so it builds on my earlier attempts to develop sociocultural frameworks that explain opportunities to learn in mathematics education (Goos, 2008, 2014). The next sections provide further details of the change and practice perspectives and how their use in research on mathematics teachers' learning has opened a window into investigating the professional work and development of mathematics teacher educators. Thus, each theoretical perspective is extended into new research domains that conceptualise mathematics teacher educators as learners. Zone theory is proposed as a framework for studying mathematics teacher educator learning as identity development, and a community of practice perspective is suggested as a means of examining mathematics teacher educator learning through boundary encounters between communities of mathematics educators and mathematicians who work in initial teacher education programs.

A CHANGE PERSPECTIVE ON MATHEMATICS TEACHER AND TEACHER EDUCATOR LEARNING AND DEVELOPMENT: VALSINER'S ZONE THEORY

Two different approaches to zone theory are evident in the mathematics education research literature, one of which defines the zones from the perspective of the student-as-learner and the other from the perspective of the teacher-as-learner.

Zone Theory Approach #1: Focus on Student-as-Learner

Teachers make decisions about what to promote and what to allow in the classroom and these decisions establish a zone of free movement/zone of promoted action complex that shapes the learning pathways available to students. This approach was taken by Blanton et al. (2005), who compared the zone of free movement/zone of promoted action complexes organised by three mathematics and science teachers in their respective classrooms as a means of revealing these teachers' understanding of student-centred inquiry. They found that two of the teachers created the appearance of promoting discussion and reasoning when their teaching actions did not allow students these experiences. Approach #1 is thus useful for explaining apparent contradictions between the types of learning that teachers claim to promote and the learning environment they actually allow students to experience.

Zone Theory Approach #2: Focus on Teacher-as-Learner

Valsiner (1997) argued that zone theory is applicable to any human developmental phenomena where the environment is structurally organised, and thus it seems reasonable to extend the theory to the study of teacher learning and development in structured educational environments. Over several years I have developed a research program that applies Valsiner's (1997) zone theory to teacher learning and development, defining all zones from the perspective of the teacher as learner (Goos, 2005, 2008, 2013).

When I consider how teachers learn, I view the teacher's zone of proximal development as a set of possibilities for their development that are influenced by their knowledge and beliefs, including their disciplinary knowledge, pedagogical content knowledge, and beliefs about their discipline and how it is best taught and learned. The zone of free movement can then be interpreted as constraints within the teacher's professional context such as students' (behaviour, socio-economic background, motivation, perceived abilities), access to resources and teaching materials, curriculum and assessment requirements, and organisational structures and cultures. While the zone of free movement suggests which teaching actions are *allowed*, the zone of promoted action represents teaching approaches that might be specifically *promoted* by prospective teacher education, formal professional development activities, or informal interaction with colleagues in the school setting. For learning to occur, the zone of promoted action must engage with the individual's possibilities for development (zone of proximal development) and must promote actions that the individual believes to be feasible within a given zone of free movement. The analytical approach developed in my previous research with teachers traces an individual's identity trajectory from past to present to (hypothesised) future (Goos, 2013).

In previous studies, I have found Approach #2 helpful for analysing alignments and tensions between teachers' knowledge and beliefs, their professional contexts, and the professional learning opportunities available to them in order to understand why they might embrace or reject teaching approaches promoted by teacher educators (Goos, 2005, 2013). One part of this research program involved investigating factors that influence how beginning teachers who have graduated from a technology-rich pre-service program integrate digital technologies into their practice. Table 3.1 maps onto each of Valsiner's zones a range of factors known to influence teachers' use of technology in mathematics classrooms.

Case study of a beginning teacher. To illustrate the analysis of mathematics teacher learning as identity development and the role of the mathematics teacher educator in

Table 3.1. Relationship of Valsiner's zones to factors influencing mathematics teachers' use of digital technologies

Valsiner's zones	*Factors influencing teachers' use of digital technologies*
Zone of proximal development (Possibilities for developing new teacher knowledge, beliefs, goals, practices)	Mathematical knowledge Pedagogical content knowledge (technology integration) Skill/experience in working with technology General pedagogical beliefs
Zone of free movement (Environmental constraints structuring teachers' access to particular areas or resources or ways of acting with resources)	Access to resources (time, hardware, software, teaching materials) Technical support Curriculum and assessment requirements Students (perceived abilities, motivation, behaviour) Organisational structures and cultures
Zone of promoted action (Activities, objects, or areas in the environment in respect of which teachers' actions are promoted)	Pre-service education (university program) Practicum and beginning teaching experience Formal professional development activities Informal professional learning with colleagues

this development, consider the case of Adam, a beginning teacher who participated in the research referred to above (more fully discussed in Goos, 2008). Adam completed his practice teaching sessions at a school that had government funding to resource all classrooms in the mathematics building with computers, Internet access, data projectors, graphics calculators and data loggers. New mathematics syllabuses additionally mandated the use of computers or graphics calculators in teaching and assessment programs. In terms of zone theory, this environment offered an expansive zone of free movement enabling integration of digital technologies into mathematics teaching. Adam's supervising teacher also encouraged him to use any form of technology that was available for promoting students' mathematics learning. The practicum environment therefore organised a zone of free movement/zone of promoted action complex that both promoted and permitted technology integration: this situation is represented by the large overlap between the zone of free movement and zone of promoted action circles in Figure 3.1. The zone of free movement/zone of promoted action complex offered by the practicum school was also likely to direct Adam's development along a pathway towards new, technology-enriched pedagogical knowledge and practices, as indicated by the large overlap between Adam's zone of proximal development and the zone of free movement/zone of promoted action complex in Figure 3.1.

After graduation, Adam was employed in the same school but experienced a different set of constraints. Because not all classes could be scheduled in the well-equipped mathematics building, Adam had to teach some of his lessons in

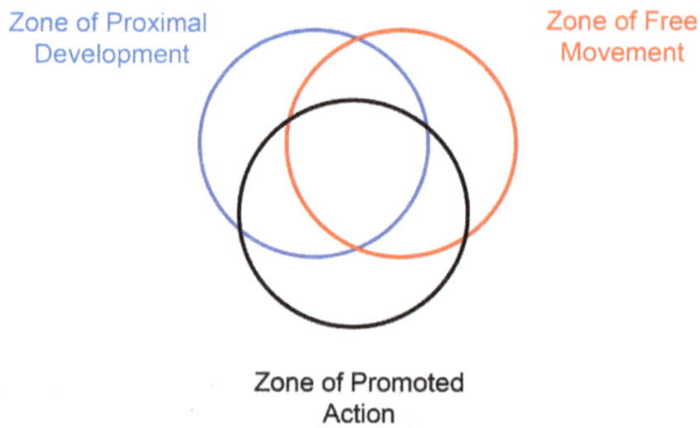

Figure 3.1. Adam's zone configuration during the practicum

other classrooms without computers, data projectors, or Internet access. Despite these difficulties he continued to integrate technology into his teaching wherever possible. The following example comes from a Year 11 lesson in which he planned to teach about the effects of the constants a, b, and c on the graph of the absolute value function $y = a|x + b| + c$. Although he clearly had specific goals in mind, the lesson was driven by the students' questions and conjectures rather than a predetermined step-by-step plan. The students first predicted what the graph of $y = |x|$ would look like, based on their knowledge of the $y = x$ function and the meaning of absolute value, and then compared their prediction with the graph produced by their graphics calculators. One student noticed that the graph involved a reflection in the y-axis and she asked how to "mirror" this effect in the x-axis. Immediately another student suggested graphing $y = -|x|$, and Adam followed this lead by encouraging the class to investigate the shape of the graph of $y = a|x|$ and propose a general statement about their findings. The lesson continued with students making similar suggestions for examining the effects of b and c. When I interviewed Adam after the lesson, he explained that he had developed a more flexible teaching approach in which the role of technology was "to help you [i.e., students] get smarter" by giving them access to different kinds of tasks that build mathematical understanding.

Nevertheless, now that Adam was a full-time staff member of the school, he discovered that many of the other mathematics teachers were sceptical about using technology. Some of these teachers accused Adam of not teaching in the "right" way. He, in turn, disagreed with their teaching approaches, which in his view betrayed negative perceptions about students:

> You do an example from a textbook, start at Question 1(a) and then off you go. And if you didn't get it – it's because you're dumb, it's not because I didn't explain it in a way that reached you.

Adam now felt that he had to defend his instructional decisions while negotiating professional relationships with other teachers, some of whom did not share his

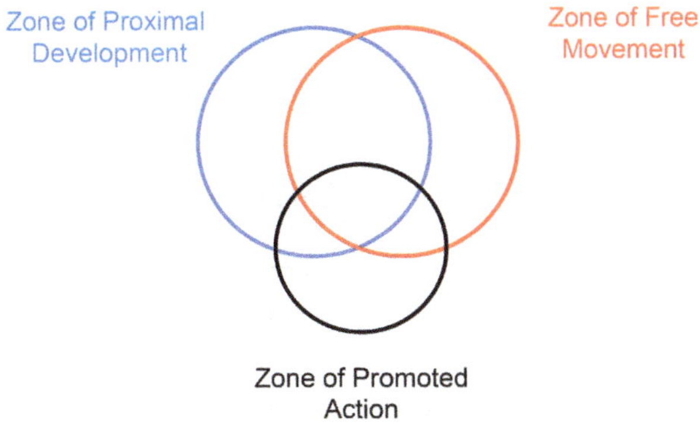

Figure 3.2. Adam's zone configuration during his first year of teaching

beliefs about teaching and learning. In these circumstances, technology-rich teaching seemed to be neither universally *permitted* (zone of free movement) nor consistently *promoted* (zone of promoted action). Nevertheless, in his first year of full-time teaching Adam continued to expand his teaching repertoire with digital technologies, often preferring to work with graphics calculators as portable tools that could be used in any classroom. He said that he saw technology as a means of giving students access to tasks that build mathematical understanding, and in this he claimed to have been influenced by my university pre-service course and the teacher who had been his practicum supervisor.

A zone theory analysis of Adam's learning would argue that he was an active agent in the development of his own professional identity, in two distinctive ways. First, he interpreted his technology-rich zone of free movement as affording his preferred teaching approach, despite subtle hindrances in the distribution of technology resources throughout the school. He also decided to pay attention only to those aspects of the mathematics department's zone of promoted action that were consistent with teaching approaches promoted by the university pre-service course. Figure 3.2 represents his situation in the year following graduation as still offering a generous zone of free movement, but with a restricted zone of promoted action that no longer overlaps so strongly with the zone of free movement because the teaching approaches Adam observed did not take full advantage of the school's technology-rich environment.

In his second year of teaching Adam was transferred to a school where there was even more limited access to computer laboratories and only one class set of graphics calculators. None of the mathematics teachers were interested in using technology, and they preferred the same kind of teacher-centred, textbook-oriented teaching approaches as some of his colleagues in his previous school. The situation in his new school is represented in Figure 3.3 by a more restricted zone of free movement and a similarly small zone of promoted action, which nevertheless overlaps considerably with the zone of free movement because the technology-free teaching actions promoted are aligned with the almost technology-free professional environment.

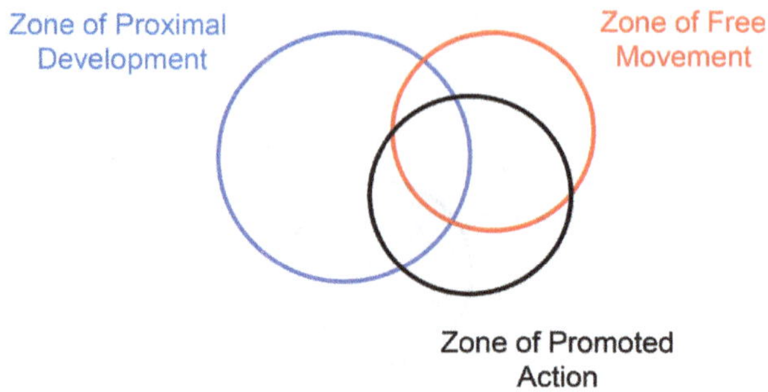

Figure 3.3. Adam's zone configuration during his second year of teaching

Zone Theory Approach #3: Focus on Teacher-Educator-as-Learner

The case study of Adam identified the zone of promoted actions provided by the school environments in which he taught – but it should also prompt questions about how it is possible for a mathematics teacher educator to intervene in these situations and offer an alternative zone of promoted action. I saw my mathematics teacher educator role as influencing Adam's interpretation of the zone of free movement/ zone of promoted action complex to maintain his sense of personal agency. For example, in his second year of teaching I encouraged him to view the single class set of graphics calculators as an opportunity he could exploit, simply because he was the only teacher who wanted to use them. I also supported him in increasing his involvement in the local mathematics teacher professional association where I hoped he would find another zone of promoted action, external to the school, that would nurture his evolving teacher identity and potential for further development. Although all of these actions provided me with opportunities for learning about my practice as a mathematics teacher educator, I now want to propose a more formal approach to studying mathematics teacher educator learning by extending the zone framework outlined above. So far, I have shown that the framework can be applied in two connected layers:

(1) when the *student is the learner*: the zone of proximal development represents possibilities for the student's development, and the teacher creates classroom zone of free movement/zone of promoted action complexes that steer the student's learning; and

(2) when the *teacher is the learner*: the zone of proximal development represents possibilities for the teacher's development, and the zone of free movement/zone of promoted action complexes that steer the teacher's learning are created by a range of factors within the teacher's professional environment. At this second layer a teacher educator may come into the picture by promoting certain teaching actions, that is, by offering a zone of promoted action for the teacher-as-learner.

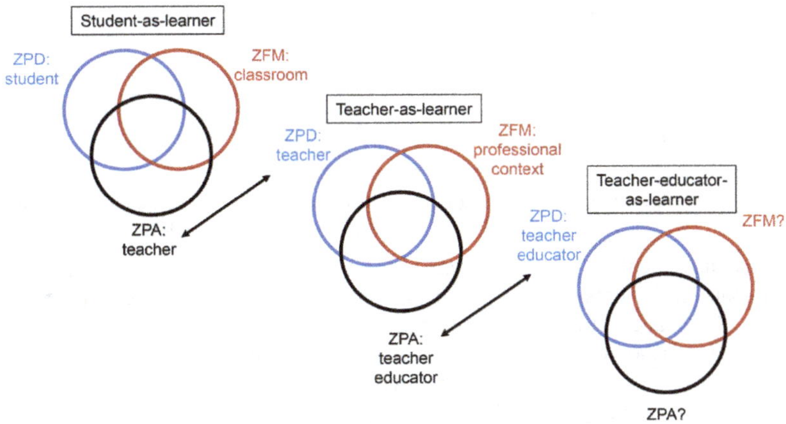

Figure 3.4. Three layers of application of zone theory

Now let us imagine a third layer, in which the *teacher educator is the learner* and the zone of proximal development represents possibilities for teacher educator development. I have represented the three layers in Figure 3.4 to show distinctions between the zone of proximal development, zone of free movement, and zone of promoted action at each layer. The arrows connecting the layers via zones of promoted action and zones of proximal development are meant to indicate that those who teach – including those who teach teachers – are also learners. This model has recursive properties similar to those describing mathematics teacher educator knowledge as containing and operating on successive layers of knowledge held by mathematics teachers and students.

Taking a zone theory perspective on mathematics teacher educator development gives rise to a new set of research questions, for example:

- How do mathematics teacher educators' professional contexts structure their interactions with prospective and practising teachers (zone of free movement)?
- What activities and areas of the professional environment do mathematics teacher educators access that promote certain approaches to educating teachers (zone of promoted action)?
- How do the zone of free movement/zone of promoted action complexes thus created shape the learning trajectories (zones of proximal development) of mathematics teacher educators, and how do mathematics teacher educators negotiate these pathways for development throughout their careers?

If the zone of proximal development represents possibilities for development as a mathematics teacher educator then we need to consider mathematics teacher educator

Table 3.2. Relationship of Valsiner's zones to influences on mathematics teacher educator (MTE) learning and development

Valsiner's zones	*Factors influencing MTE learning and development*
Zone of proximal development (Possibilities for developing new MTE knowledge, beliefs, goals, practices)	Mathematical knowledge Pedagogical content knowledge Knowledge of how new teaching practices are learned Beliefs about mathematics, teaching and learning
Zone of free movement (Environmental constraints structuring MTEs' access to particular areas or resources or ways of acting with resources)	Characteristics of teacher education students Structural characteristics of teacher education programs Organisational structures University cultures
Zone of promoted action (Activities, objects, or areas in the environment in respect of which MTEs' actions are promoted)	Reflection on practice Research with teachers Professional development Informal interactions with colleagues

knowledge and beliefs as key influences on their developing expertise (e.g., Chick & Beswick, 2017; Zazkis & Mamolo, 2018). For mathematics teacher educators, environmental constraints represented by the zone of free movement have received little research attention, but my own experience as a mathematics teacher educator suggests these might include perceptions of the knowledge and motivation of teacher education students; the structure of teacher education programs (e.g., extent of connection between courses on mathematics, general pedagogy, and mathematics teaching methods); and university organisational structures and cultures that influence timetabling, allocation of resources, and norms of what counts as "good teaching." For mathematics teacher educators, the zone of promoted action could represent teacher education approaches promoted via reflection on their practice, their research with teachers, participation in formal professional development or informal interaction with colleagues (Jaworski, 2001; Krainer, 2008; Masingila et al., 2018; Tzur, 2001; Zaslavsky & Leikin, 2004). Table 3.2 maps each of these factors influencing mathematics teacher educator learning and development onto Valsiner's zones (see Table 3.1, which represents influences on mathematics teacher learning and development in relation to technology integration).

The following sections provide brief case studies to illustrate how this extension of Valsiner's zone theory can help us understand mathematics teacher educator learning through participation in research and practice – both of which are potential elements of their zone of promoted action.

Case study of mathematics teacher educators learning through research. Mathematics teacher educators from seven Australian universities participated in a two-year project that aimed to provide a research base for improving university-based mathematics teacher education (see Callingham et al., 2011). One of the assumptions of our research team was that developing pedagogical content knowledge is central to teacher education courses, even though it was accepted that this concept is difficult to define and measure. To make it feasible to collect data from prospective teachers in all the participating universities it was decided to develop an online survey, using item formats that could be machine scored. Items were created by the project team specifically for this purpose, together with other questions collecting demographic information about respondents. We began the first project meeting by sharing information about the initial teacher education degree programs at our respective universities, as a means of introducing the task of formulating appropriate demographic items for the survey. Although all team members were known to each other and had worked together on previous research projects, we were surprised to discover the great diversity in program design both within and between universities. In other words, there was considerable unanticipated variation and complexity in the zones of free movement experienced by mathematics teacher educators team members in terms of the structural characteristics of teacher education programs, as well as student characteristics. These variations were eventually captured in our survey by seeking information such as mode of study (full-time versus part-time; on-campus vs distance vs some combination of these), type of program (1-year or 2-year postgraduate, 4-year undergraduate, 4-year or 5-year dual degree), and specialisation with respect to level of schooling or subject (early childhood, primary, middle years, secondary school; mathematics only or in combination with other subjects in secondary school teaching). Our discussions arising from this discovery caused us to wonder about the extent to which program diversity was beneficial or detrimental to the overall quality of initial teacher education in Australian universities.

The main task of the early project meetings involved designing survey items to measure prospective teachers' pedagogical content knowledge. After reviewing relevant research literature, we decided to design items to address the following aspects of mathematics teachers' pedagogical content knowledge:

a. Identifying errors and student thinking;
b. Affordances of stimulus materials;
c. Different representations of mathematical concepts;
d. Explaining mathematical ideas.

Two pedagogical content knowledge items are shown in Figures 3.5 and 3.6. Figure 3.5 illustrates pedagogical content knowledge aspects (a), (b) and (c), while Figure 3.6 illustrates aspects (c) and (d).

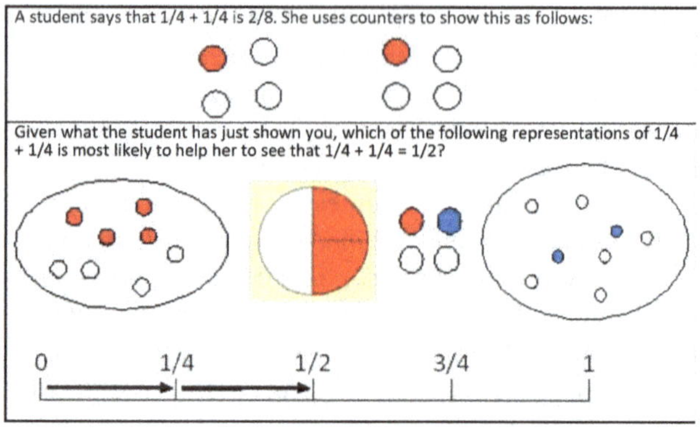

Figure 3.5. Sample pedagogical content knowledge item for primary school teachers

A teacher wants to highlight the role of the slope parameter, m, in the equation y = mx + c and its effect on the slope of the graph of the straight line. She decides to do this by using a graphing calculator to plot a sequence of linear functions, but she needs to choose a good set of functions.

For each of the following sets of functions indicate if you think it is a good choice of sequence, a poor choice of sequence, or somewhere in between.

	Poor choice	*Tolerable choice*	*Good choice*
$y = 3x - 2$, $y = 7x + 2$, $y = -3x + 4$, $y = 5x - 1$, $y = 1/5x + 4$			
$y = x - 2$, $y = 7x - 2$, $y = 1/4x - 2$, $y = -2x - 2$, $y = -1/5x - 2$			
$y = 3x + 1$, $y = 2x + 1$, $y = 1/4x + 1$, $y = 1/5x + 1$, $y = 10x + 1$			
$y = x - 1$, $y = -x - 1$, $y = 3x - 1$, $y = -3x - 1$, $y = 1/3x - 1$, $y = -1/3x - 1$			

Figure 3.6. Sample pedagogical content knowledge item for secondary school teachers

The research team of mathematics teacher educators had lengthy debates and sometimes heated arguments about what aspects of pedagogical content knowledge to incorporate into survey items, what kind of choices to include as possible answers, and which answers were "better" than others. When presenting workshops on the project for fellow mathematics teacher educators we used the tasks shown in Figures 3.5 and 3.6 to illustrate these dilemmas that we had faced in order to stimulate discussion about how mathematics teacher educators learn. These discussions not only advanced our own understanding of pedagogical content knowledge, but they also made us question the different emphases we gave to aspects of pedagogical content knowledge in our respective teacher education courses. One of the members of the research team wrote a conference paper about the dilemmas we faced in designing the surveys and the sense of exhilaration we felt from the rare experience of having conversations about our work as mathematics teacher educators (Chick, 2011). This project not only provided a glimpse of what mathematics teacher educator learning might look like, but also highlighted ways in which our research with prospective teachers intersected with structural features of our professional environments to create zone of free movement/zone of promoted action complexes that directed our learning along diverse pathways.

Case study of mathematics teacher educators learning through practice. Mathematics teacher educators from six Australian universities participated in a two-year project that aimed to improve the quality of initial mathematics teacher education by encouraging collaboration between Faculties and Schools of science, mathematics and education on course design and delivery. In this project, mathematics teacher educators were considered to include all those who taught mathematics and/or mathematics pedagogy to prospective teachers. Thus, mathematics teacher educators could be located in mathematics or education departments within institutions involved in teacher education. In each university participating in the project, interdisciplinary mathematics teacher educator teams comprising mathematicians and mathematics educators developed and implemented approaches targeting recruitment and retention strategies that promote teaching careers to undergraduate mathematics and science students, innovative curriculum arrangements that combine content and pedagogy, and continuing professional learning that builds long-term relationships with teacher education graduates. While the main purpose of the project was to create new teacher education resources and methods, it also included a research component to investigate the processes of interdisciplinary collaboration and mathematics teacher educator learning. Interviews with mathematics teacher educators provided the data for this investigation, and the transcripts were analysed using the zone theory framework shown in Table 3.2. A brief account of one case study, involving a mathematician (Leonard) and a mathematics educator (Joanne) working together, illustrates their learning during the project.

Leonard was an applied mathematician working in the School of Mathematics and Statistics. He held a PhD in physics and also a Diploma in Education (an initial teacher education qualification for secondary school teachers) that he had completed several years previously to better understand how teachers were prepared. Joanne was an experienced teacher of secondary school mathematics with a PhD in mathematics education, who worked in the School of Education in a different Faculty from Leonard. Leonard taught large undergraduate mathematics classes with students from different degree programs, including secondary teacher education, while Joanne taught mathematics pedagogy courses to future secondary school teachers: thus, both mathematics teacher educators had well-developed mathematics content knowledge and pedagogical knowledge appropriate to their teaching assignments. While they had met each other in a variety of professional contexts before the project began, both reported barriers to collaboration in the form of institutional structures and cultures that defined their zones of free movement. For example, the structure of the initial teacher education programs at this university created disciplinary "silos" that separated mathematics content (taught in the School of Mathematics and Statistics) from mathematics pedagogy (taught in the School of Education). The project represented a zone of promoted action providing funding, time and resources to support interdisciplinary collaboration. However, not even this generous support was enough to overcome institutional constraints such as Faculty budgets and academic workload policies that discouraged collaboration between different disciplines and – according to Leonard and Joanne – made it impossible to create courses that combined mathematics content and mathematics pedagogy as had been done in other universities participating in the project.

Accepting what they could not alter about their institutional context (zone of free movement), Leonard and Joanne set about making small changes to build a sense of community and collective identity as future mathematics teachers amongst their students. To do so they developed three initiatives. The first involved social networking events that brought together beginning students who were studying mathematics, but not yet any education courses, in the first year of their degree, and later years students who were studying mathematics pedagogy and had been on school placements. Leonard led the second initiative, which rearranged tutorials in his large first year mathematics course so that they were timetabled and streamed to allocate all future teachers to the same tutorial class, thus helping them to identify with peers who were aspiring to a teaching career. The final initiative, organised by Joanne, was an annual mathematics education alumni conference that connected her current students with recent graduates, experienced secondary school mathematics teachers, mathematicians, and mathematics educators. Each of these initiatives required some changes to the institutional zone of free movement and were enabled by the project zone of promoted action that promoted collaboration and provided the necessary resources. Thus, the project created a modified zone of free movement/zone of promoted action complex for Leonard and Joanne that offered possibilities for these mathematics teacher educators to develop new knowledge and practices for teacher education, thereby expanding their zones of proximal development.

A PRACTICE PERSPECTIVE ON MATHEMATICS TEACHER AND TEACHER EDUCATOR LEARNING AND DEVELOPMENT: COMMUNITIES AND BOUNDARIES

In the first edition of this *Handbook*, I discussed situative perspectives on learning to teach mathematics that drew on Lave and Wenger's (1991) concept of participation in communities of practice (Goos, 2008). I observed that the notion of community resonated with emergent ways of understanding teachers' learning through professional collaboration (e.g., Graven, 2004), and I reviewed research investigating the effects of participation in different communities on the development of a beginning teacher (Bohl & Van Zoest, 2003). However, Lave and Wenger (1991) developed the concept of community of practice from studying informal learning in apprenticeship contexts where teaching is incidental rather than deliberate, as in teacher education, and Wenger (1998) provided no discussion of teaching when he later developed the concept into a social theory of learning. At the time, then, I concluded that Valsiner's (1997) zone theory might offer a more promising pathway towards understanding the process of learning to teach mathematics. Since then, further development of my sociocultural research in mathematics teacher education and new theoretical ideas about learning at the boundary between communities have led me back to a practice perspective on mathematics teacher and teacher educator learning.

Learning to Teach in a Community of Practice

Wenger (1998) used community of practice as a point of entry into a broader conceptual framework in which learning was conceived as participating "in the *practices* of social communities and constructing *identities* in relation to these communities" (p. 4, original emphasis). These ideas proved useful to me in researching the professional socialisation of beginning teachers. One of my research questions asked how communities of practice are formed in a prospective mathematics teacher education program and sustained after graduation and entry into the profession (Goos & Bennison, 2008). This research was prompted, in part, by the unanticipated ways in which my prospective teacher education students used the course bulletin board as an online space for professional discussions during and after their university program.

A significant aspect of the study was an examination of the assumption that a "virtual" community of practice will create opportunities for teachers to learn. In teacher education research, this is a premise that is not always tested to discover whether such a community really exists or what it is actually achieving. Wenger's (1998) three dimensions of practice – mutual engagement, joint enterprise, and shared repertoire – were used to analyse more than 1500 messages posted to the Yahoo Groups bulletin boards over almost two years in order to characterise the activities of the community and trace its emergent structure.

My analysis showed that the Yahoo Groups bulletin board created emergent, rather than pre-determined, opportunities for these prospective teachers to learn in mathematics education, in keeping with Wenger's (1998) perspective on learning as an informal and tacit process. It was much more difficult to specify my own role as a mathematics teacher educator who deliberately set out to ensure that successful learning occurs. Following Wenger's argument that a community cannot be fully designed, my intention was to avoid entering into online discussions unless the prospective and beginning teachers deliberately directed a question to me. However, closer examination of bulletin board messages revealed that my role served two purposes: to ensure the effective organisation of the course and to model and encourage professional dialogue.

In this sense I participated in the community as an "old timer" who introduced newcomers to professional norms and knowledge. Although this experience helped me to think more deeply about my practice as a mathematics teacher educator, it did not provide me with theoretical tools to understand this learning. A question that arose from this study was about the communities of professional practice in which I participated as a mathematics teacher educator, and what mechanisms for learning were created by my participation. The experience of working with mathematicians on teaching-related projects then led me to wonder about opportunities for mathematics teacher educators to learn across disciplinary boundaries in mathematics education.

Learning at the Boundary between Communities of Practice

Wenger (1998) describes the three defining characteristics of communities of practice as mutual engagement of participants, negotiation of a joint enterprise, and development of a shared repertoire of resources for creating meaning. Because communities of practice evolve over time, they also have mechanisms for maintenance and inclusion of new members. Based on this description, one can accept that mathematicians, mathematics teachers, and mathematics teacher educator-researchers would claim membership of distinct, but related, communities of professional practice. Although communities of practice have "insiders" and "outsiders,"they are not completely isolated from other practices or from the rest of the world. There are various ways in which communities may be connected across the boundaries that define them.

Wenger (1998) writes of boundary encounters as distinct events that give people a sense of how meaning is negotiated within another practice. Some encounters are short-lived and transient, involving only a one-on-one conversation between individuals from two communities to help advance the boundary relationship. For example, a mathematics teacher educator might telephone a mathematics teacher who is supervising the practicum experience of one of her prospective teachers to discuss problems that the prospective teacher is encountering at the school. A more enriching instance of the boundary encounter involves immersion in another practice through a site visit. For example, a mathematician might visit a school to speak to students and teachers about careers in mathematics. However, both of these cases involve only one-way connections between different practices. A two-way connection can be established when delegations comprising several participants from each community are involved in an encounter. An example of this kind of boundary encounter can be seen when an education system is preparing a new school curriculum and organises consultation sessions that bring together delegations of school teachers, university mathematicians, and mathematics teacher educators to provide feedback on the draft curriculum document. Wenger suggests that if "a boundary encounter – especially of the delegation variety – becomes established and provides an ongoing forum for mutual engagement, then a practice is likely to start emerging" (p. 114). Such boundary practices then become a longer-term way of connecting communities in order to coordinate perspectives and resolve problems.

There is an emerging body of research on learning mechanisms involved in interdisciplinary work on shared problems. This type of work is becoming more important because of increasing specialisation within domains of expertise that requires people to collaborate across boundaries between disciplines and institutions. Akkerman and Bakker's (2011) review of this research literature emphasised that boundaries are markers of "sociocultural difference leading to discontinuity in action or interaction" (p. 133). Boundaries are thus dynamic constructs that can shape new practices through revealing and legitimating difference, translating between different world views, and confronting shared problems. As a consequence, boundaries carry potential for learning.

From their synthesis of the research literature, Akkerman and Bakker (2011) identified four potential mechanisms for learning at the boundaries between domains. The first is *identification*, which occurs when people find themselves participating in multiple overlapping communities and, as a result, the distinctiveness of established practices is challenged or threatened. Identification processes reconstruct and reinforce the boundaries between practices by delineating more clearly how the practices differ: discontinuities are not necessarily overcome. A second learning mechanism involves *coordination* of practices or perspectives via dialogue in order to translate experiences and understanding between two worlds. The aim is to overcome the boundary by facilitating a smooth movement between communities or sites. *Reflection* is the third learning mechanism, often evident in studies involving an intervention of some kind. Boundary crossing – moving between different sites – can promote reflection on differences between practices, thus enriching one's ways of looking at the world. The fourth learning mechanism is *transformation*, which, like reflection, is found in studies investigating effects of an intervention. Akkerman and Bakker stated that transformation is a learning mechanism that can lead to a profound change in practice, "potentially even the creation of a new, in-between practice, sometimes called a boundary practice" (p. 146). They suggested that transformation involved the following processes:

- Confrontation – encountering a discontinuity that forces reconsideration of current practices;
- Recognising a shared problem space – in response to the confrontation;
- Hybridisation – combining practices from different contexts;
- Crystallisation – developing new routines that become embedded in practices;
- Maintaining the uniqueness of intersecting practices – so that fusion of practices does not fully dissolve the boundary;
- Continuous joint work at the boundary – necessary for negotiation of meaning in the context of institutional structures that work against collaboration and boundary crossing.

Akkerman and Bakker (2011) noted that, although transformation is rare and difficult to achieve, it carries promise of sustainable impact. They also proposed that identification and reflection, both of which involve recognising and explicating different perspectives, are necessary pre-conditions for transformation to occur.

Case study of mathematics teacher educators learning across disciplinary boundaries. A research project described earlier in this chapter aimed to foster collaboration between mathematics educators and mathematicians for the purpose of designing new approaches to initial mathematics teacher education. One element of the project involved investigating the processes of interdisciplinary collaboration. Previously I explained how I applied a change perspective, using Valsiner's (1997) zone theory, to study the learning of mathematics teacher educators in this project. But I can also employ a practice perspective to understand how mathematics teacher educators from different disciplinary communities developed new boundary practices through their collaboration. Two rounds of interviews were conducted with participating mathematics teacher educators (mathematicians and mathematics educators) near the beginning and the end of this two-year project to probe their experiences of collaboration. One question asked about the types of exchanges and activities that were considered most successful in bringing together mathematicians and mathematics educators. Interviews were transcribed and scrutinised for evidence of the mechanisms for learning at the boundaries between domains theorised by Akkerman and Bakker (2011).

The following brief narrative presents a hybrid case constructed from all the interviews. The purpose is to illustrate what transformation can look like as a mechanism for learning at the boundary between disciplines. (Quotes have been selected from interviews. Names are pseudonyms. This material is drawn from Goos & Bennison, 2018.)

A mathematician (Carol) was working with a mathematics educator (Tess). Before the project began, they got to know each other via an externally funded teaching and learning project. Carol was then allocated to the teaching of a first-year mathematics course for initial teacher education students. She was surprised by students' apparent lack of mathematical knowledge after having completed 12 years of schooling:

> I was lamenting, "Oh my goodness me, I can't believe they don't know any maths,"like they know less that I had anticipated for someone who had come through the Australian schooling system. [Carol, mathematician]

This experience represented a *confrontation*, a kind of discontinuity between the two worlds of school mathematics and university mathematics that prompted Carol to reconsider her current practice as a teacher of university mathematics. Recognising this confrontation led both mathematics teacher educators to explore each other's worlds:

> I learned a lot about how education works and Tess learned a lot about how we function. We broke down some of the scepticism that both sides can have. [Carol, mathematician]

Carol discussed her observations with Tess, who was sympathetic and interested in exploring the differences between teaching mathematics and education in a university environment. Tess remembered "noticing that my pre-service teachers, their content knowledge was not strong,"and she pointed out to Carol the areas that she wanted her to focus on in the first-year mathematics course. Carol acknowledged that "I was teaching her [Tess's] students at the time,"and both thus recognised a *shared problem space* in which both were contributing to the mathematical preparation of future teachers.

Given this problem space, Carol and Tess were working towards a *hybridisation* of practices from their respective disciplinary contexts. The hybrid result was a new mathematics content course that was jointly planned and taught, as Tess explained:

> We're in the class together, one of us leads and the other acts as a sort of sounding board. We planned the weeks so certain weeks are Carol's weeks and certain weeks are my weeks. [Tess, mathematics educator]

There were encouraging signs that this new hybrid practice would become *crystallised,* or embedded, into institutional structures. The teacher education program was under review, and the Heads of Mathematics and Education invited Carol and Tess to design two new mathematics-specific pedagogy courses for the revised program. It was intended that the courses would be "owned" by Education, with an income sharing arrangement to recognise the teaching contribution from Mathematics.

Despite the success in creating a new hybrid practice, Carol and Tess also *maintained the uniqueness of their established practices* as a mathematician and mathematics educator. Carol acknowledged their complementary expertise when teaching the mathematics course together:

> We go to class and there are times when she says to me "That's all yours because it's beyond what I understand" and that's fine. Likewise, she'll come in and talk about the greats of education and I'm just going blank, no idea. As an educator it comes out very strongly that she's very well practised. [Carol, mathematician]

The collaboration was sustained by *continuous joint work at the boundary* between the two practices. This included weekly project meetings, attending and teaching into each other's tutorials in mathematics and mathematics education subjects, joint supervision of Honours students, and jointly conducted professional development for practising teachers.

CONCLUSION

Since the publication of the first *International Handbook of Mathematics Teacher Education* there has been growing interest in exploring the nature of mathematics teacher educator expertise as well as processes supporting mathematics teacher educators' professional formation and development. This chapter makes a theoretical contribution to the field by providing an overview of current conceptualisations of mathematics teacher educator learning and development and comparing cognitive and sociocultural frameworks that differ in their assumptions and emphases. However, the main emphasis of the chapter is on sociocultural perspectives and what aspects of mathematics teacher educator learning they can illuminate. Starting from my own practice as a mathematics teacher educator, I traced out two lines of inquiry – one based on a *change* perspective informed by Valsiner's (1997) zone theory and the other on a *practice* perspective informed by Wenger's (1998) social theory of learning within and between communities of practice. In each case I scrutinised my own actions as a mathematics teacher educator working with prospective and beginning teachers, seeking both to understand my role and to theorise my own and other mathematics teacher educators' learning. As a result, I pushed both perspectives into new territory and demonstrated their potential applicability to studying mathematics teacher educator development.

Yet these emergent analyses have given rise to even more questions that could stimulate future research. Beginning with the change perspective, we seem to know very little about the structure of mathematics teacher educators' zones of free movement and how these enable or hinder innovative teacher education practices and mathematics teacher educators' professional growth. For example, what is the effect of external regulation of initial teacher education? How do these contextual factors differ across nations and cultures? There is also scope for systematically mapping out variations in the zones of promoted action available to mathematics teacher educators, beyond those created through reflection on one's own practice or research with teachers. What formal professional development opportunities are available to mathematics teacher educators in different countries? What assumptions about the mathematics teacher educator role and preparation for this role underpin such opportunities? What are the different pathways into becoming a mathematics teacher educator, and how and why do these vary across cultures? Turning to the practice perspective, we can further extend the notion of "learning at the boundary" to explore additional aspects of mathematics teacher educators' professional work. For example, how can we understand learning at the boundary between university and school placement in initial teacher education programs? How can prospective teachers cross the epistemological boundaries between mathematics content knowledge, mathematics pedagogical content knowledge, and general educational theory – all typically taught in initial teacher education programs but with little attention to integrating these forms of knowledge so they usefully inform professional practice? These and other questions might lay a foundation for future sociocultural research into the learning and development of mathematics teacher educators.

ACKNOWLEDGEMENT

Sections of this chapter draw on previously published work (Goos, 2014; Goos & Bennison, 2018).

REFERENCES

Akkerman, S., & Bakker, A. (2011). Boundary crossing and boundary objects. *Review of Educational Research, 81*, 132–169.

Ball, D. L., Thames, M. H., & Phelps, G. (2008). Content knowledge for teaching: What makes it so special? *Journal of Teacher Education, 59*(5), 389–407.

Bansilal, S. (2011). Assessment reform in South Africa: Opening up or closing spaces for teachers? *Educational Studies in Mathematics, 78*, 91–107.

Baumert, J., Kunter, M., Blum, W., Brunner, M., Voss, T., Jordan, A., … Tsai, Y.-M. (2010). Teachers' mathematical knowledge, cognitive activation in the classroom, and student progress. *American Educational Research Journal, 47*(1), 133–180.

Beswick, K., Chapman, O., Goos, M., & Zaslavsky, O. (2015). Mathematics teacher educators' knowledge for teaching. In S. J. Cho (Ed.), *The proceedings of the 12th international congress on mathematical education: Intellectual and attitudinal challenges* (pp. 629–632). Dordrecht: Springer Science+Business Media.

Beswick, K., Goos, M., & Chapman, O. (2014). Mathematics teacher educators' knowledge (Working session 4). In S. Oesterle, C. Nichols, P. Liljedahl, & D. Allan (Eds.), *Proceedings of the joint meeting of PME 38 and PME-NA 36* (Vol. 1, p. 254). Vancouver, Canada: PME.

Blanton, M., Westbrook, S., & Carter, G. (2005). Using Valsiner's zone theory to interpret teaching practices in mathematics and science classrooms. *Journal of Mathematics Teacher Education, 8*, 5–33.

Bohl, J., & Van Zoest, L. (2003). The value of Wenger's concepts of modes of participation and regimes of accountability in understanding teacher learning. In N. A. Pateman, B. J. Dougherty, & J. T. Zilliox (Eds.), *Proceedings of the 27th Conference of the International Group for the Psychology of Mathematics Education* (Vol. 4, pp. 339–346). Honolulu, HI: PME.

Callingham, R., Beswick, K., Chick, H., Clark, J., Goos, M., Kissane, B., Serow, P., Thornton, S., & Tobias, S. (2011). Beginning teachers' mathematical knowledge: What is needed? In J. Clark, B. Kissane, J. Mousley, T. Spencer, & S. Thornton (Eds.), *Mathematics: Traditions and (new) practices* (Proceedings of the 23rd biennial conference of the Australian Association of Mathematics Teachers and the 34th annual Conference of the Mathematics Education Research Group of Australasia, pp. 900–907). Adelaide, Australia: AAMT & MERGA.

Chick, H. (2007). Teaching and learning by example. In J. Watson & K. Beswick (Eds.), *Mathematics: Essential research, essential practice* (Proceedings of the 30th Annual Conference of the Mathematics Education Research Group of Australasia, pp. 3–21). Sydney, Australia: MERGA.

Chick, H. (2011). God-like educators in a fallen world. In J. Wright (Ed.), *Proceedings of the Annual Conference of the Australian Association for Research in Education, Hobart*. Retrieved October 25, 2018, from https://www.aare.edu.au/data/publications/2011/aarefinal00667.pdf

Chick, H., & Beswick, K. (2018). Teaching teachers to teach Boris: A framework for mathematics teacher educator pedagogical content knowledge. *Journal of Mathematics Teacher Education, 21*, 475–499.

Diezmann, C., Fox, J., de Vries, E., Siemon, D., & Norris, G. (2007). Investigating the learning of a professional development team: The years 1–3 mathematics probes project. *Mathematics Teacher Education and Development, 8*, 94–116.

Even, R. (2008). Facing the challenge of educating educators to work with practising mathematics teachers. In B. Jaworski & T. Wood (Eds.), *International handbook of mathematics teacher education: The mathematics teacher educator as a developing professional* (Vol. 4, pp. 57–73). Rotterdam, The Netherlands: Sense Publishers.

Geiger, V., Anderson, J., & Hurrell, D. (2017). A case study of effective practice in mathematics teaching and learning informed by Valsiner's zone theory. *Mathematics Education Research Journal, 29*, 143–161.

Gómez, C. (2002). The struggles of a community of mathematics teachers: Developing a community of practice. In A. Cockburn & E. Nardi (Eds.), *Proceedings of the 26th Conference of the International Group for the Psychology of Mathematics Education* (Vol. 3, pp. 9–16). Norwich, UK: PME.

Goos, M. (2005). A sociocultural analysis of the development of pre-service and beginning teachers' pedagogical identities as users of technology. *Journal of Mathematics Teacher Education, 8*(1), 35–59.

Goos, M. (2008). Sociocultural perspectives on learning to teach mathematics. In B. Jaworski & T. Wood (Eds.), *International handbook of mathematics teacher education: The mathematics teacher educator as a developing professional* (Vol. 4, pp. 75–91). Rotterdam, The Netherlands: Sense Publishers.
Goos, M. (2013). Sociocultural perspectives in research on and with teachers: A zone theory approach. *ZDM Mathematics Education, 45*, 521–533.
Goos, M. (2014). Creating opportunities to learn in mathematics education: A sociocultural perspective. *Mathematics Education Research Journal, 26*(3), 439–457.
Goos, M., & Bennison, A. (2008). Developing a communal identity as beginning teachers of mathematics: Emergence of an online community of practice. *Journal of Mathematics Teacher Education, 11*, 41–60.
Goos, M., & Bennison, A. (2018). Boundary crossing and brokering between disciplines in pre-service mathematic teacher education. *Mathematics Education Research Journal, 30*, 255–275.
Goos, M., Chapman, O., Brown, L., & Novotna, J. (2010). The learning and development of mathematics teacher educator-researchers. Discussion group 4. In M. Pinto & T. Kawasaki (Eds.), *Proceedings of the 34th Conference of the International Group for the Psychology of Mathematics Education* (Vol. 1, p. 390). Belo Horizonte, Brazil: PME.
Goos, M., Chapman, O., Brown, L., & Novotna, J. (2011). The learning and development of mathematics teacher educator-researchers (Working session 5). In B. Ubuz (Ed.), *Proceedings of the 35th Conference of the International Group for the Psychology of Mathematics Education* (Vol. 1, p. 173). Ankara, Turkey: PME.
Goos, M., Chapman, O., Brown, L., & Novotna, J. (2012). The learning and development of mathematics teacher educator-researchers. Working session 2. In T. Tso (Ed.), *Proceedings of the 36th Conference of the International Group for the Psychology of Mathematics Education* (Vol. 1, p. 159). Taipei, Taiwan: PME.
Graven, M. (2004). Investigating mathematics teacher learning within an in-service community of practice: The centrality of confidence. *Educational Studies in Mathematics, 57*, 177–211.
Harré, R., & Gillett, G. (1994). *The discursive mind*. London: Sage.
Hussain, M., Monaghan, J., & Threlfall, J. (2013). Teacher-student development in mathematics classrooms: Interrelated zones of free movement and promoted action. *Educational Studies in Mathematics, 82*, 285–302.
Jaworski, B. (2001). Developing mathematics teaching: Teachers, teacher educators, and researchers as co-learners. In F.-L. Lin & T. Cooney (Eds.), *Making sense of mathematics teacher education* (pp. 295–320). Dordrecht: Kluwer Academic Publishers.
Krainer, K. (2008). Reflecting the development of a mathematics teacher educator and his discipline. In B. Jaworski & T. Wood (Eds.), *International handbook of mathematics teacher education* (Vol. 4, pp. 177–199). Rotterdam, The Netherlands: Sense Publishers.
Lave, J., & Wenger, E. (1991). *Situated learning: Legitimate peripheral participation*. New York, NY: Cambridge University Press.
Lerman, S. (1996). Socio-cultural approaches to mathematics teaching and learning. *Educational Studies in Mathematics, 31*(1–2), 1–9.
Masingila, J., Olanoff, D., & Kimani, P. (2018). Mathematical knowledge for teaching teachers: Knowledge used and developed by mathematics teacher educators in learning to teach via problem solving. *Journal of Mathematics Teacher Education, 21*, 429–450.
Shulman, L. S. (1987). Knowledge and teaching: Foundations of the new reform. *Harvard Educational Review, 57*(1), 1–22.
Tzur, R. (2001). Becoming a mathematics teacher educator: Conceptualizing the terrain through self-reflective analysis. *Journal of Mathematics Teacher Education, 4*, 259–283.
Valsiner, J. (1997). *Culture and the development of children's action: A theory of human development*. (2nd ed.). New York, NY: John Wiley & Sons.
Vygotsky, L. (1978). *Mind in society*. Cambridge, MA: Harvard University Press.
Wenger, E. (1998). *Communities of practice: Learning, meaning and identity*. Cambridge: Cambridge University Press.

Zaslavsky, O., & Leikin, R. (2004). Professional development of mathematics teacher educators: Growth through practice. *Journal of Mathematics Teacher Education, 7*, 5–32.
Zazkis, R., & Mamolo, A. (2018). From disturbance to task design, or a story of a rectangular lake. *Journal of Mathematics Teacher Education, 21*, 501–516.

Merrilyn Goos
School of Education
University of Limerick
and
School of Education
The University of Queensland

PART 2

MATHEMATICS TEACHER EDUCATORS LEARNING IN TRANSITIONS AND THROUGH COLLABORATIONS

LAURINDA BROWN AND ALF COLES

4. THEORISING THEORISING

About Mathematics Teachers' and Mathematics Teacher Educators' Energetic Learning

In this chapter, we present and illustrate our theorising of how mathematics teachers can become engaged and motivated by their own learning (theorising) about how to teach mathematics. In contrast to learning that can feel hard (for example memorisation tasks, or tasks where the learner receives consistent negative feedback), learning to teach mathematics can be experienced as something we characterise as 'energetic' when teachers themselves are involved in the process of theorising specific to their own mathematics classrooms. Our theorising, therefore, is about how to support mathematics teachers' theorising that is good enough for their own purposes. We illustrate our own theorising as mathematics teacher educators through narrative interviewing of each other. Drawing on evidence from a mathematics teachers' online discussion forum (as part of a Master's unit) we illustrate the power of teachers' theorising as part of the learning process.

INTRODUCTION

What is learning? How is learning done? There is a developing literature within mathematics education that explores these questions through an enactivist perspective. This literature covers students doing mathematics in classrooms, teachers learning about teaching mathematics to students and mathematics teacher educators' learning. In previous articles and chapters, we have discussed ideas of reflecting and developing expertise that come from an enactivist worldview in which knowing and doing are equivalent. What seems to be implicit in much of this writing is how making a new connection related to action, in a way that we call theorising, brings with it energy, what we call energetic learning. In this chapter, we explore energetic learning, including, as mathematics teacher educators, our own theorising on how to support mathematics teachers' theorising on how to teach mathematics. We illustrate our own theorising through extracts from our professional discussions after sessions whilst working together on the one-year teacher education course for prospective teachers at the University of Bristol, School of Education. Petitmengin's (2006) second-person interviewing strategies can be used as teaching strategies to support energetic learning at all levels of the system.

© KONINKLIJKE BRILL NV, LEIDEN, 2020 | DOI: 10.1163/9789004424210_005

There is a description (Brown, 1991) of Laurinda teaching mathematics that predates any engagement with enactivist ideas. We begin this chapter with extracts from that description. Later in the chapter, we will show how, over time, our language for discussing this description develops, showing theorising happening.

Initial Ways of Describing Teaching Mathematics

I walk to the board and start to draw a pair of axes. I would normally always draw four quadrants but am short of room. I plot the points (0, 0), (0, 2), (1, 5), (2, 2), (5, 2), (5, 0) and join them up (see Figure 4.1).

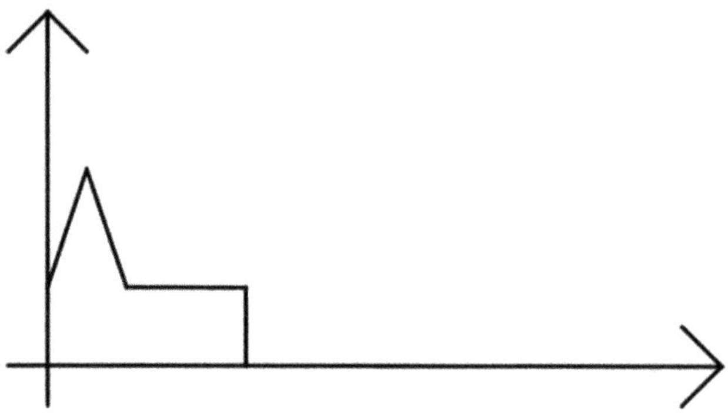

Figure 4.1. A church

Belatedly, I label the axes x and y and fill in the numbers on the axes. This has all been done quite slowly. "What does that look like?," I say, looking around expectantly. Some hands go up, and I relax.

"A church."

"Good! That's what I hoped it would look like. I'm going to make the church change shape. The process is quite strange, but you might have met something like it before."

I go through the rather strange process of multiplying two matrices, for each of the points, for instance:

$$\begin{pmatrix} 2 & 1 \\ 0 & 1 \end{pmatrix}\begin{pmatrix} 1 \\ 5 \end{pmatrix} = \begin{pmatrix} 2\times 1+1\times 5 \\ 0\times 1+1\times 5 \end{pmatrix} = \begin{pmatrix} 7 \\ 5 \end{pmatrix}$$

plotting the new points (see Figure 4.2).

I offered the students a challenge, "In five lessons' time, I will come in, write any old matrix on the board and expect you to be able to tell me what effect it would have on a shape without actually drawing it out." There were excited intakes of breath around the room: they felt challenged and were going to engage. I felt relief.

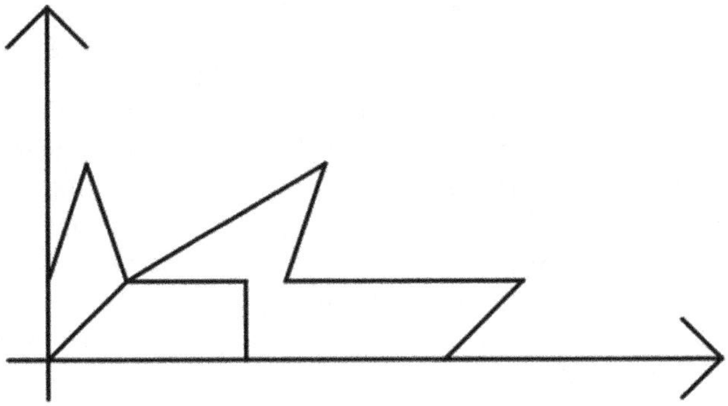

Figure 4.2. The new church

I also set up a first task, explaining:

> For a beginning, and to check that each of you can do a basic transformation, I'd like you each to take a piece of squared paper, choose your own shape and matrix, and see what happens. When you have finished, we can put them on the display board over there or on the ceiling. Then you will have a collection of a lot of information from which to work.

In 1995, as described in the Springer *Encyclopedia of Mathematics Education*, Laurinda Brown met David Reid and Alf Coles (Goodchild, 2014, pp. 209–213), becoming immersed in enactivist theories that led to her being able to become articulate about the strategies for teaching that seemed to work in her mathematics classroom and when teaching prospective teachers of mathematics. In this chapter, from an enactivist perspective, we (Laurinda and Alf) aim to explore how learning, primarily about how to teach mathematics and how to work with teachers of mathematics as mathematics teacher educators, can become energetic. The "theorising theorising" of the title will become apparent in the links we make between energetic learning and theorising of practice and has links to the work of Chapman (2008) on teacher educators researching their practice. We firstly elaborate on what we mean by 'learning' and 'energetic learning' and place our work within other contributions in the area. We will then review the literature relating to mathematics teacher education from an enactivist perspective and review some of our own previous work on developing expertise and reflecting that support the learning of teachers of mathematics. This will lead to a focus on our own energetic learning as mathematics teacher educators. We will, finally, consider similarities and differences between the energetic learning of students doing mathematics (such as in the story above), teachers of mathematics and of mathematics teacher educators.

WHAT IS LEARNING?

We begin by focusing on what we mean by learning. There is such a vast literature on learning that we focus here on how learning is viewed within enactivist writing; in the following section, where we focus on energetic learning, we will review the literature more broadly. There is a developing literature within mathematics education that explores these questions through an enactivist perspective or methodology (Reid, 1996). For instance, in the International Journal on Mathematics Education, ZDM Special Issue focused on *Enactivist methodology in mathematics education research* (2015), the articles covered research on students doing mathematics in classrooms (Abrahamson & Trninic 2015; Metz & Simmt 2015; Khan, Francis & Davis 2015; Maheux & Proulx 2015; Towers & Martin 2015; Lozano 2015; Steinbring 2015), teachers learning about teaching mathematics to students (Coles 2015; Preciado-Babb, Metz & Marcotte 2015), and mathematics teacher educators' learning (Brown 2015; Metz & Simmt 2015; Towers & Martin 2015; Simmt & Kieran 2015). This classification is from Laurinda's perspective as observer on what the main focus is, although in two cases, Metz and Simmt; and Towers and Martin, there seemed to be two main foci. In the Springer *Encyclopedia of Mathematics Education*, the entry on *Enactivist theories* (Goodchild, 2014, pp. 209–213) is distinct from that of *Embodied cognition* (Sriraman & Ke Wu, 2014, pp. 207–209) in that "[e]nactivism has biological roots [...] whereas embodied mathematics has linguistic roots" (pp. 209–210). The biological perspective is central in this chapter since we accept that our bodies learn to survive in the world through being active. As Clark (1997) says, "Minds make motions and they must make them fast" (p. 1). We "immediately cope" (Varela, 1999, p. 5) in the moment with what happens to us, bringing our patterned behaviours from the history of all our lived experience into play.

The contemporary theory of enactivism has been labelled a paradigm-shifting perspective on learning (Li, Clark & Winchester, 2010). The key authors in this area are Maturana and Varela, although it was Varela who used the term enactivist (Varela, Thompson & Rosch, 1991, p. 9; Varela, 1999, p. 12). They wrote, "All knowing is doing, and all doing is knowing" (Maturana & Varela, 1987, p. 26). We accept this equivalence of doing and knowing and take it to mean that, as we have interacted with the world, we know what to do in familiar situations and do not need to think about what to do. When we are in such moments of immediate coping our behaviours are "effective" for the purposes of maintaining that smooth functioning and relationship with the world.

Another key idea is that of the observer (Maturana, 1980; Simmt & Kieren, 2015). We inter-act with our environment, but not in the sense of representing that environment internally. Enactivism entails a radical rejection of a representationalist view of mind (as do some forms of social constructivism, see Cobb, Yackel & Wood, 1992). The environment triggers an organism and, from within the structure of the organism, a response is generated that entails a perception of that trigger. Rather than view action as resulting from processing of a passive receipt of perceptual information, enactivism suggests that organisms *select* perceptions in an active way. We are literally "bringing forth a world" (see Simmt & Kieren, 2015, p. 310, for a use of this phrase related to mathematics knowing) as we act/know. When Maturana states that "everything said is said by an observer" (Maturana, 1988, p. 9), he is pointing towards the awareness that the 'world' we have each brought forth will be different from one another. There is a world that we are all interacting with but we each make different distinctions that have come out of our personal histories of living. It could be said we live in a 'multi-verse' not universe. Any observation we make entails having made a distinction, that goes un-marked, about what kind of a thing is able to be noticed. An amoeba, from our view as an observer, is able to detect nutrients. Its membrane allows it to categorise food/not food, but it seems blind to many other dimensions. The amoeba, since it cannot tell us, may, in fact, be making distinctions that we cannot perceive.

In a mathematics classroom, experienced teachers may have trained themselves to notice elements of classroom discourse and, in so doing, will also be blind to other elements of interaction. Using a familiar technique to solve a mathematics task does not need much effort. As problems arise, however, such as prospective teachers not knowing what to do in their mathematics classroom, there is still a need to act. A problem in mathematics classrooms, or in learning to teach mathematics, can be that the immediate coping, the fast motions and doings in which students or prospective teachers engage, may not support their longer-term aims. If prospective teachers are concerned about making mathematical mistakes in front of their class, for example, their immediate coping may in part be focused on not being asked a question rather than engaging with the content of any mathematics on offer. Prospective teachers typically are overwhelmed, when they begin teaching, by trying to keep in mind school rules and systems, their own lesson plan, new student's names, targets from a previous lesson, and more. At the start of a one-year secondary Postgraduate Certificate in Education (PGCE) course in England, prospective teachers' automatic behaviours are rarely effective in the classroom. By the end of the year, if successful, their automatic behaviours support the learning of their students. How do the teachers and students learn? How do they adapt and change?

From our enactivist perspective, learning and change are defined in terms of perception and action. Perception is viewed as an active process (*i.e.*, involving action) of meaning making. Varela (1999) describes perception as "perceptually guided action" (p. 12). Learning and change are characterised by a difference in such perception/action. In other words, I recognise change (in other organisms or in myself) through noticing a change in behavior, over time, in response to the same stimulus.

WHAT MAKES LEARNING ENERGETIC?

As university teacher educators, we are interested in finding strategies to support prospective teachers in learning to change their practices. A key idea for us derives from Rosch's "basic level of categorization" (Varela, Thompson & Rosch, 1991, p. 177), "the point at which cognition and environment become simultaneously enacted" (ibid., p. 177). Basic level categorisations describe patterned ways of seeing and are linked to particular actions, in the moment, of perception. The word 'chair,' for most people, is a basic-level category linked to the action "sitting-on," but categories are not fixed and, for example, for a furniture remover, 'chair' might be part of a larger basic-level category of "objects-to-be-moved-to-and-from-houses." Categories at the basic level are the most abstract ones where we typically perform the same actions on everything within that category. Following Rosch's work, if you are sitting on a chair, you are not likely to call it furniture (a more abstract, or, superordinate category), nor call it a more precise name, which would identify just this particular chair (a more detailed, or, subordinate category) (adapted from Varela, 1999, p. 16). Words at the basic-level are tied strongly to classes of actions, these are the categories with which we make sense of the world and which allow our smooth functioning. The world we bring forth, when we are operating in a patterned way, requires no reflection on action since our perception/action is adequate to the context. This is the enactivist view of effective behavior; behavior that is good-enough (Zack & Reid, 2004) for a given purpose.

There may be a range of possible effective behaviours, in any given context, of which, one happens in the moment. We are not concerned, most of the time, with needing to act in the best way possible but rather with "immediate coping" (Varela, 1999, p. 5) and maintaining relationships to what is around us (both the animate and inanimate). At moments of breakdown, when something goes wrong, there is an opportunity to re-look at the basic-level categorisation we had been employing which, in that moment, proved ineffective (in the sense that what happened was not expected). In order to expand possibilities for action next time (a characteristic of learning) there is a need to open our habitual way of categorising the world up to question. We see this process as requiring a dwelling in the detail, via forcing our observations into the detail layer. From this detail there is the possibility that new labels can emerge (at the basic-level) which, if they prove effective in a context, can start to accrue new sets of actions. In other words, from an enactivist perspective, learning entails a change in the ways we categorise the world that we bring forth through perception/action. Although changes in basic-level categories may influence the superordinate layer, we tend not to focus attention there, since a corollary of Rosch's categorisation is that there is no direct link between the superordinate layer and action, and we see learning as evidenced primarily by action. When a new basic-level category is found, its expression is often energetic.

As an example of this energetic learning, in our own work, we have described a moment of learning for Alf, near the start of our work together (Brown & Coles, 1996). Driving in a car, Laurinda asked Alf to reflect on moments in his classroom over the last academic year when he came closest to the kind of teacher he wanted to be. This provoked two anecdotes (at Rosch's detail layer) and an energetic exclamation

"It's silence, isn't it, it's silence." At that moment, Laurinda had no idea what Alf meant, but recognised the articulation of a label that could potentially become a new basic-level category that would support him in developing his teaching. Alf was referring to a common feature of the two anecdotes he had related, which was his own deliberate silence. Laurinda's offer to focus on moments in a lesson rather than whole lessons (Alf had not been able to point to any effective lessons in their entirety) occasioned a suspension of how Alf had come to view the previous year of his own teaching and, we would interpret, through a present attention to the two stories that emerged, the creation of a new label that could start to accrue new teaching behaviours.

There are few other views of what it means for learning to become energetic, in relation to mathematics education. While there are many distinctions made about the outcomes of students' learning (for example, whether the resulting understanding is "relational" or "instrumental" (Skemp, 1976), or the widely used approaches to learning being "deep" compared to "surface" (Marton & Säljö, 1976; Marton & Booth, 1997)) there has been little work investigating qualities of the learning experience itself within mathematics education. One significant study in this area (Schmidt et al., 1996) characterised "pedagogical flow" (p. 105), drawing on a concept developed by Csikszentmihalyi (1990). For Csikszentmihalyi, flow refers to optimal experience, in contexts of high challenge and high skills. Flow moments are those moments when there is a loss of self-consciousness and a complete focus on the task at hand. These moments are often described as energetic. The notion of pedagogical flow aims to capture the way that, when things are running normally, experienced teachers handle experiences in lessons without "conscious analysis and deliberate reflection" (Schmidt et al., 1996, p. 119).

One of the few authors within mathematics education to develop the idea of flow is Liljedahl (2018) who conceptualised flow, for students in classrooms, as being a balance between the challenge presented to them and the skills needed to meet the challenge (see also, Brown, 2001). Too much challenge compared to skill leads to frustration and too little challenge compared to skill leads to boredom. When skill and challenge are well matched in a dynamical process that keeps the relationship from tipping into frustration or boredom, students can experience learning in a state of flow. We have found no other studies (apart from some of our own work that we discuss below) within mathematics education that look at the issue of qualities of learning.

Looking more broadly than mathematics education, the closest field to our concerns in this chapter is that of "transformative learning" that draws on the work of Mezirow (1978) and has now entered a second-wave of theorisation. Mezirow (2003) linked transformative learning to participation in critical reflection on experience, in order to reflect on actions and uncover insight. Gunnlaugson (2007) points to criticisms that this conception of transformative learning is overly rational and proposes (as part of the second-wave) the notion of "generative dialogue" to replace "critical reflection"

(p. 138). Central to generative dialogue is "suspending our thought processes when encountering moments of difference, dissonance, judgment" (p. 140), leading to a practice of "presencing" (Scharmer, 2000), which is a learning from attention to what is emerging in dialogue but is not yet fully formed. The moment where Alf makes the connection between the two stories could be interpreted as arising out of a generative dialogue. The idea of "flow" is also part of the characterisation of generative dialogue. The ideas around generative dialogue are influenced by Varela's three gestures of awareness, which are: suspending, re-directing and letting go (Varela & Scharmer, 2000). These gestures were Varela's synthesis of practices across phenomenological and meditative traditions. Gunnlaugson (2007) introduces the phrase "meta-awareness" (p. 145) for what is generated in the kinds of dialogue being promoted, allowing learners to become witnesses to their own learning rather than caught in dissonance or difference.

Drawing on the ideas about energetic learning above, we now expand on our own previous work, within a context of mathematics education enactivist studies, to offer examples of energetic learning.

EXAMPLES OF ENERGETIC LEARNING: ENACTIVIST PERSPECTIVES ON DEVELOPING EXPERTISE AND REFLECTING

In this section, energetic learning will be exemplified using a range of contexts: findings; interviewing; developing expertise; and reflecting. As you read through this chapter you will find yourself reading many 'ings,' such as the listing in the previous sentence. As Laurinda (Brown, 2009) wrote about her teaching of mathematics, "I offer in relation to the 'ing,' the process, and leave the mathematics to the students" (p. 8). Writing 'ings' places our attention in the actions, the process, in a range of contexts.

Find-ing(s)

Enactivist research into teaching strategies has "shed light onto the journeys that are travelled in the professional learning that takes place when developing one's teaching" (Brown, 2015, p. 193). In developing as a doer of mathematics, a teacher of mathematics, a teacher of teachers of mathematics or a mathematics education researcher, in the move from novice to expert, Laurinda has expressed her task as seeing 'nots' (p. 194). Metz and Simmt (2015) write about a process in which Metz was "able to observe aspects of learning that would otherwise have remained invisible" (an example of Laurinda's 'nots') prompting, "deeper mathematical understanding in both herself and the participants than likely would otherwise have emerged" (p. 208). In the staying with the detail of experience, we can open up the possibility of seeing some different aspects of worlds in the multi-verse of teachers and students in the classroom and, later, accrue actions linked to new basic-level categories.

Interviewing

For Laurinda, interviewing students as part of the research for her master's dissertation (Brown, 1992) was an early experience of energetic learning as she now describes the process. Having wanted to look at possible connections between the images of mathematics of teachers and their students, she wanted to get at what her interviewees thought mathematics was. Unfortunately, simply asking that question would not bring any rich statements from students, they did not seem particularly involved. Her interview protocol in the end involved each student choosing some mathematics to engage in with her, followed by each student being invited to tell stories of their classroom experiences. From these conversations, Laurinda would make a synthesis about what she thought they thought mathematics was at the end of the interview. This was followed by energetic statements such as, "Yes! And it is also …"; or "No, no, no! I think it is …" These statements were then her findings and were compared to what the teachers and other students said mathematics was in their interviews. The process could be interpreted as "presencing," learning from what is emerging in the dialogue for the interviewer, and sharing those new awarenesses of what this student thinks mathematics is produces a generative dialogue, where the student becomes energetically present to new articulations out of the experience of the interview.

Interest in energetic learning and interviewing techniques has led Laurinda currently, when interviewing, to the adoption of a protocol developed by Petitmengin (2006), a doctoral student of Varela. This protocol is a set of actions for the interviewer (what Metz & Simmt, 2015, call an "empathic second-person observer") to support the person being interviewed to give a first-person account of the detail of their experiences. It draws, similarly to generative dialogue, on phenomenological and meditative traditions. The first three of these actions were paraphrased from Petitmengin's paper, the fourth is related to work which we (authors) have done together:

1. Stabilising attention. Asking a question that brings the attention back to the experience, e.g., How did you do that? Reformulating what has been said. Asking for a recheck of accuracy (often in response to a digression or judgement).
2. Turning the attention from 'what' to 'how' (never 'why').
3. Moving from a general representation to a singular experience. A re-enactment, reliving the past as if it were present. Talking out of experience, not from their beliefs or judgements of what happened (can involve the interviewee moving to the present tense). Staying with the detail is important, a maximal exhaustivity of description that allows access to the implicit (adapted from Petitmengin, 2006, pp. 239–244).

A fourth action we have added (see, Brown & Coles, 2019), to support work with prospective teachers, is:

4. Getting to new category labels. After applying the above three actions, dwelling in the detail, telling stories from experience and exploring without judgement or digression, inviting the prospective teacher to make statements of what is being worked on. New category labels might be identified that can then be linked to new actions.

Petitmengin (2006) argues that it is "the question 'how' which triggers the conversion of the attention of the interviewee towards [their] pre-reflective internal processes, and permits [them the] awareness of these processes" (p. 241). So, the second-person interviewer is triggering a first-person account. 'Why' questions, however, "deflect the interviewee's attention [...] to abstract considerations, and must therefore be avoided" (p. 241). The actions in the protocol support us in our own professional learning in conversations with each other after teaching sessions, as we seek to trigger first-person awarenesses through the empathic second-person interviewer. These ideas will be developed in the next section, telling stories of our own energetic learning as university mathematics teacher educators.

Developing Expertise

What novices and experts can both do (see Brown & Coles, 2011) is stay with the detail of their experiences in the classroom, identifying times when they could have acted differently. At the end of a lesson, if a prospective teacher comments that the class were 'distracted' there is perhaps little space initially for thinking and acting differently. If, however, we need to act differently, then there needs to be a process of dwelling in the details to bring the possibility or new actions into play, often articulated with a new label ("expanding the space of the possible" (Davis, 2004, p. 184)). The teacher with the distracted class may, through such a process, articulate a discomfort that when they asked the class to begin a task, six hands went up asking for help and this was the point when students seemed to stop working. Importantly, at this stage, the second-person observer cannot know what the underlying issue might be for the prospective teacher (perhaps they see an issue in how they were managing behaviour, perhaps they became worried about a lack of their own subject knowledge). Further probing of the situation may lead to a recognition, from the prospective teacher, that there was a discomfort around not knowing whether the class could do the task. A new label might be articulated (by the prospective teacher or their tutor), "How do I know what the students know?."

This label can act as a basic-level category, in the sense of accruing new actions. In future, instead of experiencing a similar situation as "distracted students," actions linked to finding out what the students know may be able to be employed. A key difference between a novice and an expert in this process is that the expert is able to articulate the deliberate awarenesses that lay behind the actions that proved ineffective and do more of this kind of reflecting on their own.

The process of developing expertise, through dwelling in the detail of particular moments, to draw out inherent issues and then basic-level categories, is what we call theorising.

Reflecting

In learning to teach, prospective teachers have to think about how lessons have gone and what they can do differently in the future, as a process of developing expertise. They need to engage in reflecting on their teaching and theorising about what is effective for them. However, given our enactivist stance, we do not see reflecting and theorising as being about arriving anywhere fixed or certain. We distinguish theorising from such characterisations as Schön's (1983) theories-in-action, by the process being continually open to change. Argyris and Schön (1974) distinguish between two sorts of theories in action, ones that are espoused in response to questions and ones that are theories in use, what is actually done, often implicitly rather than explicitly known. Although not necessarily fixed as such, there is an implication that these implicit and explicit theories in action inform practice in some way from a fixed viewpoint. However, prospective teachers do not have such theories in use, even if they would be able to espouse their beliefs. From the interviewing protocol above, staying with the maximal detail in reflecting gives access to implicit theories, some of which do not fit with the images of teaching that the prospective teacher espouses, and opens them up to change.

So, energetic learning, through "presencing," pervades Laurinda Brown's work through her teaching of mathematics and the teaching of teachers of mathematics. In 2009, she wrote, "The grace of being in the moment, the energetic present, is spirituality to me" (p. 148). In 2009, she described working with matrices and transformations differently from in 1991 (see the start of this chapter) although she was talking about the same lesson. She describes knowing when she had found the "energetic present" (Brown, 2009, p. 153), her and the students being in flow:

> I knew what I was looking for, the energetic present. These students now had an intention, the affectivity into the future that would allow them to make decisions about what matrix to try next and why. I could relax because the feeling of this group who were new to me was the same as previous groups. I set them the first task to support them in checking out that they could do the procedure and [through displaying the results on a noticeboard] also give the students a whole range of examples from the space to compare and contrast, using their powers of discrimination. (2009, p. 153)

There are some accepted powers of humans (be they students or prospective teachers) from an enactivist perspective, which can always be drawn on, by the teacher or teacher of teachers, to support learning. Here, the students are active, making decisions, contributing to "a whole range of examples" that support them in making distinctions. They were in flow, responding to a challenge. Laurinda is noticing the energetic response ("excited intakes of breath"; "engagement"). Other teachers might not perceive those same things in the same way. If a classroom needs to be silent, the energy might not be met with approval. This was the classroom culture that Laurinda had grown to work with through her history of being a mathematics teacher. She is able to use all her past experiences of teaching matrices and transformations as she relaxes.

Theorising

Drawing together the concepts we will be using in the rest of this chapter, we see the story of Alf's energetic articulation of a new label 'silence' to be a paradigmatic example of what we mean by 'theorising' in the sense of arriving at new basic-level categories out of attention to the detail of experience, via deliberate analysis (see also Brown & Coles, 2012). We see a vital role for practices that allow a suspension of usual modes of making sense (getting away from existing basic-level categories) in order to bring present attention to the detail of experience. When this happens, we recognise that the awarenesses that arise tend to have an energetic quality. There is of course no guarantee that these new labels will become useful in the classroom. To become linked to new effective behaviours, the new category must be employed in practice. In the case we report in Brown and Coles (1996), 'silence' became a deliberate strategy for planning lessons and was significant in terms of allowing Alf access to strategies for developing expertise in the classroom, moving from being a novice to becoming an expert, that related to some kind of vision of how he wanted to be as a teacher.

ENERGETIC LEARNING AS UNIVERSITY MATHEMATICS TEACHER EDUCATORS

Having offered examples, in theory and in practice, of the energetic learning of students learning mathematics and teachers learning to teach mathematics, we now turn to the learning of university mathematics teacher educators. We see a self-similarity in the process of theorising of university tutors and prospective teachers on the Postgraduate Certificate in Education course, experts and novices, which is linked to presence, our adapting in the moment contingently to what arises. For us, in our professional discussions, often after teaching sessions, we use the second-person interviewing techniques discussed above, in conversation. One of us will be in 'flow,' being supported by the other in giving a detailed, first-person account or description of what was done. We seek meaning together in relation to this 'flow,' giving a second-person account of those experiences. Given that we (the authors) often teach together and have a developing language, we can share distinctions and descriptions in our conversations. Also, the exploration uncovers energetic distillations of new awarenesses. We are effectively moving between being in flow, supported by the other to stay with the detail of experiences and supporting as the conversation develops. We hope to provoke new learning through our interaction, recognised by an energetic tone, as opposed to the calm of speaking from commonly rehearsed statements. We aim to support each other to see 'nots,' what we could not see before.

Our roles when supporting are to ask questions, when we know that we do not understand; tell stories from our own experiences, to exemplify what we think is meant; and always ask for more detail from experience, such as asking 'how' something was done. We do not deal in theoretical discussions divorced from practice or experience. New labels relate to details of actions and are asking for implicit actions to be made explicit. We are interested in producing first-person accounts of what we 'do' as mathematics teacher educators, to probe our experiences. We do this, not to say that what we each do are the only or 'best' ways of being a mathematics teacher educator, but to raise possibilities for future action within ourselves and others, continuing to develop our practices.

The next section of this chapter reports on conversations that took place in April 2013, with each of us flexibly taking the roles of supporting the other to stay with the detail and being the one 'in flow.' The focus of the conversation was what we do/how we learn as mathematics teacher educators. We have used the data before in a short chapter for a book outside mathematics education (Brown & Coles, 2019) focusing on the process of "reciprocal narrative interviewing." However, in this chapter, we have revised the analysis to give a paradigmatic example of the theorising and energetic learning in conversation of mathematics teacher educators who work together on the same programme, highlighting aspects of mathematics, mathematics teaching and mathematics teacher education.

Reflecting Together

We did not have a recipe or a structure for how the conversation would emerge, but we wanted to explore some of the shifts and changes in Alf's practice as a relatively new mathematics teacher educator and the principles behind the setting up of the Postgraduate Certificate in Education mathematics course by Laurinda. So, whichever one of us began exploring some detail of experiences, the other would be supporting them to stay with the detail, using the protocol (Petitmengin, 2006) mentioned above of stabilising attention; asking what or how questions (never why); taking any general statements back to singular experiences; and using this process to support the gaining of new labels for basic-level categories. Put another way, we were looking to support each other in learning via theorising. In deciding what of the conversations to present, we were looking for energetic statements, in flow, recognised sometimes by pauses or slow speech before the new awareness can be articulated. We have selected two excerpts to analyse, one of each of us in the supporting role.

Laurinda Supporting Alf

Prior to the transcript below, the conversation had turned to focusing on the detail of teaching sessions with prospective teachers at University. As can be seen, the supporting person, Laurinda in this case, can still tell an experience or share an awareness triggered by the detailed experience of the other that itself can then trigger the articulation of new awarenesses.

> LB: In the summer it's as though we are given permission that we can teach if we want. [...] if they say they want something on discussion it can still be them sharing activities [...]
>
> AC: I think maybe in a classroom ... I thought about it in terms of giving everyone a common experience, which we then would talk about and discuss and draw out some mathematics from. And I guess what's different here is that they've all got these very different experiences from their schools and actually that's really important, that they have the space to discuss and draw that out in sessions.

Analysis – Getting to a new awareness (Alf). I had only recently joined the university from being an expert teacher of mathematics in school. So, I am acting as a novice mathematics teacher educstion out of my history of experience as a teacher. The conversation, where I was focusing on the detail of my teaching of teachers and given the trigger provided by Laurinda talking about the summer term in response to my contribution, leads to an awareness of a difference in my current teaching of teachers from being in my own classroom teaching mathematics. My speech noticeably slowed at the point in the conversation where I say, "I think maybe in a classroom." I then articulate a perhaps implicit description of a basic-level category held as a teacher, "I thought about it in terms of giving everyone a common experience." What then follows is said energetically. I make a comparison with what's happening with the university teaching, "I guess what's different here is that they've all got these very different experiences from their schools." These statements are linked closely to the actions of talking and drawing out mathematics and having space to draw out the differences in how schools operate in sessions. My label "common experience" for students (e.g., a short task for everyone, or a visualisation, or watching a short animated geometrical film) was a principle used in planning my lessons. Working with prospective teachers however, the situation is different. They already have the experiences (their time spent in school) from which they need to draw distinctions. In a sense, having too many new "common experiences" at university will only take them away from drawing out new awarenesses and learning about their time in school. What I get to, in the conversation, is the need instead to offer the prospective teachers a mechanism through which to analyse their own experiences. I am working here with the idea not of creating common experiences but giving prospective teachers "space to discuss." In this articulation, there is the potential for a new basic-level category to inform planning in this new (for me) context. This was a category I went on to use in planning and that informed new (effective) behaviours, an example of a session planned with this new idea in mind is below.

At the start of the summer term, prospective teachers identify, following small group discussions, the "school practice issues" they want to work on. I took the lead in the session on "differentiation." I began by asking for as many labels as possible that could fill in the blank: "differentiation by ______." The prospective teachers came up with: outcome, task, questioning, grouping, resource, product, and a few others. In groups of 5 or 6, I invited the prospective teachers to discuss the advantages and disadvantages of each method of differentiation, drawing on their own experiences (either via observation or in their own teaching).

I invited anyone to offer something they are going to try out that they have not done before, arising from their discussion. One teacher is going to set up two seating plans and give each student a shape and a letter. One plan will arrange students in mixed attainment groups and one plan in same attainment groups.

I then said I would offer them one mechanism for thinking about differentiation by outcome, which is the idea of setting up tasks that are 'self-generating.' An activity is self-generating because the teacher does not need to direct what to try next. One teacher, wanting to work on subtracting decimals, got students to choose a number with three digits, two of them behind the decimal point. Students then reverse the number and find the difference. I got the prospective teachers to work on this latter task for 15 minutes, sharing patterns and writing up conjectures, to get a sense of working with a self-generating activity. The challenge for the prospective teachers, still in their groups, was then to choose a topic area and come up with a self-generating activity. I suggested they imagine there has been time when students have gained the skill in question and, instead of reaching for a worksheet or textbook, they might do this.

Students had 10 minutes to plan and then each group had four minutes to give the activity and let the others work on it. Here are a few of the activities offered, generally presented as sparsely as below:

- choose a fraction, write the reciprocal, find the difference
- draw a square, split it into quarters and shade the bottom left quarter, what is left unshaded? … then split each unshaded quarter into quarters, and shade the bottom left section … what is left unshaded? … continue this process
- choose any 3-digit number. Multiply it by 7, then by 11, then by 13, what do you notice?

Alf Supporting Laurinda

The context of the short excerpt below was that Laurinda was discussing what she does during selection interviews for the Postgraduate Certificate in Education course.

> AC: My sense is that what you're talking about isn't necessarily captured in writing about enactivism.
>
> LB: (laughs) (60 second pause) I think what enactivism speaks to me on is that my awareness in the moment in that interview is not on what he's actually talking on [ ] what I've been asking myself in the last couple of minutes when I went silent is well what the hell am I [doing] if I'm not listening to what he's saying, which I suspect I've always thought I was but I'm not of course.

Analysis – Getting to a new awareness (Laurinda). My spontaneous laughter is an indicator that I recognised the potential for a new awareness in the question from Alf. It is energised and followed by a pause. One purpose of our conversations is to find out about connections we would not otherwise think about, to uncover our 'nots.' The long wait time following the laughter (60 seconds) we see perhaps re-living an experience. This pause was not interrupted by Alf. He recognised something in my focus of attention that indicated important work was taking place. I then articulate my own learning/awareness in the articulation of difference between what I always thought I had done and what I realise now that I actually do. My attention is on 'what' I do in interviews, provoked by Alf's sense of an incompatibility between what I had been saying and other articulations about enactivism. My implicit and unquestioned behaviours during interviews are brought to present attention and what I am actually doing is capable of being articulated.

There are two discussions that we have during interviews for the Postgraduate Certificate in Education course, one in relation to "Why teaching?" and the other in relation to "Why mathematics?" I interview in a pair, almost always, the other interviewer being either a teacher from a local school who works with us, say, a mentor, or, sometimes, Alf. I am an experienced interviewer and would have thought that I was listening to the content of what the interviewee is saying. However, what became apparent to me in the interview with Alf was that I don't! I am listening for something else. What?

When interviewing, we are looking for people whom we can work with. When things are going well, the interviewee is able to talk in detail about their experiences (*e.g.*, visiting a school) and also able to shift the level of the discussion to be about their learning. One particular question is important in our decision-making, "What have you learned about yourself from that visit to the school?." At the start of the interview, we say that if there is something that is blocking our offering a place we will feed that back and there will be another chance. No two answers are ever the same but, in the majority of cases, there is no issue, there is an answer that is good-enough for us to know this person can learn from experience, for example:

> That's a good question. (long pause) I learnt that I can't just tell them. I was talking with a girl who'd called me over because she was stuck. I told her what to do and she said she did not understand. I asked her to show me how far she'd got and this worked better. She sorted out where the problem was for herself.

> I felt uncomfortable because I didn't know what to do when one of them misbehaves. The next time it happened I decided to try to distract him back into the mathematics, "Show me how to do that one?" and it worked. I know I don't want to end up shouting.

These are answers that I have made up, fictions if you like. But they are a distillation of my experiences. The terrain across which the variations happen is large.

What about when there are issues? I know then that it is hard to know when to ask the question. The issues have arisen before the question is asked, one example of which is when asked to describe a lesson that they had observed when visiting a school, the interviewee talks in terms of judgements. There were bad teachers, shocking low achievement in the students, the students talked! We feedback that it is best not to make judgements. It is important to try to focus on what they can learn from this experience. Were all the students misbehaving? It is not until the interviewee begins to describe their experiences, rather than sharing judgements that it feels like asking the question about self is a possibility. Even so, if, "No-one's ever asked me a question like that before" followed by anger or occasionally bursting into tears from the frustration of not being able to get in touch with their learning are the responses, then no place is offered.

In some cases, the question is asked at the point where judgements are put aside along with negative emotions and I realise that I am not listening to the context of these messages, but to the process, the meta-messages. There might be the story of an individual child's learning, told with energy, linked for me with the idea of presence. Being here, now, "no memory or desire" (Bion, 1970, p. 34) and having heard this shift, I would feel able to ask the question. The bombast disappears, along with the person who arrived being who they thought we might want them to be and they answer openly, sometimes crying, and what they really fear comes out. The place is offered. I am not listening to the content of what is being said but to presence. There is a self-similarity in Alf's question that provoked the awareness/learning in me and what can potentially happen in my asking the question of the interviewee. For us both we are in what Maturana has called the praxis of living.

REFLECTING ON THEORISING: THEMES ACROSS MTE ENERGETIC LEARNING

We see, in Alf's analysis, a description of his shifting from teaching in school to teaching prospective teachers. In the story of a session on differentiation, Alf offers the prospective teachers a way of reflecting on their (different) experiences, using the notion of self-generating activities. This allows the prospective teachers to interrogate their experience in what for some of them may have been a new way. The different experiences of the prospective teachers can be shared through a common focus. We still see the echo here of Alf's previous thinking about generating 'common experiences' in a school context. As we learn, we do not leave behind who we were and the ways we acted. However, what Alf generates now is a common focus, not to reflect on a common experience, but to help reflect on experiences in school. The example of Alf's learning is offered to illustrate how finding the time and space to engage in theorising (starting with dwelling in the detail of events) can support the development of a mathematics teacher educator.

Laurinda's analysis articulates ideas around the question of what does she listen to in interviews, if not the content of what is being said. What Laurinda gets to, is that she is listening for 'presence,' or another way of saying this might be that Laurinda has attuned herself to noticing moments or possibilities for energetic learning – and she is listening for this in interviews. If someone is able to get to a moment of energetic learning in an interview, we will be able to work with them during the Postgraduate Certificate in Education year. Again, this example is offered, not to suggest others should use the same interview process, but to point to the process of theorising.

Our theorising of theorising concerns what happens when we do not know what to do, or when we set out, as in the narrative interviews above, to learn, to see or act differently in the future. The theorising technique of staying with the detail of our experiences for a while, in flow, prompted through inter-action with another or others in conversation, allows the space in this moment, now, for us to become aware of implicit actions to see or act differently in the future. Theorising involves extending our range of possible behaviours, not always acting in this different way, but allowing the awareness of that behaviour and consequences to emerge, mapping out the domain of variation as part of the praxis of living.

CONCLUSION

In this chapter we have set out an enactivist perspective on what is learning, linking learning to changes in effective behaviours, prompted by a breakdown in our (more typical) smooth functioning in the world. Looking across the mathematics education literature, there has been little attention given to the question of different qualities of learning. Our concern is with learning that we label 'energetic' since our belief is that when learning becomes energetic then new behaviours can quickly accrue, making the journey towards effective action quick. We link what it is that makes learning energetic to Rosch's basic-level categories. Our experience as teachers and mathematics teacher educators points to a pattern, that learning becomes energetic when we are forced out of our typical functioning, our normal way of relating to the world, focus on the detail of our experience and are able to get to a new category (at the basic-level) that has the potential, then, to become a focus for attention and accrue new and effective behaviours (of course there is no guarantee that these new behaviours will be effective and there may be a need for further dwelling in the detail to arrive at new labels). The process of exploring experience at the point of a breakdown in effective functioning is a characteristic of expert behaviour. What we have set out in this chapter is a series of mechanisms by which novices can learn in the same way as experts, in a process we call theorising (relevant to becoming an expert mathematician, teacher of mathematics, or mathematics teacher educator).

We have pointed to the way that the process of energetic learning, seen as the development of new basic-level categories, can operate for students learning mathematics, teachers learning to teach mathematics and mathematics teacher educators learning to work with teachers of mathematics. As mathematics teacher educators ourselves, the techniques of narrative interviewing inform our work with teachers (for example, with prospective teachers). Both for ourselves, and in our work with teachers, what seems important is that there is a movement between dwelling in the detail (of experiences, events, readings) and then considering the issue at a more general level, in order to arrive at new categories for description, linked to the possibility of new ways of acting. The processes we have described are supported by an 'other' who is able to attend to what is being said, notice judgements (in order to dissipate them) and recognise the kind of theorising that is characteristic of energetic learning. We suggest that such processes would potentially be applicable in any teaching and learning context, where there is a second-person observer.

The process of reflecting on experience, exemplified in this chapter, we have described as theorising. This is the process in which we engage prospective teachers of mathematics. In this chapter we have been theorising about theorising, dwelling in the detail of our experiences of theorising and then drawing out commonalities and new labels (including the label 'theorising'). There are significant questions that follow on from our writing. For example, how might mathematics teacher educators learn to recognise energetic learning? Are there any patterns linked to when a new label or category does or does not lead to new behaviours in practice?

When learning is energetic, our experience suggests it is meaningful and learners become committed and engaged in the process. We believe this is true across the system, whether for students learning mathematics, prospective teachers learning to teach mathematics, and mathematics teacher educators learning to work with teachers. For example, it is possible to go back to the classroom example of mathematics teaching from the start of this chapter and see that the students, in being offered the new space of matrices and transformation to explore, are bringing forth a world through actions, trying out matrices and seeing what they do. Comparing and contrasting their examples with those of others they are energetic as they create new basic-level categories of transformations and patterns. The teacher is engaged and asks what and how questions, being a second-person empathic observer, supporting students to become articulate and aware of their first-person accounts. The interview strategies that we described can be used to support teachers' learning, in flow; and, in turn, their energetic learning can occasion the energetic learning of us, in conversation, as mathematics teacher educators as we theorise about their theorising.

REFERENCES

Abrahamson, D., & Trninic, D. (2015). Bringing forth mathematical concepts: Signifying sensorimotor enactment in fields of promoted action. *ZDM Mathematics Education, 47*(2), 295–306.

Argyris, C., & Schön, D. (1974). *Theory in practice: Increasing professional effectiveness.* San Francisco, CA: Jossey-Bass.

Bion, W. (1970). *Attention and interpretation.* London: Rowman & Littlefield Publishers, 2004.

Brown, L. (1991). Stewing in your own juice. In D. Pimm & E. Love (Eds.), *Teaching and learning school mathematics* (pp. 3–15). London: Hodder and Stoughton.

Brown, L. (1992). The influence of teachers on children's image of mathematics. *For the Learning of Mathematics, 12*(2), 29–33.

Brown, L. (2001). The story of mathematics. *International Journal of Educational Policy, Research and Practice, 2*(2), 187–208.

Brown, L. (2009). Spirituality and student-generated examples: Shaping teaching to make space of learning mathematics. In S. Lerman & B. Davis (Eds.), *Mathematical action & structures of noticing: Studies on John Mason's contribution to mathematics education* (pp. 147–160). Rotterdam, The Netherlands: Sense Publishers.

Brown, L. (2015). Researching as an enactivist mathematics education researcher. *ZDM Mathematics Education, 47*(2), 185–196.

Brown, L., & Coles, A. (1996). The story of silence. In L. Puig & A. Gutierrez (Eds.), *Proceedings of the 20th Conference of the International Group for the Psychology of Mathematics Education (PME20)* (pp. 145–152). Valencia: University of Valencia.

Brown, L., & Coles, A. (2011). Developing expertise: How enactivism re-frames mathematics teacher development. *ZDM Mathematics Education, 43*, 861–873.

Brown, L., & Coles, A. (2012). Using "deliberate analysis" for learning mathematics and for mathematics teacher education: How the enactive approach to cognition frames reflection. *Educational Studies in Mathematics, 80*(1–2), 217–231.

Brown, L., & Coles, A. (2019). Reciprocal narrative interviewing. In C. Comanducci & A. Wilkinson (Eds.), *Matters of telling: The impulse of the story* (pp. 177–184). Leiden, The Netherlands: Brill Rodopi.

Chapman, O. (2008). Mathematics teacher educators' learning from research on their instructional practices. In B. Jaworski & T. Wood (Eds.), *The international handbook of mathematics teacher education: The mathematics teacher educator as a developing professional: The mathematics teacher educator as a developing professional* (Vol. 4, pp. 115–136). Rotterdam, The Netherlands: Sense Publishers.

Clark, A. (1997). *Being there: Putting brain, body, and world together again.* Cambridge, MA & London: The MIT Press.

Cobb, P., Yackel, E., & Wood, T. (1992). A constructivist alternative to the representational view of mind in mathematics education. *Journal for Research in Mathematics Education, 23*, 2–33.

Coles, A. (2015). On enactivism and language: Towards a methodology for studying talk in mathematics classrooms. *ZDM Mathematics Education, 47*(2), 235–246.

Csikszentmihalyi, M. (1990). *Flow: The psychology of optimal experience.* New York, NY: Harper and Row.

Davis, B. (2004). *Inventions of teaching: A genealogy.* New York, NY: Lawrence Erlbaum Associates.

Goodchild, S. (2014). Enactivist theories. In S. Lerman (Ed.), *Encyclopedia of mathematics education* (pp. 209–213). Dordrecht, The Netherlands: Springer.

Gunnlaugson, O. (2007). Shedding light on the underlying forms of transformative learning theory: Introducing three distinct categories of consciousness. *Journal of Transformative Education, 5*(2), 134–151.

Khan, S., Francis, K., & Davis, B. (2015). Accumulation of experience in a vast number of cases: Enactivism as a fit framework for the study of spatial reasoning. *ZDM Mathematics Education, 47*(2), 269–279.

Li, Q., Clark, B., & Winchester, I. (2010). Instructional design and technology with enactivism: A shift of paradigm? *British Journal of Educational Technology, 41*(3), 403–419.
Liljedahl, P. (2018). On the edges of flow: Student problem solving behavior. In S. Carreira, N. Amado, & K. Jones (Eds.), *Broadening the scope of research on mathematical problem solving: A focus on technology, creativity and affect* (pp. 505–524). New York, NY: Springer.
Lozano, M.-D. (2015). Using enactivism as a methodology to characterize algebraic learning. *ZDM Mathematics Education, 47*(2), 223–234.
Maheux, J.-F., & Proulx, J. (2015). Doing mathematics: Analysing data with/in an enactivist-inspired approach. *ZDM Mathematics Education, 47*(2), 211–221.
Marton, F., & Booth, S. (1997). *Learning and awareness.* Mahwah, NJ: Lawrence Erlbaum Associates.
Marton, F., & Säljö, R. (1976). On qualitative differences in learning. 1: Outcome and process. *British Journal of Educational Psychology, 46*, 4–11.
Maturana, H. (1980). The biology of cognition, In H. Maturana & F. Varela (Eds.), *Autopoiesis and cognition: The realization of the living.* Boston, MA: D. Reidel.
Maturana, H. (1988). Ontology of observing: the biological foundations of self-consciousness and the physical domain of existence. In R. Donaldson (Ed.), *Texts in cybernetic theory: An in-depth exploration of the thought of Humberto Maturana, William T. Powers, and Ernst von Glasersfeld* (conference workbook). Felton CA: American Society for Cybernetics. Retrieved September 17, 2016, from http://ada.evergreen.edu/~arunc/texts/cybernetics/oo/old/oo.pdf
Maturana, H., & Varela, F. (1987). *The tree of knowledge: The biological roots of human understanding.* Boston, MA & London: Shambala.
Metz, M., & Simmt, E. (2015). Researching mathematical experience from the perspective of an empathic second-person observer. *ZDM Mathematics Education, 47*(2), 197–209.
Mezirow, J. (1978). Perspective transformation. *Adult Education Quarterly, 28*, 100–110.
Mezirow, J. (2003). Transformative learning as discourse. *Journal of Transformative Education, 1*, 58–63.
Petitmengin, C. (2006). Describing one's subjective experience in the second person: An interview method for the science of consciousness. *Phenomenology and the Cognitive Sciences, 5*, 229–269.
Preciado-Babb, A. P., Metz, M., & Marcotte, C. (2015). Awareness as an enactivist framework for the mathematical learning of teachers, mentors and institutions. *ZDM Mathematics Education, 47*(2), 257–268.
Reid, D. (1996). Enactivism as a methodology. In L. Puig & A. Gutierrez (Eds.), *Proceedings of the twentieth Annual Conference of the International Group for the Psychology of Mathematics Education (PME20)* (Vol. 4, pp. 203–209). Valencia, Spain: University of Valencia.
Scharmer, O. (2000). *Presencing: Learning from the future as it emerges.* Paper presented at the Conference on Knowledge and Innovation, Helsinki School of Economics, Finland, and at the MIT Sloan School of Management. Retrieved from http://www.ottoscharmer.com/sites/default/files/2000_Presencing.pdf
Schmidt, W. H., Jorde, D., Cogan, L. S., Barrier, E., Gonzalo, I., Moser, U., Shimizu, K., Sawada, T., Valverde, G. A., McKnight, C., Prawat, R. S., Wiley, D. E., Raizen, S. A., Britton, E. D., & Wolfe, R. G. (1996). *Characterizing pedagogical flow: An investigation of mathematics and science teaching in six countries.* Dordrecht: Kluwer Academic Publishers.
Schön, D. (1983). *The reflective practitioner: How professionals think in action.* New York, NY: Basic Books.
Simmt, E., & Kieren, T. (2015). Three "moves" in enactivist research: A reflection. *ZDM Mathematics Education, 47*(2), 307–317.
Skemp, R. (1976). Relational understanding and instrumental understanding. *Mathematics Teaching, 77*, 20–26.
Sriraman, B., & Wu, K. (2014). Embodied cognition. In S. Lerman (Ed.), *Encyclopedia of mathematics education* (pp. 207–209). Dordrecht: Springer.
Steinbring, H. (2015). Mathematical interaction shaped by communication, epistemological constraints and enactivism. *ZDM Mathematics Education, 47*(2), 281–293.
Towers, J., & Martin, L. C. (2015). Enactivism and the study of collectivity. *ZDM Mathematics Education, 47*(2) 247–256.

Varela, F. (1999). *Ethical know-how: Action, wisdom, and cognition.* Stanford, CA: Stanford University Press.

Varela, F., & Scharmer, O. (2000). *Three gestures of becoming aware. Conversation with Francisco Varela January 12, 2000, Paris.* Retrieved April 23, 2019, from https://studylib.net/doc/18288861/three-gestures-of-becoming-aware

Varela, F., Thompson, E., & Rosch, E. (1991). *The embodied mind: Cognitive science and human experience.* Cambridge, MA: The MIT Press.

Zack, V., & Reid, D. (2004). Good-enough understanding: Theorizing about the learning of complex ideas (Part 1). *For the Learning of Mathematics, 24*(1), 43–50.

Laurinda Brown
School of Education
University of Bristol

Alf Coles
School of Education
University of Bristol

JUDY ANDERSON AND DEBORAH TULLY

5. MATHEMATICS TEACHER EDUCATOR COLLABORATIONS

Building a Community of Practice with Prospective Teachers

Prospective teachers frequently struggle to feel they belong to the teaching profession during their university preparation program. While they participate in multiple professional experience placements, the outcomes of these experiences rely on being mentored by a committed, knowledgeable and supportive mathematics teacher who is willing to provide opportunities to implement the kind of practices advocated in the program. At the University of Sydney, a community of practice was formed joining defined networks of prospective teachers, practising teachers and mathematics teacher educators with the aim of building connections through shared experiences, and for prospective teachers to grow in partnership with those already in the teaching profession without being judged for assessment purposes. To investigate ways of merging these communities, three distinctive strategies were designed, implemented and evaluated to develop our understanding of the best ways to connect and nurture this emergent mathematics educators' community. This chapter describes this community, its networks and the process of assessing their value from the prospective and the practising teachers' perspectives as well as from the perspectives of the participating mathematics teacher educators.

INTRODUCTION

Retention of early career teachers is a critical issue internationally (e.g., De Jong & Campoli, 2018), as well as in Australia (e.g., Weldon, 2018). Recent studies have suggested that between 30% and 50% of novice teachers leave the profession within their first five years of teaching (Buchanan et al., 2013; Paris, 2010). In Australia, the demand for qualified mathematics teachers outpaces the supply and this critical shortage may in part explain why secondary school students choose not to enrol in the highest mathematics curriculum levels (Marginson, Tytler, Freeman, & Roberts, 2013). In recent years, the Australian government has encouraged initiatives to improve K-12 student success in STEM subjects (Office of the Chief Scientist, 2016) and to improve teacher education programs through more collaboration between education and science faculties in universities. Through collaborations, new ways of integrating the pedagogical and content expertise of mathematics education lecturers

© KONINKLIJKE BRILL NV, LEIDEN, 2020 | DOI: 10.1163/9789004424210_006

and mathematicians have the potential to enhance recruitment and retention into teaching careers while also developing platforms for continued professional learning after graduation (Goos, 2015). By recruiting practising teachers as partners in the enterprise of prospective teacher education, there is potential to further enhance existing teacher education programs to develop but also retain new teachers. As Zeichner suggests (2010, p. 97)

> This work in creating hybrid spaces in teacher education where academic and practitioner knowledge and knowledge that exists in communities come together in new less hierarchical ways in the service of teacher learning represents a paradigm shift in the epistemology of teacher education programs.

This chapter describes the development of a community of practice of mathematics teacher educators comprising practising teachers who are alumni of the university, mathematics education lecturers and mathematicians, together with prospective teachers at the University of Sydney.

Designed and developed by the authors of this chapter, three distinctive strategies were implemented and evaluated over a three-year period – an Alumni Conference, a Teaching in Practice day, and Mentoring Mosaics. These strategies were implemented sequentially by using data from the first to inform the design of the second and so forth. Our research uses the description of a community of practice as a group that shares a practice with the specific features of mutual engagement of participants, a joint enterprise, and the development of a shared repertoire of resources (Wenger, 1998). Nested within the concept of the community is the notion of networks – we use the term networks to refer to the connections formed between participants in each of the three distinctive strategies. Wenger, Trayner, and de Laat (2011, p. 10) define a network as a "set of relationships, personal interactions, and connections among participants who have personal reasons to connect." Providing opportunities through the networks to develop and strengthen the professional identity of both prospective and alumni teachers are key to developing the community of practice, and improving our teacher education program.

The purpose of our research is to ascertain how a community of practice linking together prospective teachers and mathematics teacher educators enhanced identity formation for prospective teachers and created professional value for alumni teachers who contributed to this community as mathematics teacher educators. Our research also aims to learn how the networks described in this chapter may have built connections and facilitated shared understandings between prospective teachers and the mathematics teacher educators. Additionally, we seek to discover how initial teacher education programs may sustainably embed such strategies into their ongoing program delivery. The chapter begins with a review of relevant literature followed by a detailed description of the three strategies, the development of networks and the community of practice. Interviews with participants, all of whom have engaged in at least two of the three strategies, enabled a judgement about its value, its potential to promote identity formation, and the efficacy of the approach as an integral component of the university initial teacher education program.

LITERATURE REVIEW

Professional learning communities may be uniquely situated to contribute positively to teacher identity formation and hence retention by creating a supportive platform for mathematics teachers through expanding a shared repertoire of complementary expertise between experienced and beginning teachers (Grossman, Wineburg, & Woolworth, 2001; LeCornu & Ewing, 2008). Although time consuming to build, fostering participation in educator-based learning communities may prove beneficial for the professional development and career commitment of both those studying to become teachers and those already engaged in the profession (Campbell & Brummet, 2007). The collaborative professional environments in which teachers seek and gain mutual encouragement facilitate not only teacher support for continued retention but also reduce the negative experience of isolation often felt by beginning teachers (Buchanan et al., 2013; Le Cornu & Ewing, 2008). For many early career teachers, a lack of ongoing collegial support and a sense of isolation within the profession are contributing factors to their attrition (Buchanan et al., 2013). Finding strategies to enhance the development of positive teaching identities for both prospective and early career teachers informs mathematics teacher educators' efforts to better prepare mathematics teachers for the profession. This literature review examines research pertaining to the development of learning communities for teachers, teacher mentoring, teachers as teacher educators, and building teacher identities through communities of practice.

Professional Development within the Context of Teacher Learning Communities

Teachers' professional development often depends upon quality interactions with other members of their community. "From a community-of-practice perspective, one's work and one's professional development are inextricably entwined with those with whom one works" (Schlager & Fusco, 2003, p. 204). As teachers collectively engage through "talking about the work from inside the practice" (p. 203), they learn valuable shared knowledge and resource building through collaboration and observation. Communities encourage teachers to ask questions focused on their practice, which in turn facilitate growth in teachers' professional identities as they expand their repertoire of knowing how to work and support their colleagues (Lieberman, 2009).

Communities of practice play significant "direct and catalytic roles in teacher learning" (Schlager & Fusco, 2003, p. 206). Since participation in a community may vitally support the professional growth of teachers, it is essential that prospective teachers are exposed to the value of such communities and how to effectively participate in them, even while still in their teacher preparation programs. Le Cornu and Ewing (2008) suggest that the climate of teacher education in Australia calls for an increase in emotional and intellectual support between all teachers, including those who are new to the profession.

Mentoring in Learning Communities

Participating in a collaborative community of practice and reflecting on one's own learning are important practices for both prospective and practising teachers in which to engage. When this exercise merges the voices of both prospective and practising teachers into a collective narrative, a form of symbiotic mentoring emerges (Le Cornu & Ewing, 2008). Universities in Australia have begun incorporating mentoring programs as part of each student's professional experience to provide support and guidance for these novice teachers as they enter the profession. However, not all mentor roles in teacher education programs are shaped alike. In some universities, mentor teachers serve primarily as supervisors who oversee the practicum or professional experience of novice teachers. In this supervisory role, they may be the gatekeepers of prospective teachers' development. The hierarchy of roles of supervisor and novice teacher have the potential to be fraught with negative connotations, a place where judgment, evaluation and criticism may define essential aspects of the relationship (Sanford & Hopper, 2000). However, a more reflective and less critical approach anchored in a shared relationship between mentor and mentee may prove more beneficial in meeting the needs of affirming and encouraging prospective teachers as they begin their professional journeys (Ambrosetti & Dekkers, 2010).

In describing mentoring as an "holistic process," Ambrosetti and Dekkers (2010, p. 31) argue that while supervisors may be responsible for assessing prospective teachers, the role of a mentor should be more centred on a personal relationship in which trust underpins the relationship, and where both the mentor and mentee navigate the relationship and their professional development through dialogue and reflection. For practising teachers, mentoring a prospective teacher requires more than focused discussions centred exclusively on content knowledge and subject competency (Ambrosetti, 2014). Successful mentors possess high-level interpersonal skills as well as professional competencies; they exhibit sensitive and nonjudgmental tendencies to facilitate the growth of mutual trust and confidence with their mentees (Inzer & Crawford, 2005). In many ways, mentoring is a complex activity, encompassing both emotional and cognitive dimensions (Ambrosetti, 2014; Hudson, 2013). For all teachers, developing emotional competence and resiliency is a worthwhile pursuit as it not only impacts their well-being but also influences how they interact with students and others in their school environment (Collie, Martin, & Frydenberg, 2017; Morgan, Ludlow, Kitching, O'Leary, & Clarke, 2010).

One of the potential benefits of mentoring in teacher educator programs is the way in which collaborative engagement between practising and prospective teachers enhances the professional development of both mentor and mentee (Campbell &

Brummet, 2007). Effective teacher mentors may come to recognize that the role they play is significant in their own development, not simply an added task or responsibility to assist prospective teachers. While many teachers who serve as mentors for novice practitioners describe a renewed passion for the profession as an outcome of their mentoring experience (Inzer & Crawford, 2005), the benefits of engaging in a mentoring relationship appears to be "profound" for beginning teachers (Buchanan et al., 2013, p. 119). Finding ways for teachers to work collaboratively with mathematics education lecturers as a community of mathematics teacher educators in teacher education programs, has the potential to develop shared engagement in developing teachers of the future.

Teachers as Teacher Educators

Teachers as teacher educators is not a new concept, as classroom teachers are often called upon to mentor prospective teachers during their practicum experience. University teacher education programs often rely on teachers' contextualized knowledge and skills of practice to guide and prepare those who are about to embark upon a teaching career. However, even as they work alongside novice teachers guiding them and helping them learn to teach, classroom teachers rarely see themselves in the role of teacher educator (Feiman-Nemser, 1998), even though the teacher assumes a supervisory or mentoring role during a prospective teacher's field placement.

The notion that classroom teachers should play a "substantial role" (Feiman-Nemser, 1998, p. 65) in the education of novice teachers gained significant attention in the United States as seen through the establishment of professional development schools (Holmes Group, 1990). While this model may not be universally feasible, it does signal the need to rethink models of teacher preparation programs and the way prospective teachers are mentored through respecting and utilising the knowledge of classroom teachers. In this way, practising teachers and university education lecturers form a collective enterprise of teacher educators focused on assisting prospective teachers to develop in their craft as emerging teachers (Feiman-Nemser, 1998).

Through forming links between school and university, teachers who act as teacher educators may be considered "boundary spanners" who embody the idea that the "dynamics and cultures of both worlds are vital in linking schools and universities in viable collaboration" (Sandholtz & Finan, 1998, p. 24). Given their unique context of being grounded within a school context, practising teachers who have bridged the boundary from teacher to teacher educator (Bullough, 2005) guide and inspire novice teachers through providing "powerful images of good teaching and strong professional commitment" (Feiman-Nemser, 1998, p. 1017).

While the professional development of prospective teachers may be enhanced through "co-learning partnerships" formed between teachers and university education lecturers (Jaworski, 2001; Wagner, 1997), novice teachers are not the only beneficiaries of this partnership. By working together to enrich the learning experiences of prospective teachers (Bullough, 2005), teacher educators from both school and university can learn deeply from the experience of the other, expanding their appreciation for the unique contributions to their own growth as educators while also helping to shape the identity of novice practitioners (Bullough, 2005; Jaworski, 2001).

Building Teacher Identity through Communities of Practice

Prospective teachers begin to develop their identity as a 'teacher' during their university initial teacher education program and this development continues as it shifts and reshapes depending on context and connections with new communities (Beauchamp & Thomas, 2009; MacLure, 1993). Beauchamp and Thomas (2009) suggest that because their identities may only be tentative, prospective and early career teachers will feel the impact of new communities in more powerful ways. Building a strong teacher identity helps to instil confidence, power and agency (Williams, 2010) and supports retention in the profession (Izadinia, 2015). Since having both a professional and a personal identity has been recognised, Wenger (1998) links identity with practice and suggests the same five characteristics apply to both types of identity – it is a negotiated experience; involving community membership; has a learning trajectory; combines different forms of membership; and involves local and global contexts. Further embedded within these described characteristics of community are networks whose attributes include personal interactions among participants and a focused connection between members who have a vested interest in coming together (Wenger et al., 2011).

Prospective and practising teachers who engage in shared networks not only gain a space to dialogue with others, but also to reflect authentically on their practice, further facilitating growth in their professional identities. As Wenger (2000) asserts

> … in the generational encounter between newcomers and established members, the identities of both get expanded. Newcomers gain a sense of history. And old-timers gain perspective as they revisit their own ways and open future possibilities for others …. (p. 241)

Networks provide prospective teachers entering the profession with access and insight into the patterns of current teaching practice through the voices of those already engaged in the profession. Lieberman (2009) found that as teachers learn within a community, their questions gradually shift from 'what do I know about teaching' to 'who am I as a teacher'? This becomes an important dialogue for new teachers as competency in subject knowledge, in tandem with developing confidence as a teacher, informs practice and teacher identity. These findings affirm Wenger's (2000) assertion that communities of practice facilitate growth in professional identities as these communities are "a place where a person can experience knowing as a form of social competence" (p. 241). Learning within a community of practice allows teachers to experience growth in their professional selves since identity and learning are intricately woven together (Lave & Wenger, 1991).

Identity development is influenced by many experiences. For example, identity creation is enhanced as prospective teachers observe and work with mentor-teachers during professional or field experience placements in schools (Izadinia, 2015; Walkington, 2005). However, not all school-based experiences expose prospective teachers to the types of practices university programs advocate with evidence that some mentor teachers actively discourage prospective teachers from the 'impractical ideas' recommended by teacher education lecturers (Feiman-Nemser, 2001; Izadinia, 2015). The disconnect between the 'theories' delivered during university programs and the 'practicalities' of what is experienced in school settings can be challenging and frequently disconcerting (Zeichner, 2010).

Ideally, universities should select and develop supervising teachers' mentoring capabilities before professional experience (Izadinia, 2015) but at our university, we struggle to find sufficient placements in schools for the large number of prospective teachers in our programs. Without the benefit of a mentor development program or other approaches such as the successful professional development schools model in the United States (Darling-Hammond, 2010), we sought to build networks and ultimately, a community of practice, to expose prospective teachers to a larger number of more experienced teachers, and to acknowledge the importance of teachers' practical knowledge of teaching through inviting our alumni teachers to collaborate with us as mathematics teacher educators within these networks and the community of practice we hoped to develop. Our challenge as university lecturers was to design strategies which would facilitate community building between the prospective teachers in our programs and the collective of mathematics teacher educators, which included the university mathematics education lecturers, the mathematicians, and our former secondary mathematics education students, who we refer to as 'alumni teachers.'

BUILDING A COMMUNITY OF PRACTICE BETWEEN PROSPECTIVE TEACHERS AND MATHEMATICS TEACHER EDUCATORS: OUR STRATEGIES

To develop a community of practice and create a greater connection between schools and the university and between the group of mathematics teacher educators as described above, we developed a "third space" (Gutierrez, 2008). We use this term to describe the various boundary crossings between universities and schools (Zeichner, 2010). The notion of a third space rejects the binaries of academic and practitioner, between theory and practice, and

> … is concerned with the creation of hybrid spaces in preservice teacher education programs that bring together school and university-based teacher educators [mathematics teacher educators], and practitioner and academic knowledge, in new ways to enhance the learning of prospective teachers …

> in less hierarchical ways to create new learning opportunities …. (Zeichner, 2010, p. 92)

We wanted to strengthen our initial teacher education program by acknowledging that practitioner knowledge and experiences were as valuable as university lecturer knowledge but we also wanted to provide opportunities for our alumni mathematics teachers to reconnect with the university campus, and to reconnect with each other. In contrast to the approaches described by Zeichner (2010) of hiring experienced teachers to deliver methods programs on campus, employing a teacher-in-residence, or delivering university programs in a local school, we developed three strategies to bring alumni teachers of varying experience (from 'early career' to those leading faculties) back on campus. We wanted to provide opportunities for the mathematics teacher educators to collaborate so that we could build a community of practice with the prospective teachers and help them develop their identities as teachers.

Over a three-year period, we sequentially instituted three networks that joined together prospective teachers and mathematics teacher educators within specific social spaces for learning to form a broader community of practice focused on teacher development – an Alumni Conference, Teaching in Practice (TIP) Day, and Mentoring Mosaics (Tully, Poladian, & Anderson, 2017). Informed by research, these opportunities for connected learning encouraged the sharing of resources, perspectives and stories between and amongst the prospective teachers and our collectively defined mathematics teacher educators (Graven, 2004; Mason, 2013; McGraw, Lynch, Koc, Budak, & Brown, 2007). After each network was implemented, feedback from participants informed the development of the next networking opportunity.

The Alumni Conference

The initial approach to connecting prospective teachers and mathematics teacher educators involved the creation of an annual one-day Alumni Conference beginning in 2014. Building on the connections that already existed between groups of alumni who had studied together, the aim of the conference was to provide an opportunity for them to reconnect and to interact with the prospective teachers by sharing their experiences and practical advice about teaching and learning (Mason, 2013). The first author invited all practising mathematics teachers with whom she had worked over the past 12 years at the university. To facilitate teachers' attendance, the conference was held during school holidays, at no cost to participants, and the current group of prospective teachers were required to attend as a compulsory component of their university course.

The program for the day involved a combination of keynote presentations, panels of speakers, and short presentations about practical teaching ideas. Each year, the conference has a theme focusing on a current issue in mathematics education (e.g., catering for diversity of students' needs, motivation and engagement of students, inquiry-based learning, assessment practices and purposes), with an invited keynote speaker to begin the day's deliberations. Our decisions about the focus of the event and the purpose of supporting and connecting with the alumni teachers as members of the mathematics teacher educator group, is grounded in the notion of accepting teaching as a challenging profession – that there is no one model of 'expert teaching,' that each of these teachers works in a complex context with many challenges and that each has their own purpose for attending the event each year. Acknowledging their role as mathematics teacher educators, we consult with the alumni teachers to seek their advice on suitable topics for discussion, for potential inspiring keynote speakers, and for suggestions about topics for debate. When inviting teachers to present, we accept a wide range of practical ideas for teaching, as well as topics such as managing challenging students, coping with parents, etc. While we may not promote one explicit model of teaching, we do aim to promote teaching as inquiry, with opportunities for participants to share "knowledge-for-practice" and "knowledge-in-practice" as well as "knowledge of practice" (Cochran-Smith & Lytle, 1999, p. 250). Like Hong, Day and Greene (2018), we notice that early career teachers tend to be more focused on survival and want practical, ready to use advice while more experienced teachers are more interested in the debates and discussions of 'bigger picture' issues.

Popular program inclusions have encompassed panels of experts (leaders in schools, mathematicians, university education lecturers) discussing aspects of school life, and short presentations of two or seven minutes called 'TeachMeets.' These short presentations are given by volunteers although we often need to encourage both alumni teachers and prospective teachers to 'volunteer.' To build further connections, the program included dedicated time and space for pairs of prospective teachers and alumni teachers to exchange ideas and share experiences. These pairs typically connected with other pairs over lunch with alumni teachers frequently offering professional learning field placements to their prospective teacher colleagues. This has been an unexpected bonus since we are always seeking school placements to fill the quota required each year.

With between 120 and 150 participants each year (approximately 45% alumni teachers, 45% prospective teachers and 10% university education lecturers, mathematicians, and university tutors), the energy and enthusiasm for mathematics teaching is palpable. Some alumni teacher participants have been teaching for just three months and often arrive feeling quite overwhelmed by their work, seeking reassurance and support from others, and wanting to share their experiences with peers. Other alumni teachers are quite experienced (8 to 12 years of teaching), have leadership positions in their schools or education systems, and are keen to offer support and encouragement. Feedback is often very positive, with particularly favourable comments from both alumni teachers and prospective teachers about the short, practical presentations about effective teaching ideas which we have referred to as TeachMeets. As noted by one alumni teacher, "the TeachMeet presentations were absolutely amazing, so many new and innovative ideas."

While building the network may be challenging given the size of the event, feedback from alumni teacher participants indicated they feel connected with the university and with the university staff who attend the day, and connected with peers from their university graduating class – they feel accepted as members of the collective mathematics teacher educator group. One teacher indicated he felt "valued," particularly when invited to be on a panel of experienced teachers, and that he appreciated the opportunity to "give back" to the profession. The prospective teachers appreciate the opportunity to hear from experienced teachers and seek out teachers during morning tea and lunch time to ask for advice. Feedback indicated there was interest from alumni teachers to continue their engagement with the university by volunteering to attend further professional networking opportunities – this led to the development of the next strategy, the Teaching in Practice Day.

Teaching in Practice Day

Beginning in 2015, we designed the next network, a one-day forum for one class of prospective mathematics teachers held early in second semester before they began their first professional experience placement. Facilitated by the authors (two university mathematics education lecturers) and one mathematician, the purpose of this smaller experience was for prospective and alumni teachers to work together on classroom issues and curriculum design, thus providing a platform for alumni teachers to serve in the capacity as teacher educators, to share experience and inspire the prospective teachers, while the prospective teachers raised and discussed issues of concern. Typically, there were 15 mathematics teacher educators in attendance, of which 12 were alumni teachers, working with 24 prospective teachers. The 12 alumni teachers and 24 prospective teachers were placed in mixed groups of three to work on a range of tasks throughout the day.

Knowing that prospective teachers worry about practical aspects of teaching including classroom management, managing workloads, preparation time and creating innovative lessons, prior to the Teaching in Practice day we asked them to contribute issues and questions which we reframed into scenarios to use as discussion triggers throughout the day. One example follows:

> You may have a different teaching style to your mentor/supervising teacher during professional experience. Your mentor teacher may want you to do things or 'be' a teacher that you are not. They may teach in a certain way and are not open to alternative practices. They may give you feedback that you don't necessarily agree with.
>
> What are some ways to manage this situation given the mentor/supervising teacher is tasked with assessing your competence at the end of professional experience?

> Should you stay on the 'good side' of the mentor/supervising teacher by doing what they want from you, or bring your own style and strategies into the classroom?

These scenarios promoted much discussion and debate about the best strategies to use to address the power imbalance in such situations.

The program also included sessions dedicated to identifying aspects of quality mathematics lessons, group lesson planning, strategies to combine topics from curriculum documents, working with mixed-ability classes, designing rich tasks from newspaper articles, and classroom management. Many of the sessions promoted teaching as a dynamic and challenging profession but sought to enhance the notion that by working with colleagues, planning lessons and coping with challenges helps to make the tasks more manageable. All the mathematics teacher educators helped to reinforce images of teaching as collaborative, collegial and connected since the prospective teachers would be joining mathematics departments led by an experienced head teacher who would mentor and support them as they developed their teaching identity. During morning tea and lunch breaks participants continued informal conversations and built upon prior connections from the Alumni Conference. Like findings from Murray, Mitchell and Dobbins (1998), data from participant feedback indicated alumni teachers particularly valued the shared conversations with the other mathematics teacher educators during the day, and the opportunities to mentor and offer professional guidance to the prospective teachers.

Mentoring Mosaics

The final network was also based on feedback from the Alumni Conference and Teaching in Practice Day, with most alumni teachers and prospective teachers indicating a desire and willingness to participate in further mentoring relationships. Considering this level of demonstrated interest, a pilot mentoring program was instituted for second semester 2016 linking prospective mathematics teachers who were preparing for their first professional experience placement, with alumni teachers.

In considering how to make the most efficient use of time for those participating in our mentoring pilot program, we explored the feasibility of designing an online mentoring model. Internet technologies are being used to sustain the learning and collegial interactions amongst teachers. Although teachers may prefer to collaborate in person rather than in a virtual space (Stephens & Hartmann, 2004), online professional learning communities provide a viable means of communication, collaboration, learning and reflection for educators, especially when time and space constraints make face to face meetings impractical (Macià, & García, 2016; McConnell, Parker, Eberhardt, Koehler, & Lundeberg, 2013; Trust, Krutka, & Carpenter, 2016). While developing effective and sustained professional learning communities is challenging in and of itself (Booth, 2012), Goos and Bennison

(2008) found that emergent online communities contribute to the sustainability of a community of practice for teachers.

Critical factors in building an online professional community include having a clear task focus, with participants sharing a commitment of purpose and a shared identity (Booth, 2012; Goos & Bennison, 2008). However, community facilitators face the dynamic challenge of designing and structuring an online community while also allowing participants to manage and construct their own learning space. In their study, Goos and Bennison (2008) used minimal structure in the online platform designed for continued discussion and support for new mathematics teachers. They found that an online community of practice as defined by mutual engagement, joint enterprise, and shared repertoire (Wenger, 1998), did in fact emerge. However, the participants' shared repertoire was less about the resources useful for teaching mathematics, and more geared towards the ways in which teachers make sense of the different experiences they faced as new practitioners. Goos and Bennison (2008) found participants in this online community of practice for prospective teachers regularly initiated communication with each other discussing themes related to their professional lives, with the community remaining together after graduation and the "official" conclusion of the program. There is potential to build communities within and between prospective and practising teachers worthy of further investigation, particularly in relation to building identity.

After exploring a variety of mentoring models that could be adopted for this pilot program, we organised 'mentoring mosaic' groups of two to three prospective teachers with two to three practising teachers. This arrangement allowed for a diverse range of sharing experiences as well as an efficient use of the mentor's time since the group mentoring experience would take place "online." Also, mosaic groupings permitted a more informal style of mentoring hinged on the relational component of shared trust and the potential of peer mentoring between practising teachers in each of the groups (Ambrosetti, 2014; Inzer & Crawford, 2005). To facilitate open and honest sharing within groups, none of the alumni teachers served in a supervisory role for the prospective teachers during their professional experience.

Overall, ten prospective and ten alumni teachers participated in the mentoring pilot program. Participants completed a preference questionnaire and groupings were organised based on these preferences. For example, one prospective teacher preferred to be in a single-gender mentoring group, while others preferred to be grouped with alumni teachers from public schools, as opposed to private schools. The program began with a two-hour workshop where participants met each other, engaged in discussions on meaningful mentoring and designed their online platform which served as the medium for ongoing 'mentor-type' discussions. Most groups chose to create a secret Facebook group as their platform for group discussions. Before the completion of the workshop, each group's online platform was established and participants sent multiple exchanges to test their platform. Within each mentoring mosaic group, the minimum requirement for each participant was to post one original comment/question a week and to respond to at least one other post during the week. Each group designated a 'point-person' who would encourage on-going participation of group members. It was the responsibility of the "point-person" to track group participation and privately message each person who had not fulfilled their weekly posting requirement.

To maintain transparency and confidentiality of each group's discussions, the online mentoring forums were limited in membership to just those in each mentoring mosaic group. Each month, one of the university-based mathematics teacher educators would check in with members to see how their groups were progressing. As this mentoring experience was set up as an "online" model, face-to-face mentoring meetings during the semester was not a requirement of the program. However, some prospective and alumni teachers forged meaningful connections and took it upon themselves to set up face-to face meeting times for more personal mentoring.

After the semester pilot program was completed, we hosted a dinner where participants shared their stories of their mentoring experiences. While the expectation of commitment to this community was one semester, one of the mosaic groups decided to remain together due to the perceived benefit to all participants. Although only a small number of prospective and alumni teachers participated in this pilot program, the responses from their feedback were consistently positive with almost all participants indicating they felt valued, connected and had an increased level of commitment to the teaching profession. All participants found the mentoring mosaics experience encouraging and worthwhile. To further investigate the efficacy of these strategies to develop prospective teachers' identities as teachers, we designed an approach to collecting and analysing richer data to inform the collective work of the mathematics teacher educators.

METHODOLOGY

Considering that the professional life of a teacher is experienced within a dynamic social milieu, our framework centres upon Wenger's (1998) social theory of learning and his concept of communities of practice. More specifically, we further draw upon the Wenger et al. (2011) definition of a community of practice as:

> … a learning partnership among people who find it useful to learn from and with each other about a particular domain. They use each other's experience of practice as a learning resource. And they join forces in making sense of and addressing challenges they face individually or collectively. (p. 9)

We use the concept of a learning partnership in analysing teacher learning where both face-to-face and online communities were developed to enhance the prospective teachers' developing identities, through the collaborative efforts of the mathematics teacher educators. Feedback from all three strategies indicated the alumni teachers also developed their sense of belonging to both the collective group of mathematics teacher educators as well as to the overall community of practice of secondary mathematics teachers.

The *third space* offered through the three strategies of Alumni Conference, the Teaching in Practice Day and the Mentoring Mosaic groups provided a flexible platform through which these interconnected networks created a community of practice centred upon connections, shared understanding and professional development. The research questions for this study were:

- How has a community of practice linking prospective teachers and the collective of mathematics teacher educators enhanced identity formation and created professional value for prospective and alumni teachers (who also act as teacher educators)?
- How do the networks build connections and shared understandings between prospective teachers and the mathematics teacher educators?
- How can prospective education programs sustainably embed such strategies into their ongoing program delivery?

To explore the first two questions, and assess how the described networks built connections and shared understanding, and how professional value was created through this community of practice, we adopted the Wenger et al. (2011) conceptual framework. This framework was initially developed as a guide to assist professionals in giving voice to narratives centred on the stories of those participating in communities and networks that are specifically aimed at advancing teaching practice. The focus of this framework centres upon the value created by networks and communities used for learning activities within a socially derived context such as teaching:

> … sharing information, tips and documents, learning from each other's experience, helping each other with challenges, creating knowledge together, keeping up with the field, stimulating change and offering new types of professional development opportunities. (Wenger et al., 2011, p. 7)

The framework is premised on the notion that to appreciate the depth of the value created by networks and communities, it is beneficial to reflect upon these values in a more cyclical manner.

In their framework, Wenger et al. (2011) propose five cycles for consideration when assessing the professional value offered through networks and communities:

- Immediate value – activities and interactions – this is the most basic cycle where value is assigned. In assessing aspects of value in this cycle, one would consider things such as simple connections made with others, giving input, being with others who challenge one's thinking, or receiving a good tip from a colleague. Events and collaborations can offer significance in and of themselves.
- Potential value – knowledge capital – in this cycle one would evaluate the acquisition of different forms of potential knowledge that may be stored and ready for access for future use. This knowledge can be in the form of a useful

teaching skill or technique or a shared understanding between participants, or the potential for future collaborations.

- Applied value – changes in practice – while Cycle 2 may focus on the potential value of knowledge capital, this cycle focuses on how acquired knowledge is applied in specific situations. In teaching this may be using a lesson idea or implementing a technique shared by the network or community participants.
- Realised value – performance improvement – the application of new teaching techniques doesn't necessarily imply an improvement in performance. This cycle allows one to consider the influence that changes in practice may have on determining professional growth outcomes for teachers such as growth in identity formation.
- Reframing value – redefining success – in this cycle, one should consider how participation in networks and communities may have proven influential in helping to shape new metrics of success through reframing previously held goals or values. Any redefinition of success can be applied to both an individual as well as to a group.

Employing this framework as a guide, we used these cycles to analyse teacher learning where both face-to-face and online communities were developed linking the mathematics teacher educators with the prospective mathematics teachers. While these cycles can be independently analysed, there may be some causal relationship assumed between the cycles, such as potential value-knowledge (Cycle 2) leading to changes in practice (Cycle 3). Additionally, it should be noted that each cycle offers a unique structure in which various values of the community of practice are assessed through the lenses of the prospective teacher and mathematics teacher educator participants. One should also not assume that a community is only successful when teachers focus their reflections on multiple attributes from the fifth cycle. These cycles provide a means to assist us in appraising how and in what ways the community of practice developed between the mathematics teacher educators and the prospective teachers offered focused value to those who participated.

We conducted semi-structured interviews with a volunteer sample group of 10 prospective teachers and 12 alumni mathematics teachers who were involved with more than one of the network learning spaces described in this chapter. Prospective teacher participants were in their final year of their teacher education program and the range of professional experience of alumni teacher participants spanned between one and ten years since graduation. During the interviews, we asked participants to discuss their overall impressions and personal experience of the Alumni Conference, the Teaching in Practice Days and their Mentoring Mosaic group, as well as any of their experiences with other professionally based communities. We also asked teachers to comment on if and how their involvement with these networks may have influenced their professional growth. Interviews ranged from 20–60 minutes in length, were audio-recorded and transcribed verbatim. It was our intention to collect and weave together a combined narrative of the teacher voices as it is in the "interplay between personal and collective narratives" (Wenger, et al., 2011, p. 16) that we begin to analyse shared understandings and the value attributed through participating in the network experiences and the potential for developing a community of practice.

The analysis of the interview data was initially guided through explicitly using the five cycles of the Wenger et al. (2011) framework. Through using this deductive approach in analysing the qualitative interview data, we independently assigned various excerpts from the teacher narratives and reflections to the relevant cycle as guided by this framework, compared our analyses and collaboratively agreed on the final coding. After this initial coding process was completed, we then further parsed the data by collectively defining subthemes within each cycle. After multiple iterations, this analytic process produced a matrix that summarised, in multi-dimensional form, the value that participants ascribed to the network experiences and their evolving professional identity (Miles & Huberman, 1994).

THE PROFESSIONAL VALUE OF SHARED CONNECTIONS WITHIN A COMMUNITY OF PRACTICE

In this next section of the chapter, we present our understanding of the shared connections and value offered by the strategies by presenting the prospective (PS) and alumni teachers' (AT) reflections and stories, weaving together themes through the five cycles of the Wenger et al. (2011) framework.

Cycle 1: What Immediate Value Did Prospective and Alumni Teachers Place on Their Participation in These Networks and in This Community?

Networking. One of the main benefits noted by the alumni teachers was the ability to link with other teachers, to strengthen prior ties, and to feel connected to a larger network of alumni. For many of the prospective teachers, the Alumni Conference provided a welcome opportunity to network with other prospective and practising teachers. Having the occasion to talk with others about their own practice and teaching experiences provided participants with both the chance to learn from others, as well as to express their own ideas and thoughts. One prospective teacher noted that the event helped him feel a "part of the group" and "networking was the highlight" (PS 10). The ability to connect with others was also "unexpectedly helpful," remarked one participant, who came out of the day "feeling really, really great" because of her conversation with a first-year alumni teacher (PS 7). The benefits of networking were equally expressed by the alumni teachers. As one noted,

> For me, I have gained a lot from that community, and the Alumni Conference has kept me connect, has kept me looped back into that. Just personally having those connections and just knowing someone to ask a question to, I feel is incredibly powerful. (AT 1)

The networking opportunities also helped teachers to make unanticipated connections. As an alumni teacher expressed, "I've linked with other people who I wouldn't have" (AT 10).

Learning and showcasing. Prospective teacher participants in the Alumni Conference all reported being impressed with the variety of ideas they gathered from the presentations throughout the day. Activities and presentations showcasing creative teaching concepts exposed prospective teachers to new ways of thinking and the opportunity to build their own database of resources through the "short, little titbits of tips" (PS 8). For some prospective teachers, this may have been their first exposure to the value that a community of practice may offer for their ongoing professional development. Prospective teachers also noted the importance of peer teaching. A participant expressed that he loved "the fact that it was peers teaching peers, like teachers being peers. Teachers teaching teachers" (PS 1).

Unlike prospective teachers who are also current students and immersed in university life, alumni teachers appreciated the opportunities that these networks offered for them to return to the university and to be in a space of learning again.

> I think part of having that alumni day or the mentoring program helps because there is a lot of teacher talk and exchanging of anecdotes and advice and stuff and I think that is quite helpful in itself. (AT 6)

Inspiration and encouragement. Prospective teachers felt their participation in these networks was helpful because of the opportunity to engage with and hear about the experiences of current teachers.

> It was really good hearing from past students that were in this course and see where they've gone, and then – it's good to see where they've gone from this course. (PS 6)

Another participant noted that making connections with teachers of varied experience and hearing about their journeys was helpful and provided, "emotional [support]" (PS 5). One of the immediate benefits to prospective teachers was being able to discuss creative teaching ideas with alumni teachers. Prospective teachers found this aspect of the community inspiring as they could see first-hand, how novel approaches to teaching could be connected to the curriculum.

First year teachers can often find their new professional life challenging. An early career teacher commented, "Gosh, I found it really energetic and inspiring and I thought as a new person, fantastic" (AT 8). Aside from the connections, sharing, and value placed on learning, alumni teachers expressed how these networks personally inspired and re-invigorated them towards their profession. "I love the ideas, but also the energy. Still it's quite inspiring. I just love to have that and get the reinvigoration" (AT 11). Alumni and prospective teachers alike were inspired not only through the sharing of teaching ideas, but also through the collegiality amongst this diverse set of educators.

Cycle 2: What Resources Were Produced through Participation in This Community?

Networking conversations that contribute to learning. While networking was an immediate value to both prospective and alumni teachers, the productive conversations that took place between colleagues produced a variety of resources for the teachers in these networks. Prospective teachers who interacted with other teachers felt particularly encouraged by the direction of their career and the connections made with alumni teachers "who were already there [and could] tell you how it is" (PS 4). One participant was intrigued by a presentation about inquiry-based learning at the Alumni Conference, and could further discuss this idea with the presenter during the day. Others made connections with teachers with whom they would be eager to work for their professional placement and appreciated forming "another network outside what I've already got now" (PS 6).

Additionally, many prospective teachers appeared fearful of issues involving classroom management. However, having the opportunity to discuss this issue with alumni teachers helped them realize that while managing a classroom of 25 to 30 school students might be especially challenging for new teachers, they could still confidently learn techniques to employ in their own classrooms in the future. A participant remarked that having a conversation during that day, "helped me to think that I'll get better at that when I've got my own class" (PS 10). One of the extended values of networking is the feeling of being a part of something bigger. A prospective teacher explained that, "I find basically whenever you feel like you are not alone as a teacher is usually a good experience" (PS 1).

Resources. Sharing resources and teaching ideas were central to each of the networks. During the Teaching in Practice day, alumni teachers encouraged prospective teachers to think about new ideas and exposed them to their own methods of practice and teaching during the roundtable discussions. Prospective teachers were also exposed to new and practical ways of generating lesson plans and activities, which were "much more valuable from what you can see in a [mathematics] textbook" (PS 9). Another commented on the value of hearing from various teacher experiences so that "you got to see a bit of insight into what happens. You get experience without necessarily having to experience it yourself" (PS 5). Alumni teachers placed a high value on the variety of resources they acquired through their participation in these various networks as well. Teachers spoke favourably of the generation of new ideas, and immediate ideas to implement in their classrooms; "Everyone was talking about these great ideas; I think that's fantastic. I think as teachers there is no time to come up with all this information for yourself" (AT 7).

Emotional encouragement. Alumni teachers, though, expressed even greater valuing of the people resources that were now at their disposal, someone to not only share teaching resources, but teaching stories as well. As reflected by an alumni teacher,

> We're sharing stories which is nice and that kind of support is really nice to have for one another, just to share teaching ideas, 'I did this. It was really great,' 'That's awesome. It's really awesome. I did this this week.' I'm just being really encouraging. I think just having that positivity and that recognition is something we kind of don't get as much on a day-to-day basis. So just having that online [mentoring mosaic group] with a group is quite nice. (AT 5)

Prospective teachers also expressed the helpfulness of the online mentoring community. A prospective teacher explained that after having a bad lesson during his practicum, he sought advice from the group, and received feedback and counsel that was "really supportive and beneficial to keeping in a positive mindset" (PS 10). Many prospective teachers remarked that having access to alumni teachers who were a few years into their teaching career was extremely valuable in terms of giving perspective as well as support because "sometimes you need an outside perspective" (PS 5). They also felt that the mentoring program was valuable for personal reflection, getting practical classroom advice, and emotional support. The mentoring program specifically "made teaching more real to me … I felt really supported by people" (PS 10). Some teachers reflected that they will likely continue relationships with others they met during the Alumni Conference, which "will build a sense of community."

Cycle 3: How Have Prospective and Alumni Teachers Applied What They Have Learned through This Community?

Curriculum development. Sharing creative curriculum ideas was one of the hallmarks of these mathematics teacher networks. Several teachers commented on how they had not only acquired fresh resources but could apply these new teaching ideas in their classrooms.

> I thought that was really good [Alumni Conference]. I was astounded by how good those students [prospective teachers] were presenting when they had the TeachMeets. That was so good. Some of that stuff I was going, "That's awesome." One pre-service [prospective] teacher presented on Kahoot and I was like, "Well, yes. That's really good," so I've used that in the classroom. (AT 7)

Also, building on the theme of people resources, one of the alumni teachers forged a working relationship with another alumni teacher from a different school to work on designing assessment tasks together.

Support and persistence. Participants in the mentoring mosaic groups also found the online mentoring community instructive in assisting teachers to reflect honestly on their practice. The mentoring program was beneficial particularly for prospective teachers who needed extra encouragement and support. One participant even said, "if I didn't have that mentor, I would have dropped out [of the university program]" (PS 6).

Cycle 4: What Has Been Achieved through Prospective and Alumni Teacher Community Participation?

Personal value and self-worth. Considering that many teachers leave the profession due to a lack of collegial support (Buchanan et al., 2013), in this cycle we highlight teachers' perceptions of personal value and sense of self-worth as it relates to their participation in these communities of practice and their legitimacy as a member of the collective of mathematics teacher educators. As an alumni teacher shared, "It's certainly inspired me to feel that I am valued" (AT 7). These data also confirm feedback received from the various network events.

> One of the things that has engaged me in what Sydney Uni [university] is doing, is feeling like I matter. I think that could go some way [reaching out to alumni teachers] to make teachers feel more valued in a field where they quite often don't get very much of that. (AT 4)

> The sense of community that came from the set up [mentoring mosaics]. I am part of a few different teacher Facebook groups, but they are more about sharing news that affects teachers on a whole, to be part of this where the purpose was to share your own experience and feelings was really cathartic. (AT 5)

For several alumni teachers, participating in this community promoted opportunities to reflect on their personal growth as teachers. As an alumni teacher noted,

> It's also nice for me to come into an environment and just realise how much I know about teaching that I didn't know when I was an undergraduate. It's like a marker of how far I have come. (AT 4)

Growth in identity formation. Hearing the thoughts and experiences of others through their network connections was particularly reassuring for prospective teacher participants. Prospective teachers had some common fears assuaged and gained confidence to continue the path as future educators. They realized it is "okay to be junior … you'll grow and you will not only grow on your own, you'll grow because these other experts are there to help you" (PS 1). When reflecting on a connection made with an alumni teacher, a prospective teacher commented, "I want to strive to be as good as you" (PS 4). Network interactions between prospective and alumni teachers enabled the novice teachers to be able to envision themselves as practising classroom teachers, dedicated and excited about their future career trajectory. One prospective teacher felt that her participation in these programs gave her a "strong connection to the profession," and she developed "a sense of community [which] makes you feel stronger in your ability to do your job and to do it well" (PS 8). Yet another prospective teacher reflected on what he believed to be the shifting nature of teacher identity and how his authentic connections with alumni teachers through the mentoring mosaics and the Alumni Conference shaped his emerging identity as a teacher:

> Seeing some of the variety of teachers that can be successful was – yes, I guess it let me compare my ideas with their ideas and that sort of thing. So, it's probably contributed to my identity as a teacher … my identity is very much in a changing place at the moment. It's likely to change a lot in the first couple of years of teaching. But seeing a variety of teachers and what their ideas are and what they seem to do helps me to compare what I'm like as a teacher with other teachers. (PS 10)

For prospective teachers, the process of forming a professional identity is a natural yet dynamic process and is often informed by those a few steps ahead in their professional journeys. One disheartened prospective teacher said her mentors "were encouraging because I would think that I wasn't good enough. They would allow me to think in a different way" (PS 6). The trust shared through these affirming conversations with alumni teachers guided prospective teachers into a space where they began to test their own pre-conceived notions of what it means to be a teacher.

Cycle 5: How Has Participation in This Community Changed Teachers' View of What Matters?

Several of the teachers described a shift in their views about what mattered to them professionally after participating in the community and networks. Responses were more individually aligned, with teachers' responses focusing upon their future contribution as educators. Motivated by discussions on the future of mathematics teaching that occurred during the network gatherings, one of the practising teachers has since enrolled in a PhD program focusing on STEM education. Through seeing teachers mentor others and develop a community through their sharing, one prospective teacher noted that he changed his thinking about the scope of teachers' influence: "So I added that to my definition of being a successful teacher – to help other teachers and upcoming teachers in the future" (PS 10).

After experiencing the benefit of shared learning in our mathematics teachers' community of practice, an alumni teacher asserted that this type of learning platform should be formalised within the teacher education program.

> I know I personally would love the opportunity to come in, perhaps during school holidays, to assist in some of the tutorials in a similar way to how I did at the Teaching in Practice Day. If expert teachers were paid to contribute their time, perspective and insights to the students, I think it would contribute a huge amount to developing the kind of community of practice that we're seeking. (AT 1)

These networking experiences affirmed this alumni teachers' role as an mathematics teacher educator, but he wanted the opportunity to take it even further and to make a greater time commitment to supporting prospective teachers' development and identify formation.

REFLECTIONS ON SHARED UNDERSTANDING, AUTHENTIC CONNECTIONS AND PROFESSIONAL VALUE

Exploring ways in which communities of practice support teachers' professional growth is considered an under-researched area of study (Schlager & Fusco, 2003). In this chapter, we sought to discover how a community of practice linking prospective and alumni mathematics teachers created shared understanding, authentic connections and professional value using a multi-focal viewpoint incorporating the perspectives of those preparing to teach and those already engaged in the profession. Applying the "cycles of value" as outlined in Wenger et al.'s framework (2011) allowed us to decipher how and in what ways the community and networks described in this chapter elevated the professional experience of prospective and alumni mathematics teachers while furthering the identity formation of those soon to embark upon the profession.

This community of practice offered participants not only access to practical teaching ideas and novel curriculum approaches, but also inspired emotional resilience, professional encouragement and a deepening commitment to the teaching profession. Through collaborative connections, critical reflection and meaningful mentoring, alumni teachers expressed feeling affirmed and valued, often necessary components for continued commitment to the profession (Hudson, 2013; Loughran, 2002; Morgan et al., 2010). Through contributing to the educational experience of prospective teachers as mathematics teacher educators, alumni teachers felt recognised as valued members of a mathematics educators' community. For alumni teachers, this sense of personal value was further reinforced through the personal invitation they received from the university welcoming their participation, contribution and influence in these networks and community, which in turn promoted a deeper connection to the profession.

As they are still immersed in their initial teacher education program, prospective teachers may have difficulty envisioning themselves as future classroom practitioners. However, through their participation in these networks, prospective teachers could place themselves within a wider community of experienced teachers and could grow in their identity formation through making "sense of themselves in relation to others" (MacLure, 1993, p. 311). Through facilitating opportunities to affect growth in teacher identity we hope to influence continued retention in the profession (Izadinia, 2015), which was a catalyst for the formation of the three unique strategies, the subsequent networks and the overall community of practice described in this chapter.

In defining a community of practice, Wenger et al. (2011) use descriptors such as "a learning partnership," "use each other's experience of practice" and "join together in making sense" (p. 10); each of these phrases aptly describes the community of practice developed through our approach to improving the initial teacher education program at our university and using the expertise of our alumni teachers as fellow mathematics teacher educators. Applying a framework to assess the professional value offered through interconnected networks of prospective and alumni teachers allows us to move forward to further enhance professional learning, but also to seek ways to embed these networks into our prospective mathematics teacher education program.

EMBEDDING PROFESSIONAL NETWORKS INTO A PROSPECTIVE TEACHER EDUCATION PROGRAM: WHAT HAVE WE LEARNT?

The first author of this chapter is the coordinator of the secondary mathematics initial teacher education program at the University of Sydney and during the 16 years she has been in that role, she has tried several strategies to include practising mathematics teachers in program delivery. This has included prospective teachers visiting schools to observe mathematics lessons, and one or two teachers visiting campus to talk to the prospective teachers about their experiences or to facilitate tutorial sessions. While prospective teachers have always valued the experiences, these strategies have relied on the good will of teachers to give up valuable time in their classrooms and were liable to last minute cancellation because of other priorities at the teacher's school.

The strategies described in this chapter have enabled connections with a much larger pool of alumni teachers who engaged in one or more of the three strategies depending on their commitments and availabilities. One strategy was conducted in school holidays (the Alumni Conference) and another after school (the Mentoring Mosaics) thus having little impact on the school day and teachers' work in classrooms. The only strategy to occur during school time was the Teaching in Practice day. So pragmatically, the Alumni Conference and the Mentoring Mosaics strategies are more sustainable. We have now held five Alumni Conferences and they are growing in popularity. Many of our prospective teachers look forward to returning to the campus once they begin teaching and the alumni teachers look forward to engaging once again in a familiar learning space forging deeper connections with those they studied with while forming new bonds with those still enrolled at the University.

The real benefit in creating this "third space" (Guitierrez, 2008) was to encourage 'boundary crossing' as the collective of mathematics teacher educators from universities and schools come together in less hierarchical ways to enhance the learning of prospective teachers. We wanted to create networks to promote teacher identity formation for our prospective teachers while at the same time expanding our boundary definition of mathematics teacher educators to include those alumni teachers who offer their valuable experience as an asset to our mathematics teacher education program. We wanted our prospective teachers to feel as though 'they belong' before they begin teaching. We wanted them to have a safe space to ask difficult questions and not be judged. We believed that through bringing our alumni teachers back on campus as mathematics teacher educators, they would be willing to join us in these endeavours. They were and they did!

We recognise our conclusions are based on the reflections of a small group of participants in these three networks but as mathematics teacher educators, we have learnt a great deal about connecting and creating a community of practice. We plan to continue to implement these strategies and to conduct further research. We also plan to follow some of the participants as they finish their initial teacher education program and continue into the profession, collecting longitudinal data about their journey and identity formation.

DEDICATION

Judy and Debbie would like to dedicate this chapter to the memory of Associate Professor Leon Poladian who sadly passed away on 13th February 2018 during the development and early drafting of this chapter. A mathematician, Leon was our collaborator, our friend, and an insightful colleague, dedicated to supporting his mathematics teacher educators colleagues and to developing new ways to support our prospective teachers.

REFERENCES

Ambrosetti, A. (2014). Are you ready to be a mentor? Preparing teachers for mentoring pre-service teachers. *Australian Journal of Teacher Education, 39*(6), 30–42.

Ambrosetti, A., & Dekkers, J. (2010). The interconnectedness of the roles of mentors and mentees in pre-service teacher education mentoring relationships. *Australian Journal of Teacher Education, 35*(6), 42–55.

Beauchamp, C., & Thomas, L. (2009). Understanding teacher identity: An overview of issues in the literature and implications for teacher education. *Cambridge Journal of Education, 39*(2), 175–189.

Booth, S. E. (2012). Cultivating knowledge sharing and trust in online communities for educators. *Journal of Educational Computing Research, 47*(1), 1–31.

Bullough Jr., R. V. (2005). Being and becoming a mentor: School-based teacher educators and teacher educator identity. *Teaching and Teacher Education, 21*(2), 143–155.

Buchanan, J., Prescott, A., Schuck, S., Aubusson, P., Burke, P., & Louviere, J. (2013). Teacher retention and attrition: Views of early career teachers. *Australian Journal of Teacher Education, 38*(3), 112–129.

Campbell, M. R., & Brummet, V. M. (2007). Mentoring preservice teachers for development and growth of professional knowledge. *Music Educators Journal, 93*(3), 50–55.

Cochran-Smith, M., & Lytle, S. L. (1999). Relationships of knowledge and practice: Teacher learning in communities. *Review of Research in Education, 24*, 249–305.

Collie, R. J., Martin, A. J., & Frydenberg, E. (2017). Social and emotional learning: A brief overview and issues relevant to Australia and the Asia-Pacific. In E. Frydenberg, A. Martin, & R. J. Collie (Eds.), *Social and emotional learning in Australia and the Asia-Pacific* (pp. 1–13). Singapore: Springer.

Darling-Hammond, L. (2010). Teacher education and the American future. *Journal of Teacher Education, 61*(1–2), 35–47.

De Jong, D., & Campoli, A. (2018). Curricular coaches' impact on retention for early-career elementary teachers in the USA: Implications for urban schools. *International Journal of Mentoring and Coaching in Education, 7*(2), 191–200.

Feiman-Nemser, S. (1998). Teachers as teacher educators. *European Journal of Teacher Education, 21*(1), 63–74.

Feiman-Nemser, S. (2001). From preparation to practice: Designing a continuum to strengthen and sustain teaching. *Teachers College Record, 103*(6), 1013–1055.

Goos, M. E. (2015). Learning at the boundaries. In M. Marshman, V. Geiger, & A. Bennison (Eds.), *Mathematics education in the margins* (Proceedings of the 38th Annual Conference of the Mathematics Education Research Group of Australasia, pp. 269–276). Sunshine Coast, Australia: MERGA.

Goos, M. E., & Bennison, A. (2008). Developing a communal identity as beginning teachers of mathematics: Emergence of an online community of practice. *Journal of Mathematics Teacher Education, 11*(1), 41–60.

Graven, M. (2004). Investigating mathematics teacher learning within an in-service community of practice: The centrality of confidence. *Educational Studies in Mathematics, 57*, 177–211.

Grossman, P., Wineburg, S., & Woolworth, S. (2001). Toward a theory of teacher community. *The Teachers College Record, 103*, 942–1012.

Gutierrez, K. D. (2008). Developing sociocultural literacy in the third space. *Reading Research Quarterly, 43*(2), 148–164.

Holmes Group. (1990). *Tomorrow's schools. Principles for the design of professional development schools*. East Lansing, MI: Holmes Group.

Hong, J., Day, C., & Greene, B. (2018). The construction of early career teachers' identities: Coping or managing? *Teacher Development, 22*(2), 249–266.

Hudson, P. (2013). Mentoring as professional development: Growth for both mentor and mentee. *Professional Development in Education, 39*(5), 771–783.

Inzer, L. D., & Crawford, C. B. (2005). A review of formal and informal mentoring: Processes, problems, and design. *Journal of Leadership Education, 4*(1), 31–50.

Izadinia, M. (2015). A closer look at the role of mentor teachers in shaping preservice teachers' professional identity. *Teaching and Teacher Education, 52*, 1–10.

Jaworski, B. (2001). Developing mathematics teaching: Teachers, teacher educators, and researchers as co-learners. In F. Lin & T. J. Cooney (Eds.), *Making sense of mathematics teacher education* (pp. 295–320). Dordrecht: Kluwer Academic Publishers.

Lave, J., & Wenger, E. (1991). *Situated learning: Legitimate peripheral participation.* Cambridge: Cambridge University Press.

Le Cornu, R., & Ewing, R. (2008). Reconceptualising professional experiences in pre-service teacher education: Reconstructing the past to embrace the future. *Teaching and Teacher Education, 24*(7), 1799–1812.

Lieberman, J. (2009). Reinventing teacher professional norms and identities: The role of lesson study and learning communities. *Professional Development in Education, 35*(1), 83–99.

Loughran, J. J. (2002). Effective reflective practice: In search of meaning in learning about teaching. *Journal of Teacher Education, 53*(1), 33–43.
Macià, M., & García, I. (2016). Informal online communities and networks as a source of teacher professional development: A review. *Teaching and Teacher Education, 55*, 291–307.
MacLure, M. (1993). Arguing for yourself: Identity as an organising principle in teachers' jobs and lives. *British Educational Research Journal, 19*(4), 311–323.
Marginson, S., Tytler, R., Freeman, B., & Roberts, K. (2013). *STEM: Country comparisons*. Melbourne: The Australian Council of Learned Academies.
Mason, K. O. (2013). Teacher involvement in pre-service teacher education. *Teachers and Teaching, 19*(5), 559–574.
McConnell, T. J., Parker, J. M., Eberhardt, J., Koehler, M. J., & Lundeberg, M. A. (2013). Virtual professional learning communities: Teachers' perceptions of virtual versus face-to-face professional development. *Journal of Science Education and Technology, 22*(3), 267–277.
McGraw, R., Lynch, K., Koc, Y., Budak, A., & Brown, C. A. (2007). The multimedia case as a tool for professional development: An analysis of online and face-to-face interaction among mathematics pre-service teachers, in-service teachers, mathematicians, and mathematics teacher educators. *Journal of Mathematics Teacher Education, 10*, 95–121.
Miles, M. B., & Huberman, A. M. (1994). *Qualitative data analysis: A sourcebook*. Beverly Hills, CA: Sage Publications.
Morgan, M., Ludlow, L., Kitching, K., O'Leary, M., & Clarke, A. (2010). What makes teachers tick? Sustaining events in new teachers' lives. *British Educational Research Journal, 36*(2), 191–208.
Murray, S., Mitchell, J., & Dobbins, R. (1998). An Australian mentoring program for beginning teachers: Benefits for mentors. *Australian Journal of Teacher Education, 23*(1), 22–28.
Office of the Chief Scientist. (2016). *Australia's STEM workforce: Science, technology, engineering and mathematics*. Canberra: Commonwealth of Australia.
Paris, L. (2010). Reciprocal mentoring residencies … better transitions to teaching. *Australian Journal of Teacher Education, 35*(3), 14–26.
Sandholtz, J. H., & Finan, E. C. (1998). Blurring the boundaries to promote school-university partnerships. *Journal of Teacher Education, 49*(1), 13–25.
Sanford, K., & Hopper, T. (2000). Mentoring, not monitoring: Mediating a whole-school model in supervising preservice teachers. *Alberta Journal of Educational Research, 46*(2), 149.
Schlager, M. S., & Fusco, J. (2003). Teacher professional development, technology, and communities of practice: Are we putting the cart before the horse? *The Information Society, 19*(3), 203–220.
Stephens, A. C., & Hartmann, C. E. (2004). A successful professional development project's failure to promote online discussion about teaching mathematics with technology. *Journal of Technology and Teacher Education, 12*(1), 57–73.
Trust, T., Krutka, D. G., & Carpenter, J. P. (2016). "Together we are better": Professional learning networks for teachers. *Computers & Education, 102*, 15–34.
Tully, D., Poladian, L., & Anderson, J. (2017). Assessing the creation of value in a community of practice linking pre-service and in-service mathematics teachers. In A. Downton, S. Livy, & J. Hall (Eds.), *40 years on: We are still learning!* (Proceedings of the 40th Annual Conference of the Mathematics Education Research Group of Australasia, pp. 522–529). Melbourne, Australia: MERGA.
Wagner, J. (1997). The unavoidable intervention of educational research: A framework for reconsidering research-practitioner cooperation. *Educational Researcher, 26*(7), 13–22.
Walkington, J. (2005). Becoming a teacher: Encouraging development of teacher identity through reflective practice. *Asia-Pacific Journal of Teacher Education, 33*(1), 53–64.
Weldon, P. (2018). Early career teacher attrition in Australia: Evidence, definition, classification and measurement. *Australian Journal of Education, 62*(1), 61–78.
Wenger, E. (1998). Communities of practice: Learning as a social system. *Systems Thinker, 9*(5), 2–3.
Wenger, E. (2000). Communities of practice and social learning systems. *Organization, 7*(2), 225–246.
Wenger, E., Trayner, B., & de Laat, M. (2011). *Promoting and assessing value creation in communities and networks: A conceptual framework*. Heerlen: Ruud de Moor Centrum.

Williams, J. (2010). Constructing a new professional identity: Career change into teaching. *Teaching and Teacher Education, 26*(3), 639–647.

Zeichner, K. (2010). Rethinking the connections between campus courses and field experiences in college- and university-based teacher education. *Journal of Teacher Education, 61*(1–2), 89–99.

Judy Anderson
Sydney School of Education and Social Work
The University of Sydney

Deborah Tully
Sydney School of Education and Social Work
The University of Sydney

MAHA ABBOUD, ALINE ROBERT AND JANINE ROGALSKI

6. EDUCATING MATHEMATICS TEACHER EDUCATORS

The Transposition of Didactical Research and the Development of Researchers and Teacher Educators

Based on our work, a theoretical framework is presented addressing design and analysis of educating mathematics teacher educators as a preliminary step towards teacher education. The double approach is extended to articulate a didactical stance centered on students' mathematical task analysis and an ergonomical one oriented toward analyzing teachers' practices from the mathematics teacher educator standpoint. The approach involves both a transposition of didactical research from didacticians to mathematics teacher educator and a process of co-development between them. Another line of extension takes into account the tensions the introduction of technology triggers or emphasizes. The conclusion underlines research issues concerning the relationship between being a mathematics teacher and becoming a teacher educator.

INTRODUCTION: CONTEXT AND BACKGROUND

This chapter addresses the design and analysis of education for mathematics teacher educators as introduced earlier by Abboud-Blanchard and Robert (2013). Henceforth we use the term *didacticians* to designate researchers working in the domain of mathematics education (mathematical didactics in the French context), whether or not they are involved in the education of mathematics teachers or the education of mathematics teacher educators. Teachers who are being engaged in mathematics teacher educator education programs are referred to as *prospective* mathematics teacher educators.

We examine two processes involving didacticians and prospective mathematics teacher educators: the transposition of key didactical elements and co-development. We illustrate our work by drawing upon a French mathematics teacher educator education program. In this case, prospective mathematics teacher educators are experienced secondary school teachers who (only) teach mathematics. To help the reader understand the two processes, we describe a current example of a mathematics teacher educator education program.

© KONINKLIJKE BRILL NV, LEIDEN, 2020 | DOI: 10.1163/9789004424210_007

The Context

In the French education system, mathematics teacher educators intervene at various points in teacher education (initial, continuous, with respect to specific mathematical content, etc.). However, until recently, there was no dedicated education program for them, and they are usually selected by pedagogical authorities on the basis of the quality of their teaching.

Here we present a university-level mathematics teacher educator education program that was designed and conducted by the two first authors of this chapter. This professional master's degree runs over two years. The first year is generalist (taught by Aline Robert), while the second is partly focused on using technology to teach mathematics (taught by Maha Abboud). The course has run for more than ten years. The first year has been shown to be reproducible by another didactician (from the same research team). Prospective mathematics teacher educators must have at least five years' experience as secondary school mathematics teachers and continue teaching during the program.

It should be noted that the program was not designed as a research project or an experiment, which is why an indirect assessment method was adopted. This method is often used in ergonomic psychology when there are no empirical data and consists of *a posteriori* questioning of the designers and educators. Here, we apply the insider/outsider distinction introduced by Goodchild (2007) to research on educating mathematics teachers. The 'outsider,' this chapter's third author (Janine Rogalski), who was not involved in the program, questioned the 'insiders' (the first two authors).

Educating Teachers and Teacher Educators

We argue that prospective teacher education must help teachers to understand, outside the classroom environment, what a teacher really does with their students, and support their investment in the teaching function. Education should also be extended to practising teachers, as it enriches their practices by exposing them to new notions or the use of new instruments. It can also look at alternative approaches, or notions that can be implicit, and which are often ignored at the beginning of their career.

In our practice, we have often observed that teachers find it difficult to address these points by themselves, because of the need to take a step back. Another complication comes from the fact that there is no 'gold standard.' Some common threads are: adaptability to the context; consideration of students' diversity; and the transposition of mathematical content. These features are the fruit of experience, but experience alone cannot highlight similarities and differences. In other words, providing tools that can be adapted to specific needs seems to be a more useful approach than precise prescriptions.

The Crucial Role of Mathematics Teacher Educators

Mathematics teacher educators play a crucial role in educating mathematics teachers. However, to be fully effective, they themselves need to be educated. We therefore designed an initial mathematics teacher educator education program based on our research. Prospective mathematics teacher educators who take the program are experienced teachers who bring practical experience that enriches the formal, general, and decontextualized knowledge taught during the program. Their experience gives them a legitimacy that didacticians, who may not necessarily have experience of teaching at secondary level, may lack. Participants also develop the capacity to adapt to different practices, based upon the tools presented during the program.

Our experience suggests that a key issue in mathematics teacher educator education is to be able to design and deliver a program that enriches teachers' practices.[1] It is, therefore, strongly oriented toward analyzing teaching practices, particularly when teachers interact with students in the classroom. In fact, we consider the relation between teachers' practices (tasks and their implementation in the classroom) and students' potential learning as the bedrock of mathematics teacher educator education. Prospective mathematics teacher educators are expected to be able to integrate this into their work.

The tools offered to mathematics teacher educators are crucial, and we believe that they should be co-developed by didacticians and prospective mathematics teacher educators to ensure that they are relevant. A key challenge is the need for mathematics teacher educators to understand and take ownership of didacticians' discourse. To achieve this, it is essential to select appropriate *operating modalities* – and not only *content*. In particular, the practical experience of prospective mathematics teacher educators should play a part in selecting the tools used in the analysis of practices. A *shared working environment* needs to be created, which merges didacticians' mathematical and theoretical knowledge, and the experience of prospective mathematics teacher educators. The didactician presents general analytical tools that gain meaning as mathematics teacher educators are able to relate them to their experience. Through this process, tools continue to be modified in ways that enrich teachers' knowledge.

International Perspectives

Our approach to mathematics teacher educator education does not tackle the issue of development through practice (Zaslavsky & Leikin, 2004), nor does it see mathematics teacher educators as observers of their own practices through "self-reflective analysis" as expressed by Tzur (2001). Furthermore, we do not directly deal with the issue of "mathematics knowledge for teaching," which is a specific line of research on mathematics teacher practices and education as expressed in Beswick and Chapman (2012, 2013) or in Voogt, Fisser, Pareja Roblin, Tondeur, and van Braak (2013) for "technological pedagogical content." The relationship with mathematics teacher educators' own practices is only indirectly considered through the issues of co-development and transposition[2] from research to mathematics teacher educator education and mathematics teacher education and does not directly address "reflective practitioners" and "researchers" (as in Chapman, 2009, or Tzur, 2001).

In general, the literature on teacher education is more extensive than for mathematics teacher educator education (Krainer & Llinares, 2010). This is also true in France (Robert, Roditi, & Grugeon, 2007). Nevertheless, the growing interest in teacher education over the past decade has increased interest in how to educate mathematics teacher educators. Research is often based on case studies of the transition from classroom teacher to university teacher educator (Zeichner, 2005). Another approach relates to collaborations between researchers and teachers who consequently become mathematics teacher educators (Lin & Cooney, 2001). Within this field some studies have sought to use research and theory to enrich teachers' practices (Lin et al., 2018). Another line of investigation is teacher inquiry groups, and communities of practice which now play an important role in teacher education. The relationship between researchers working in the domain of mathematics education and mathematics teachers has often been noted as a possibility of development (Krainer, 2014). This diversity in approaches to teacher education can be found, to some extent, in mathematics teacher educator education.

This chapter is structured as follows. Our theoretical perspectives are presented in the second (next) section. The design of the mathematics teacher educator education program is developed in the third section and illustrated by some examples. The fourth section discusses the two focal points of our work: the transposition of didactical research to mathematics teacher educator education and the co-development, of didacticians and prospective mathematics teacher educators. The conclusion returns to the main issues of our study, addresses some open questions and examines future research perspectives for teacher education.

THEORETICAL PERSPECTIVES ORIENTING MATHEMATICS TEACHER EDUCATOR EDUCATION

Like many other researchers, our starting point is the teaching goal: students' mathematical learning. From a constructivist Piagetian stance, the acquisition of mathematical notions is the product of students' activities when performing mathematical tasks, or in reflecting upon their performance (reflective abstraction). However, in classroom contexts, students are not left alone to face a task. Their activities depend not only on the teacher's choice of content (the cognitive dimension of the 'cognitive route'), but also on the implementation of the chosen tasks (the mediative dimension), including times when the teacher addresses the class (the teacher's telling moments).

Hence, our analysis focuses not only on the type of mathematical tasks teachers choose for their students, but also on the way they implement the cognitive route in the classroom and provide assistance to students, via mathematical aids that are oriented toward task performance or developing the concepts needed to perform the task (in a *scaffolding* process).

While many other factors affect students' learning (such as their social origins or interactions within the class), the teacher's practice is at the heart of their learning opportunities. These opportunities are closely linked to the relation between tasks, students' activities, and the teacher's interventions, which are more or less involving the students' Zone of Proximal Development (ZPD) (Vygotsky, 1986, chap. 6).[3]

Analyzing teachers' practices in the classroom, and taking into account their complexity, is at the heart of our conception of mathematics teacher educator education. In the case we analyze below, we highlight both the tools used for this purpose, and some types of classroom analyses. This approach constitutes a possible answer to the "*challenge of educating educators*" discussed some years ago by Even (2008), as it aims to help prospective mathematics teacher educators to take a step back from their own teaching experience and opens up a wider range of possibilities for educating teachers.

Frameworks to Analyze Teachers' Practices

The Double Approach (Robert & Rogalski, 2005; Robert & Hache, 2013) encompasses a didactical stance that is centered on students' mathematical task analysis, and an ergonomics approach that is oriented toward analyzing teachers' practices in relation to students' mathematical knowledge and activities. The ergonomic stance is extended to teaching practices from the standpoint of teacher education in the context of prospective mathematics teacher educators' practices. Teachers' choices are not simple, and they are not only determined by learning purposes; they also involve external institutional and social constraints, such as the curriculum, the classroom atmosphere, and even the social context. It is clear that personal determinants also play a role in teachers' practices. For individual teachers, these dimensions are interlinked with cognitive and mediative ones, in a way that may explain the stability of their practices (Paries, Robert & Rogalski 2013).

With respect to Schoenfeld's model of mathematics teachers' activity (Schoenfeld, 2010), a main difference with our work lies in how social and institutional dimensions can be considered: for us, they are determinants that interact with the individual dimension, while for Schoenfeld the personal dimension of "beliefs" integrates these social and institutional elements.

Teachers Practices as the Target of Mathematics Teacher Educator Education

We believe that classroom practice and the corresponding choices lie at the heart of mathematics teacher educator education. It is necessary to consider not only cognitive and mediative dimensions, but also professional factors that impact these practices (Robert & Hache, 2013, pp. 50–51). Mathematics teacher educator education must make participants aware of the need to prepare teachers to make inevitable adaptations.

Our earlier research shows that teachers' practices quickly become stable – particularly the mediative dimension (Chappet-Pariès, Robert, & Rogalski, 2013) – making them difficult to modify. The best we can hope for is to enrich them. For this reason, educating mathematics teacher educators (and consequently teachers), by simply presenting and discussing their teaching practices seems to be insufficient, as what emerges is the result of cognitive, individual, social and institutional choices that are not easily visible. Other theories about learning or teaching – whether drawn from cognitive psychology or the didactics of mathematics[4] – also seem insufficient, as many of the transpositions/adjustments that are at the heart of activities in the classroom have to be made by teachers themselves.

Schematically, our position is that prospective mathematics teacher educators have to develop key elements of knowledge about this complexity from a pragmatic perspective: They must be given '*analytical tools*' and be equipped with '*words to express it*' in order to go beyond their individual experience, and transpose what they know into a more depersonalized and decontextualized approach. In the course of our earlier research, and unlike other studies, we have developed tools that can be used by mathematics teacher educators either within or outside a research context. Indeed, generally the findings of theoretical research are limited, and not intended to have a practical use or to be directly exported into practice. Often, the scope and limits are not identified by researchers, nor are students and teacher' difficulties identified during task implementation.

In our work, mathematics teacher educator education does not aim to change learners into "reflexive practitioners," even if a certain degree of reflexivity may be the product. Neither is it intended to transform them into researchers. Instead, the aim is to equip them to enrich teachers' practices, not only by working with teachers on mathematical content, but also by working with them on day-to-day practice (Abboud-Blanchard & Robert, 2013). Prospective mathematics teacher educator education has to take into account the personal and collective Zone of Proximal Professional Development (ZPPD) – and the same is true for educating teachers.

From an ergonomic point of view, if mathematics teacher educators are seen as enriching teaching practices, we must consider another level of analysis. When mathematics teacher educators educate teachers, both the content and the situation are contextualized. With a greater range of tools, other possibilities open up. These interventions aim to expand teachers' awareness, by asking them to focus not only on their students (the 'productive' dimension of their activity), but also to pay attention to their own actions (the 'constructive' dimension).[5] Educating mathematics teacher educators constitutes a third level, as the didactician encourages prospective mathematics teacher educators to think about their own teaching activities.

These three levels are not embedded in the way Zavlasky and Leikin (2004) presented Jaworski's embedded triad for educators of mathematics teachers (p. 8). In fact, teachers' practices in mathematics are a sort of "common object" that researchers, didacticians, prospective mathematics teacher educators, and teachers have different attitudes to.

The Zone of Proximal Professional Development

Like other didacticians involved in teacher education, we consider that teaching students in order to develop their mathematical conceptualization requires taking into account their zone of proximal development.

We borrow this notion and adapt it to mathematics teacher educator education practices. The zone of proximal professional development refers to what teachers have experienced, without necessarily being able to talk about, or even identify it. As the mathematics teacher educator is able to put this experience into words (based on research), prospective mathematics teacher educators become able to use the same notions appropriately, leading to a more general reflection on teaching practices. This proximity between the concepts used by didacticians who are familiar with teachers' practices and teachers' own experience is a 'go-between' between experience and knowledge. At the same time, the didactician has to adapt his/her knowledge to the reality described by the prospective mathematics teacher educators (cf. co-development). For example, a group analysis of videos can provide a link between the actual needs of prospective mathematics teacher educators and what the didactician thinks they need. In turn, the prospective mathematics teacher educators may introduce new elements about observed practices depending on what their students have said. For instance, options include a specific question such as, "Why do you think that this question is important?," or a more general observation such as, "What alternative would you suggest?" In practice, the question depends on the goal: prospective mathematics teacher educators are interested in general principles, while this is not the case for teachers.

THE MATHEMATICS TEACHER EDUCATOR EDUCATION PROGRAM

In this section, we present an outline of our mathematics teacher educator education program, then illustrate the approach with an example that describes in detail some of the tools and the methods taught to prospective mathematics teacher educators.

The Modules of the Program: Principal Components

The program spans two years. The first, 'Generalist,' year is divided into three modules. The second, 'Technology,' year begins with a 'Technology module' designed according to the same theoretical perspectives as in the first year and continues with two other, more general modules.

The first module of the Generalist year begins with an introduction to task analysis. Video excerpts provided by the didactician are analyzed. As these excerpts focus on a lesson in progress, they enable participants to make real-time observations and simulate partial immersion in the classroom. The exercise therefore examines both the work done by the teacher in the classroom, and upstream preparations. Prospective mathematics teacher educators are asked to consider alternatives and a range of possible practices in relation to the notion to be learnt, the different contexts, and the different personal points of view.

The second module (of the first year) consists of each prospective mathematics teacher educator presenting their analysis of a video excerpt from his/her own class and leading a discussion on what he/she would do in a teacher education session, following the format presented in the first module. This is fundamental, as it enables everyone to follow the process and ask questions, helped by the other members of the group and the didactician. Furthermore, each prospective mathematics teacher educator interviews some colleagues who previously attended other teacher education programs. The summary of the results of these surveys helps to clarify the expectations of their future 'teachers to educate' and anticipate any weak points, particularly with regard to educating modalities.

The third module (of the first year) is devoted to the design of a hypothetical teacher education scenario. Prospective mathematics teacher educators work in small groups, each of which chooses a theme, spends two months working on it, and presents the results at the end of the year. The group presentation includes the key elements of the scenario, the main choices and their justification, and a full-scale animation of a specific moment. This exercise is designed to complement the surveys carried out during module 2. It helps prospective mathematics teacher educators to become aware of the diversity of teacher education processes, move away from the myth that there is one good way to teach, and move towards the idea of generating teaching scenarios.

The technology module runs over one semester during the second year. Participants are asked to create a collaborative environment where bottom-up and top-down dynamics work together. The bottom-up dynamic consists of prospective mathematics teacher educators sharing their experience, practice, vision, and knowledge related to the use of technology. The top-down dynamic consists of the didactician presenting the results of research that can help in critically considering the issues being debated. Coursework is organized into topics that are related to the use of specific technology tools. The latter are chosen at the beginning of the course and are related to the expertise of participants (dynamic geometry, algorithmic software, interactive whiteboards, etc.). For each topic, one or two sessions are organized based on the two dynamics presented above. First, a group prepares a summary of their own experience of the use of the technology tool in their classroom: issues include institutional constraints, physical conditions, the benefits of using the tool, the range of tasks proposed to students, difficulties encountered by the teacher and their solutions. Secondly, the didactician presents an overview of relevant research that takes account of a variety of issues, including potentialities and limits of the tool, classroom management, tensions in the classroom, and possible disturbances in the students' cognitive route. This work is based on reading and discussing papers, or on observing video recordings provided by the didactician that show different practices.

As in the first year, prospective mathematics teacher educators have to work in pairs and design a scenario that can be used in teacher education focused on technology uses. Unlike the first year, this scenario must draw upon their own technology-based lessons. They are asked to draft a paper presenting this scenario and discuss it with the whole group. Overall, the exercise consists of preparing a lesson and implementing it in their classroom. Then, they use the video recording of the lesson as a basis to design a scenario and prepare a paper that explains the theory underlying their work. The result is that participants clarify the implicit elements they encounter during the creative phase, and project them into a real teacher education context.

A Typical Example from the Beginning of the First Year

The module requires participants to understand task analysis and its implementation. The didactician informs prospective mathematics teacher educators that the course will begin with a detailed presentation of these concepts and their application in practice, before addressing issues related to teacher education. She then presents the exercises shown in Figures 6.1 and 6.2, which are part of the assessment of 10th grade students who have learnt the conditions for two triangles to be similar. She shows a video in which the teacher corrects students' answers after they have received their marks.

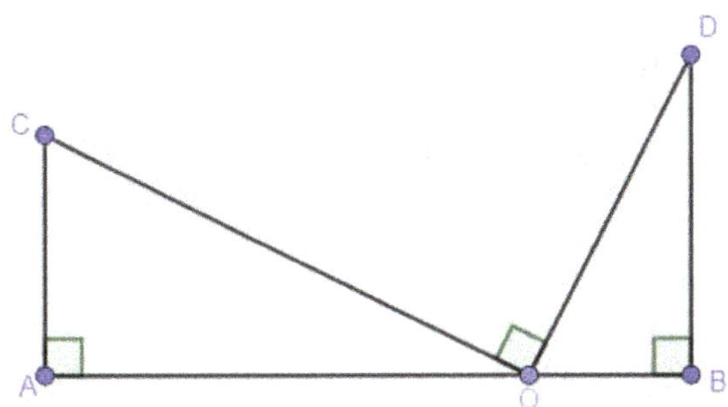

Show that the triangles CAO and OBD are similar

Figure 6.1. Exercise A

The first question the didactician asks is: "Which exercise seems easier to you?" Usually, prospective mathematics teacher educators quickly answer that exercise A is easier because the figure in exercise B is complicated. Then the didactician reveals

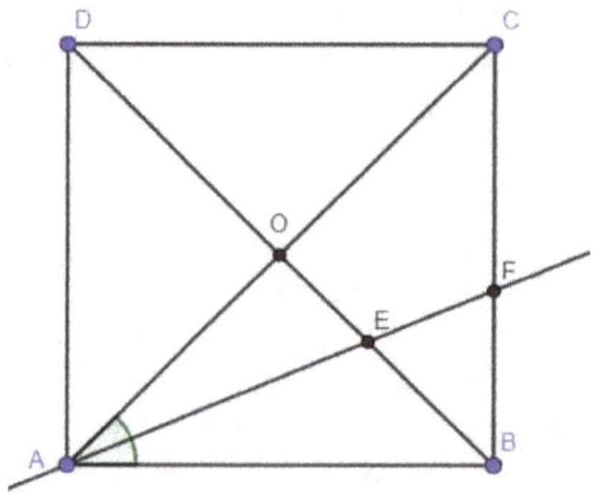

ABCD is a square; (AF) is the bisector of the angle CAB.
Show that the triangles AOE and ABF are similar.

Figure 6.2. Exercise B

that almost all students completed the second exercise successfully, unlike the first. So, her second question is: "Why?" Then, she introduces an analysis of the task that may explain the result. Students have to prove that the two angles of the triangles are respectively equal. But, as the triangles are right angled, it is sufficient to show that another angle of the first triangle is equal to one of the angles of the second. Therefore, the first question can be resolved using something like: AOC + COD + DOB = 180°, so AOC + DOB = 90°, but AOC + OCA = 90°, hence COA = DOB. As the values of the angles are unknown, the task requires the adaptation of knowledge and the use of algebraic transitivity. In the second question, students have only to understand that the diagonals of a square are perpendicular, and the property of the bisector.

The video highlights the teacher stating that the difference in success rates is due to a difference between forgotten, 'old' knowledge about complementary angles, and current knowledge about the properties of a bisector. There is no reference to the adaptation that is made evident by the task analysis. The issue becomes clear: if he had referred to this adaptation maybe more students would have understood the difference between the two exercises and would have been more successful when working on the first. Finally, however, it should be noted that the didactician nuances this conclusion by stressing the complexity of teaching practices: it is possible that, at this stage, the teacher wanted his students to focus on reviewing their 'old' knowledge.

The A Priori Analysis of Mathematical Tasks as a Key Tool

The following sessions of the first module are devoted to the analysis of exercises, and then the tools to be used in such analyses are presented. However, as these tools are designed for prospective mathematics teacher educators, participants have to transpose them into their own teacher education.

Knowledge adaptation. Solving a mathematical task requires the use of old and new knowledge. What we are interested in here is the way in which students use their knowledge. Consequently, task analysis does not directly refer to the potential learning benefits of an exercise, rather the aim is to identify the activities students will undertake in the context of the exercise, given their assumed level of knowledge (Horoks & Robert, 2007). Clearly, the analysis depends on the level of education and the class to be taught. For tasks that are not simple and isolated (such as the immediate application of new knowledge) we determine the adaptations that students must implement (Robert, 1998) (cf. Appendix).

This analysis requires us to examine certain features of the notions to be learnt. More precisely, we need to identify the different ways to perform the task, identify whether the required knowledge has already been learnt or not, and anticipate students' difficulties. Therefore, we merge these three issues into one, called the relief study. This consists of a mathematical analysis of the notion in question, an analysis of the curricula, and the classification of students' difficulties. The aim is to list expected task activities and design some ways to approach the notion.

Level of knowledge and adaptions. For simple isolated tasks, students' work is examined at a 'technical' level. When tasks seem to require a certain level of knowledge adaptation, we draw upon the notion of a 'mobilizable' level of knowledge. When students are asked to use knowledge they already have, we refer to the 'available' level of knowledge. In our work, students must be able to reach the third level for most tasks, and the design of scenarios deduced from the relief study has to take this into account (Robert & Rogalski, 2002).

Tools for the Analysis of In-Class Events

As we explain below, the task analysis does not provide comprehensive information about possible student activities, and it must be supplemented by an analysis of the implementation of tasks in the classroom. First, the didactician shows videos to prospective mathematics teacher educators; here, the aim is to compare the *a priori* task analysis and the implementation shown in the video. This study is followed by an explicit analysis of the local, *a posteriori*, implementation. These modalities are specific to prospective mathematics teacher educator education. Teacher education gives less time to such an analysis, and there is no direct link made with the actual teaching of participants. The analysis of the relationship between expected activities and their implementation in the classroom highlights potential student activities. The following, systematic questions are thus addressed.

The form and nature of students' work and the didactic contract. These elements clarify students' activities and lead to the identification of the role of all interactions in the classroom. The didactic contract, along with the habits and memory of the class, also plays a role. Students may, for example, engage in a task because they have understood that this is what the teacher expects, and not necessarily for mathematical reasons.

Teacher interventions. One important source of observable elements relates to the classroom discussion and what is said by the teacher, whether through soliciting students' responses, responding to them, or developing a didactic project. These elements are analyzed in terms of their assumed influence on students' activities. Some relate to the format of interactions with students, while others concern content (assistance, assessment, reminders, explanations, corrections and evaluations, presentation of knowledge, mathematical content, etc.). One important variable is the proximity of these interventions to students' acquired knowledge (i.e., their zone of proximal development).

Modalities of teacher assistance. Here, we define the type of assistance, the moment when it is given, and its format. We distinguish two types as a function of whether they modify expected activities or whether they add something to students' actions.

The first type is procedural. It corresponds to statements made by the teacher before or during students' work, and includes open-ended questions such as "What theorem can you use?" It can lead to subdividing the task into explicit subtasks, or having students choose a contextualized method. In general, it changes task adaptations. The second type is constructive. It intervenes between specific student activities and the construction of the desired knowledge. It may consist of a simple summary of what was done, even immediately after it was done, or reminders, partial generalizations, assessments and the like. This assistance can partly decontextualize students' work, for example by presenting the corresponding generic case. Constructive interventions help students put what they have done into perspective, find a (slightly) more general method, and discuss results. For a given student, procedural assistance can become constructive if the student extracts a generalization. In other cases, constructive assistance can remain procedural.

By extending this typology to technological environments, we add a dimension that takes into account the instrumental nature of assistance (Abboud-Blanchard, 2014; Abboud & Rogalski, 2017b). We distinguish assistance that refers to the use of software (and is unrelated to the associated mathematical content), and assistance that links these tools to mathematics. The first type is termed 'manipulative instrumental' assistance. The second combines the tool and mathematical objects and is termed 'mathematical instrumental' assistance. Note that the procedural/ constructive distinction also applies to instrumental assistance.

Another type of instrumental assistance concerns the articulation between different environments, either physically present or referred to. This occurs when the teacher asks students to say or do something that they have already done in a more familiar environment. For example, if a formula is to be entered on a spreadsheet, the teacher can ask the student to do it as if they were using a paper-and-pencil, but not forget to put the symbol "=" at the beginning. They can also remind students that the spreadsheet syntax must be respected, as they are used to doing with scientific calculators. This type of assistance is called 'inter-instrumental' assistance.

Corrections: Zooming in on teachers' 'meta' comments. We differentiate several types of correction (oral, written, continuous, at the end of an exercise, by the teacher, by students, etc.). This classification can be compared with the role of errors in learning. Errors can reveal false or incomplete representations that can remain undetected if nothing specific is said on the subject, or if the correction phase does not bring them to light. Simply presenting a model solution can mean that they are unnoticed by the teacher and their students.

Several variables are important in the knowledge presentation phase. Within the overall dynamics of the contextualization/de-contextualization process, issues include: the order in which the different phases unfold, relationships between exercises (contextualization), knowledge exposition (de-contextualization), or even the format of the course (lecture-based, interactive, dialogue-based). Here, we pay particular attention to 'meta' comments on the mathematics. We determine whether such comments exist, if they only refer to mathematical knowledge (or address it more broadly), or present alternative worked solutions. Meta comments may also explicitly present the structure of the lesson, notably proofs, or summarize the new knowledge that has been learned in light of the problems students have encountered. This is found, for example, in the case of FUG[6] (Robert, 1998, p. 64) where no introductory problem is given because the distance between what is known and what is to be learned is too great. Meta comments may also include a presentation of future applications of the concept.

What Characterizes the Design of the Mathematics Teacher Educator Education Program?

It is important to remember that the ultimate goal is students' learning. Over the past decade, our mathematics teacher educator education program has been designed and developed with this goal in mind. The program is not only *extensive* and *collaborative* as we explained above, but also inductive, random, and holistic.

Inductive, because it is based on the response of prospective mathematics teacher educators when they are exposed to our theoretical tools. Their practice as teachers enables them to give meaning to the theory that we present, which enriches their analyses and discussions. *Random,* because there is a degree of randomness in the contents of the videos that are analyzed in the first year, and also in the technologies chosen to be analyzed in the second year. *Holistic,* because these analyses and discussions focus on the entire teacher's activity, and not on single tasks or their management. It places students' activities at the center. All of these characteristics contribute, in turn, to the development of the didacticians themselves, both as researchers and as mathematics teacher educator educators.

TRANSPOSITION AND CO-DEVELOPMENT

The method of investigating the mathematics teacher educator education program used below was developed as an 'outsider' view at each of the two years of the program. The non-educator didactician, Rogalski, placed herself in an 'external' position, taking the point of view of ergonomic psychology and professional didactics: Semi-directed interviews were held with the two insider didacticians (Abboud and Robert), and addressed two specific issues. The first was the transposition of research from didacticians to prospective mathematics teacher educators. The second was co-development between, on the one hand, didacticians and prospective mathematics teacher educators and, on the other hand, within the group of prospective mathematics teacher educators. The results are presented for each year separately, beginning with the Generalist year. The findings relating to the Technology year require the extension of the Double Approach to technology environments (Abboud-Blanchard, 2014) taking into account the *tensions* and *disturbances* that these environments trigger or exacerbate (Abboud-Blanchard, Clark-Wilson, Jones, & Rogalski, 2018).

Transposition and Co-Development in the First (Generalist) Year

Transposition. Here, the question to be answered is, "What has mathematics didactics to offer to prospective mathematics teacher educators, who do not intend to become researchers in mathematics education?" Transposing academic practices to prospective mathematics teacher educators is not the same as educating future researchers. In general, it is widely accepted (if not proven) that educating teachers consists in working with them to improve their mathematics teaching. While didacticians' work is oriented by current research challenges, prospective mathematics teacher educators are focused on an analysis of existing teachers' practices. The goal is to start from existing practice and improve it (in-service education), or to build upon the teacher's initial experience including those acquired when being students or pupils, and – obviously – during their pre-service education.

It should be noted that in educating prospective mathematics teacher educators, didacticians as academic researchers have legitimacy for the theoretical point of view and suffer a lack of legitimacy for the practical one.[7] Nevertheless, didactical research has developed tools for practices analysis that can link theoretical and practical perspectives. These tools are the focus of the analysis of teaching practice and include the "*words-to-say-it*" – in other words, the operational language that is required to refer to, and to discuss practices.

Another aim is to encourage prospective mathematics teacher educators to replace their initial references to their own practice with new references to the practices of other teachers. We believe that this helps them to adopt a depersonalized posture and be able to listen to the teachers they will be educating in order to understand rather than to assess. Our underlying conviction is that there is no such thing as 'expert practice,' and that a variety of determinants have to be taken into account which are a function of each teacher's position and professional context.

Although the tools that are used to analyze mathematical contents and their implementation in the classroom are inspired by research in mathematics didactics, they are not used in the same way. Classroom analyses do not seek to develop a better understanding of a particular concept but are used to identify examples of effective implementation. This involves a transposition of the notion of task analysis (Robert & Rogalski, 2006), and its implementation. This transposition is based on an analysis of multiple exercises and examples selected from classroom videos. The first aim is to encourage prospective mathematics teacher educators to compare expected student activities with what actually happens, which helps them to understand the importance of their teaching practice. A second point they must understand is that practices are not only determined by students' learning needs, but also by personal and institutional considerations.

With respect to tools, transposition mainly consists in adapting them to the specific usefulness for them as future mathematics teacher educators. The breadth of analyses is more limited in mathematics teacher educator education than in research, and the order also differs: task analysis and issues related to implementation lead them to question the relief of the targeted notion in general, while researchers begin with the study of the relief.

In the case of the analysis of practices, transposition concerns both a method for examining practices (Robert & Rogalski, 2002; Vandebrouck, 2013), and research findings chosen as relevant for future mathematics teacher educators (Robert et al., 2012). Two findings are particularly important: first, the fact that teachers initially suffer from a lack of resources at the micro (classroom) level; second, the stability of practices of experienced teachers, particularly in the mediative dimension (Chappet-Pariès, Robert, & Rogalski, 2013).

In the research context, tools are part of a well-defined theoretical framework and used to address current issues; in the education context, these theoretical questions are not fully developed. Nevertheless, our experience shows that the lack of theoretical debate does not concern prospective mathematics teacher educators, although taking a more critical approach could be a future goal in their education. A first step in this direction is to encourage them to read and discuss papers directly addressing professional issues.

Co-development. What is the impact of mathematics teacher educator education on didacticians? Two effects are observed: first, they learn more about prospective mathematics teacher educators – their beliefs, expectations, responses to education; and second, they learn more about the education situation itself.

Didacticians learn more about how prospective mathematics teacher educators feel about their profession, for instance the importance given to students as individuals in their analysis of what is happening in the classroom. Prospective mathematics teacher educators require many months of experience to be able to develop a task analysis that is sufficiently independent of their students. Initially task analysis is contextualized with reference to familiar student activities, and the relative importance of curriculum and institutional requirements. This understanding is underpinned by their initial beliefs about the role of the mathematics teacher educator. Another example is that the didactician can identify differences in representations (beliefs) regarding teaching and learning mathematics, particularly the issue of transfer. Some prospective mathematics teacher educators place great value on the potential transfer of reasoning and rigor, and are less interested in the mathematical conceptualization of their students.

With respect to the program as such, two aspects were modified. Initially, prospective mathematics teacher educators were asked to analyze and summarize books. This was replaced by reading articles from the professional literature, first with the help of a "reading guide," then with help from the didactician in how to use the reading guide. The second concerns the task given to prospective mathematics teacher educators. Over the course of a term, they were asked to design a teacher education scenario, and identify and make explicit the research hypotheses that underpinned this scenario. In practice, the task could not be completed. Nevertheless, it was retained because it was considered that it was meaningful and helped to foster distancing, even if prospective mathematics teacher educators took a more pragmatic approach to their work. It became apparent that hypotheses emerged from the scenarios that were developed, and that they were consistent with students' experience (for instance, with respect to their temporal composition: questionnaires for teachers, task analysis, classroom video analysis, and so on).

Transposition and Co-Development in the Second (Technology) Year

The first semester of the second year focuses on technology-mediated lessons. It is deliberately designed to follow on from the first year, and the analysis tools that have been acquired in the first year are reused and adapted to the context of technologies. Before the tools are introduced, prospective mathematics teacher educators are made aware of several important points regarding how teachers can work with technologies in their classroom. Initially, they are asked to work in groups and reflect on the difficulties they have encountered in their technology-based sessions. The aim is to identify areas of difficulty when educating teachers in how to use technology. The didactician seeks to develop the awareness of prospective mathematics teacher educators within the general framework of the three forms of awareness distinguished by Mason (1998, p. 33). Nevertheless, there are some differences in focus. The first point we draw attention to is awareness of students' behavior when performing tasks using technology (Mason's *awareness-in-action*); the second is awareness of the implementation of tasks in the classroom (this differs slightly from *awareness-in-discipline* as defined by Mason); while the third focuses on future teachers' education (*awareness-in-counsel*).

Next, mathematics teacher educators analyze existing technology resources. It is pointed out to the group that such resources are often focused on the design of tasks. Little thought is given to what happens when they are implemented in the classroom, and neither the actual activities of students nor those of teachers are taken into account. Through this exercise, mathematics teacher educators gradually become aware of the importance of analyzing how technology is implemented. This analysis is one of the main objectives of the didactician. Mathematics teacher educators must understand the complexity of technology-based classroom practices in order to be able to design teacher education that takes this complexity into account. This understanding is developed through group work, for two reasons: first, prospective mathematics teacher educators have worked together during the previous year and second, they are able to take an analytical approach (which they could not do at the beginning of the generalist year).

Transposition. In the second year, pragmatism dominates, and there is an emphasis on increasing awareness of teaching practices based on technological tools. This is mainly due to the fact that there are currently not many analyses in the field of research on technology mediated lessons, on which educating knowledge could be rooted and developed. Still, certain concepts and tools resulting from the research are presented to mathematics teacher educators. These include instrumental orchestrations (Drijvers et al., 2010), and tensions, hiccups and disturbances (Abboud et al., 2018). These concepts can be useful tools for prospective mathematics teacher educators – whether they emerge from the teacher's own actions (Clark-Wilson & Noss, 2015) or via the researcher who works with video recordings (Abboud & Rogalski, 2017). Other concepts are presented in less detail, such as the structuring features of classroom practice (Ruthven, 2009).

In this second year, prospective mathematics teacher educators are exposed to more specialized tools. The aim is that they understand that the analysis of the mathematical task is not sufficient; it is also necessary to analyze its implementation, taking into account both the teacher's instrumented activity and that of students (Abboud & Rogalski, 2017). It is important to look at technological tools in more detail and be able to analyze resources, evaluate their adaptability (Gueudet et al., 2012), or even go so far as to experiment with them in the classroom.

For this reason, prospective mathematics teacher educators are asked to design a scenario based on a real classroom session that they have actually experienced. The task is consistent with a desire to go beyond simple task analysis and move towards an understanding of the 'non-transparency' of technologies. In this case, the scenario design differs from the equivalent task mathematics teacher educators are set in the first year: instead of having to construct a (fairly long) scenario, they have to prepare a short teacher education scenario that must be based on a real-life lesson and reflect on how they could use it in teacher education.

It should also be noted that there is a difference in emphasis between the two years. In the first year, the focus is on tools that enable teachers to become aware of what is happening in the classroom. In the second year, the main focus is prospective mathematics teacher educators' awareness of their own experience of using technology in the classroom.

Moreover, the transposition from mathematics teacher educator education to teacher education raises several issues, notably that of temporality. A tool that is developed during the two-year mathematics teacher educator education program is difficult to transpose to in-service teacher education that only lasts a few days (in the French context). The scenarios that are developed therefore highlight a shift that is observed during the course. Rather than presenting real-life situations where technologies bring added value to mathematics teaching, mathematics teacher educators move towards taking into account the conditions and constraints that may affect their implementation, even if all of the tools that they learnt about are not used.

Co-development. Usually, by the beginning of the second year, prospective mathematics teacher educators have bonded, notably as a result of having in the first year to develop in groups teacher education scenarios. The second year helps to strengthen a community that uses the same tools and vocabulary, building on their acquired knowledge.

As the didactician is not a secondary school teacher, the question of legitimacy arises. For this reason, there is a two-way process of, "I learn from them and they learn from me," based on a collective approach to learning. It is here that the process of sharing experience and knowledge becomes most apparent.

This sharing of learning brings the community even closer together. Both the didactician (as a researcher) and practitioners bring their respective expertise, and prospective mathematics teacher educators become both learners and teachers. Working together helps to facilitate discussions on the constraints that are encountered during technology-based lessons, and how they can be dealt with (e.g., machine/student interactions and possible aids).

Feedback from prospective mathematics teacher educators has generated, and continues to generate, ideas for the didactician. One example emerged when mathematics teacher educators highlighted the important, but not obvious point that the teacher "never thinks about what they say or how they help students when they switch between different technological tools used for the same task." This observation led to a group discussion and was taken up by the didactician in her own research.

At the end of the program many prospective mathematics teacher educators are more confident about their practices. Some have already found jobs as mathematics teacher educators but continue to participate in the group's work in order to expand their resources and continue their professional development. As several members noted a need to continue the group's work the didactician created a working group within the institutional frame of the Institute of Research in Mathematics Education (IREM) and invited mathematics teacher educators who had completed the program to join it on a voluntary basis. The aim is to continue to work together and explore themes related to the use of technologies in mathematics education in more detail. There is also a practical goal – to produce resources for teacher education.

Evolution of the Program

The mathematics teacher educator education program is neither fixed nor unvarying. Its evolution has three drivers: research, the needs of mathematics teacher educators, and institutional changes. As it is run by researchers, it has evolved as research has progressed, both in terms of concepts and tools, and with respect to research findings. As it is aimed at professional development it has evolved as a result of interactions with prospective mathematics teacher educators, as participants express their needs, or they are identified by the didactician. Mathematics teachers work in an institutional context, and their education has to take into account changes at this level (e.g., mathematical content, curricula, requirements).

One example is the introduction of an analysis of knowledge exposition. This was requested by participants and reflected new research findings. A second example is the introduction of the notion of proximities (Robert & Vandebrouck, 2014; Bridoux, Grenier-Boley, Hache & Robert, 2016), which only followed new advances in research. Institutional changes, such as the introduction of new mathematical content into the curricula or new teaching modalities (e.g., the flipped classroom), create new demands for teacher education and, consequently, mathematics teacher educator education.

A third example is the shift from learning to use technology at the beginning of the second year, to analyzing resources that use these tools. Over time, and under institutional pressure, many software packages have become familiar to teachers and are frequently used in the classroom. This was not the case when the program was launched. Both prospective mathematics teacher educators and the didactician agree that it is no longer relevant to devote time to perfecting the use of such software, and it is more useful to study on-line resources that use them. Moreover, new themes that address the use of technology have been introduced, following the introduction of new topics into school curriculum (e.g., coding and programming).

What about the transposition to teacher education? Teachers are concerned about the tasks they give their students and seek a better understanding of what occurs in the classroom, including students' learning, and their own explanations and assessment of their students' work. Mathematics teacher educators must be able to help, by going beyond a simple categorization of tasks as 'easy' or 'difficult.' They can draw upon the more developed tools they have learned about, depending on whether they are offering prospective or practising teacher education. The breadth and depth of knowledge regarding these tools differs as a function of the situation (educating teachers or educating mathematics teacher educators), and the transposition should also differ. Mathematics teacher educators' attention is drawn to this point throughout the program.

CONCLUSION AND DISCUSSION

This chapter aimed to contribute to knowledge on educating mathematics teacher educators. We started from the widely-shared postulate that students' learning develops from performing mathematical tasks that are 'well' chosen and 'correctly' implemented by teachers. The feasibility of a mathematics teacher educator education program consistent with this approach was shown through a detailed description of an example from France, centered on providing prospective mathematics teacher educators with the tools they need to observe and understand teaching practices. These tools can be used to analyze both tasks and events that occur during task implementation in the classroom, along with 'the words-to-say-it.' A detailed example was given to enable a better understanding of how research can be transposed and the co-development that occurs between didacticians and prospective mathematics teacher educators.

Our approach assumes that there are two key dimensions relating to mathematics teacher educator education: experience, on the one hand, and knowledge, on the other. There is a difference between education that is closely linked to practice, and education that is more closely based on theoretical aspects of learning and teaching. In both cases it is important that prospective mathematics teacher educators change their point of reference and move away from notions that are based on their own practices.

Transposition and Co-Development

In both years of the program, tools for analyzing teaching practices are transposed from research to practice. These tools are not designed to be used as they are in research, but as a medium for understanding specific teaching practices and identifying possible alternatives. In other words, we emphasize that mathematics teacher educators are not being educated to become researchers. From the point of view of co-development processes, didacticians learn about prospective mathematics teacher educators' professional needs, and adapt how they present their tools. Moreover, the feedback from prospective mathematics teacher educators about their own experience in using technologies in the classroom has alerted didacticians to needs that have been overlooked (for students, teachers, and educators).

It is important to note that the transposition requires didacticians to move from using tools for research purposes (an epistemic use) to prospective mathematics teacher educators using them to analyze teaching practice (a pragmatic use). Mathematics teacher educators are not expected to learn didactical knowledge, but they should be able to take a step back when observing teachers in the classroom. This choice of modalities enables prospective mathematics teacher educators to participate and contribute, provides didacticians with opportunities for co-development, and offers more opportunities for adaptations to be made. These features of the program could be exported to other contexts outside France.

Variation in Mathematics Teacher Educator Education

The differences between the two years of the program also depend on prospective mathematics teacher educators' needs (expressed or implicit) and requests (which are not necessarily consistent with their needs), and the didactician's perception of these needs. The challenge for the didactician is to interpret and transform this feedback, given what she anticipated would happen, and the available research resources. This is a source of tension in the didactician's work in both years, as it raises issues related to transposition and co-development, and requires them to adapt their representations.

In the first year, the didactician considers both the knowledge available to prospective mathematics teacher educators, and the impact of their personal experience of practice. In the second year, mathematics teacher educators' experience is initially given priority. They gradually become more aware of the need to have access to didactical tools and concepts that can help them go beyond technology-based task design and begin to think about the complexity of the implementation of these tasks. Prospective mathematics teacher educators initially develop an awareness of their own practices and discuss them. This is a first step in de-contextualization.

Moreover, the approach to prospective mathematics teacher educators differs in the two years. It is more 'distant' in the first year; there is more guidance from the didactician, and it is more oriented towards building awareness of students' tasks and activities. In the second year, participants' practices are given priority, and the focus is on building 'internal' awareness. This could be explained by differences in the scope of research findings, as studies of classroom practices with technologies are more recent.

The Method

We return to the expectations of the 'outsider' didactician. The interview method is focused on two specific points: the transposition of knowledge and tools from research; and the co-development between didacticians and prospective mathematics teacher educators. The evaluation did not attempt to undertake a fine-grained analysis of what happened in the long term as the program unfolded. The answers highlighted both the main aims and structure of the program, and particularly the fact that the general education component was centered on the practices of mathematics teachers in terms of what they do for and with students to develop their understanding and learning.

Further Discussion and Perspectives

The hurdles to analyzing and evaluating such mathematics teacher educators educating situations are linked to a deeper-rooted phenomenon: an intricate, fourfold research project. Students' learning is the endpoint for evaluating how much teachers have learned (in both pre-service and in-service education). This evaluation of teachers is, itself, a key element in evaluating mathematics teacher educators' performance and, finally, evaluating whether mathematics teacher educators' education meets its goals. The question is, how can we design research on mathematics teacher educator education with such limited feedback? A first step seems to be an investigation of mathematics teacher educator practices when educating mathematics teachers, taking into account the variables related to specific contexts. In particular, in France, it should be possible to compare the practices of educated mathematics teacher educators with the practices of other mathematics teacher educators.

NOTES

1 In this text, the words 'teachers' practices' refer to the singular dimension of teaching practices.
2 Transfer with some adaptation.
3 Arguments for referring both to Piaget and to Vygotsky were developed by Shayer (2003) in a paper with the provocative title "Not just Piaget; not just Vygotsky, and certainly not Vygotsky as alternative to Piaget." A similar point of view to the Double Approach is developed in Rogalski (2013).
4 In the French context, the Théorie des Situations Didactiques (TSD) and the Théorie Anthropologique du Didactique (TAD), did not develop research on mathematics teacher educator education, and adopted an approach for teacher education centered on the teacher as a function in the didactical triangle, rather than as an individual with the mission of teaching mathematics to individual students and classes.
5 This distinction was introduced by Samurçay and Rabardel (2004) for the general case of professional activity; it is applied to mathematics education in Vandebrouck (2018).
6 A notion is expressed with a new Formalism Unifying previous formulations and Generalizing them.
7 However, both of the didacticians are teachers, but not at the secondary level.

REFERENCES

Abboud, M., Clark-Wilson, A., Jones, K., & Rogalski, J. (2018). Analysing teachers' classroom experiences of teaching with dynamic geometry environments: Comparing and contrasting two approaches. *Annales de Didactique et de Sciences Cognitives, 23*, 93–118.

Abboud, M., Goodchild, S., Jaworski, B., Potari, D., Robert, A., & Rogalski, J. (2018). Use of activity theory to make sense of mathematics teaching: A dialogue between perspectives. *Annales de Didactique et de Sciences Cognitives, 23*, 61–92.

Abboud, M., & Rogalski, J. (2017a, February 1–5). Real uses of ICT in classrooms: Tensions and disturbances in the mathematics teacher's activity. *TWG 15 Teaching Mathematics with Technology and Other Resources, CERME 10.*

Abboud, M., & Rogalski, J. (2017b). Des outils conceptuels pour analyser l'activité de l'enseignant "ordinaire" utilisant les technologies. *Recherches en Didactique des Mathématiques, 37*(2–3), 161–216.

Abboud-Blanchard, M. (2013). *Technology in mathematics education. Study of teachers' practices and teacher education. Syntheses and new perspectives.* Synthesis for HDR, University of Paris Diderot.

Abboud-Blanchard, M. (2014). Teachers and technologies: Shared constraints, common responses. In A. Clark-Wilson, O. Robutti, & N. Sinclair (Eds.), *The mathematics teacher in the digital era: An international perspective on technology focused professional development* (pp. 297–318). London: Springer.

Abboud-Blanchard, M., & Robert, A. (2013). Strategies for educating mathematics teachers. The first step: Educating the trainers. In F. Vandebrouck (Ed.), *Mathematics classroom: Students' activities and teacher's practices* (pp. 229–245). Rotterdam, The Netherlands: Sense Publishers.

Abboud-Blanchard, M., & Robert, A. (2015). Former des formateurs d'enseignants de mathématiques du secondaire: un besoin, une expérience et une question d'actualité [Educating trainers of secondary school mathematics teachers: a need, an experience and a current question]. *Annales de didactique et de sciences cognitives, 20*, 181–206.

Beswick, K., & Chapman, O. (2013). Mathematics teacher educators' knowledge. In A. M. Lindmeier & A. Heinze (Eds.), *Proceedings of the 37th Conference of the International Group for the Psychology of Mathematics Education (PME)* (Vol. 1, p. 215). Kiel, Germany: PME.

Bridoux, S., Grenier Boley, N., Hache, C., & Robert, A. (2016). Les moments d'exposition des connaissances en mathématiques, analyses et exemples. *Annales de Didactiques et de Sciences Cognitives, 21*, 187–233.

Chapman, O. (2008). Mathematics teacher educators' learning from research on their instructional practices: A cognitive perspective. In B. Jaworski & T. Wood (Eds.), *The international handbook of mathematics teacher education: The mathematics teacher educator as a developing professional* (Vol. 4, pp. 115–134). Rotterdam, The Netherlands: Sense Publishers.

Chapman, O. (2009). Educators reflecting on (researching) their own practice. In R. Even & D. L. Ball (Eds.), *The professional education and development of teachers of mathematics* (pp. 121–126). New York, NY: Springer.

Chappet-Pariès, M., Robert, A., & Rogalski, J. (2013). Stability of practices: What 8th and 9th grade students with the same teacher do during a geometry class period. In F. Vandebrouck (Ed.), *Mathematics classrooms. Students' activities and teachers' practices* (pp. 91–116). Rotterdam, The Netherlands: Sense Publishers.

Clark-Wilson, A., & Noss, R. (2015). Hiccups within technology mediated lessons: A catalyst for mathematics teachers' epistemological development. *Research in Mathematics Education, 17*(2), 92–109.

Drijvers, P., Tacoma, S., Besamusca, A., Doorman, M., & Boon, P. (2013). Digital resources inviting changes in mid-adopting teachers' practices and orchestrations. *ZDM – The International Journal on Mathematics Education, 45*(7), 987–1001.

Drijvers, P., Doorman, M., Boon, P., Reed, H., & Gravemeijer, K. (2010). The teacher and the tool: Instrumental orchestrations in the technology-rich mathematics classroom. *Educational Studies in Mathematics, 75*(2), 213–234.

Even, R. (2014). Challenges associated with the professional development of didacticians. *ZDM – The International Journal on Mathematics Education, 46*(2), 329–333.

Goodchild, S. (2007). Inside the Outside: Seeking evidence of didacticians' learning by expansion. In B. Jaworski, A. B. Fuglestad, R. Bjuland, T. Breiteig, S. Goodchild, & B. Grevholm (Eds.), *Learning communities in mathematics* (pp. 189–203). Norway: Caspar Forlag AS.

Gueudet, G., Pepin, B. & Trouche, L. (Eds.). (2012). *From text to 'lived' resources*. Dordrecht: Springer.

Gueudet, G., & Trouche, L. (2011). Mathematics teacher education advanced methods: An example in dynamic geometry. *ZDM – The International Journal on Mathematics Education, 43*(3), 399–411.

Horoks, J., & Robert, A. (2007). Tasks designed to highlight task-activity relationships. *Journal* of *Mathematics Teacher Education, 10*(4–6), 279–287.

Jaworski, B. (2008). Mathematics teacher educator learning and development. In B. Jaworski & T. Wood (Eds.), *The international handbook of mathematics teacher education: The mathematics teacher educator as a developing professional* (Vol. 4, pp. 335–361). Rotterdam, The Netherlands: Sense Publishers.

Krainer, K. (2008). Reflecting the development of a mathematics teacher educator and his discipline. In B. Jaworski & T. Wood (Eds.), *International handbook of mathematics teacher education: The mathematics teacher educator as a developing professional* (Vol. 4, pp. 177–199). Rotterdam, The Netherlands: Sense Publishers.

Krainer, K. (2014). Teachers as Stakeholders in mathematics education research. *The Mathematics Enthusiast, 11*(1), 49–60.

Krainer, K., & Llinares, S. (2010). Mathematics Teacher Education. In P. Peterson, E. Baker, & B. McGaw (Eds.), *International encyclopedia of education* (Vol. 7, pp. 702–705). Oxford: Elsevier.

Lee, H. S., & Hollebrands, K. (2008). Preparing to teach mathematics with technology: An integrated approach to developing TPACK. *Contemporary Issues in Technology and Teacher Education, 8*(4), 326–341.

Lin, F.-L., & Cooney, T. (Eds.). (2001). *Making sense of mathematics teacher education.* Dordrecht: Kluwer Academic Publishers.

Mason, J. (2008). Being mathematical with and in front of learners. Attention, awareness and attitude as sources of differences between Teacher educators, teachers and learners. In B. Jaworski & T. Wood (Eds.), *International handbook of mathematics teacher education: The mathematics teacher educator as a developing professional* (Vol. 4, pp. 31–55). Rotterdam, The Netherlands: Sense Publishers.

Pastré, P., Mayen, P., & Vergnaud, G. (2006). La didactique professionnelle. *Revue Française de Pédagogie, 154*, 145–198.

Robert, A. (1998). Outils d'analyse des contenus mathématiques à enseigner au lycée et à l'université. *Recherches en Didactique des Mathématiques, 18*(2), 139–190.

Robert, A. (2005). Des recherches sur les pratiques aux formations d'enseignants de mathématiques du second degré: un point de vue didactique. *Annales de Didactique et Sciences Cognitives, 10*, 209–250.

Robert, A., & Hache, C. (2013). Why and how to understand what is at stake in a mathematics class. In F. Vandebrouck (Ed.), *Mathematics classroom: Students' activities and teacher's practices* (pp. 23–74). Rotterdam, The Netherlands: Sense Publishers.

Robert, A., & Rogalski, M. (2002). Comment peuvent varier les activités mathématiques des élèves sur des exercices – le double travail de l'enseignant sur les énoncés et sur la gestion de la classe. *Petit x, 60.*

Robert, A., & Rogalski, J. (2005). A cross-analysis of the mathematics teacher's activity. An example in a French 10th-grade class. *Educational Studies in Mathematics, 59*(1–3), 269–298.

Rogalski, J. (2003). Y a-t-il un pilote dans la classe? Une analyse de l'activité de l'enseignant comme gestion d'un environnement dynamique ouvert. *Recherches en Didactique des Mathématiques, 23*(3), 343–388.

Rogalski, J. (2013). Theory of activity and developmental frameworks for an analysis of teachers' practices and students' learning. In F. Vandebrouck (Ed.), *Mathematics classroom: Students' activities and teacher's practices* (pp. 3–23). Rotterdam, The Netherlands: Sense Publishers.

Rogalski, J., & Robert, A. (2015). De l'analyse de l'activité de l'enseignant à la formation des formateurs. Le cas de l'enseignement des mathématiques dans le secondaire. *Raisons Éducatives, 19*, 95–114.

Ruthven, K. (2009). Towards a naturalistic conceptualisation of technology integration in classroom practice: The example of school mathematics. *Education et Didactique, 3*(1), 131–149.

Sakonidis, C., & Potari, D. (2014). Mathematics teacher educators'/researchers' collaboration with teachers as a context for professional learning. *ZDM – The International Journal on Mathematics Education, 46*, 293–304.

Samurçay, R., & Rabardel, P. (2004). Modèles pour l'analyse de l'activité et des compétences, propositions. In R. Samurçay & P. Pastré (Eds.), *Recherches en didactique professionnelle* (pp. 163–180). Toulouse, France: Octarès.

Schoenfeld, A. H. (2010). *How we think: A theory of goal-oriented decision making and its educational applications.* New York, NY: Routledge.

Trgalova, J., Soury-Lavergne, S., & Jahn, A. P. (2011). Quality assessment process for dynamic geometry resources in Intergeo project. *ZDM – The International Journal on Mathematics Education, 43*(3), 337–351.

Tzur, R. (2001). Prospective a mathematics teacher educator: Conceptualizing the terrain through self-reflective analysis. *Journal of Mathematics Teacher Education, 4*(4), 259–283.

Vandebrouck, F. (Ed.). (2013). *Mathematics classrooms. Student's activities and teachers' practices.* Rotterdam, The Netherlands: Sense Publishers.

Vandebrouck, F. (2018). Activity theory in French didactic research. In G. Kaiser, H. Forgasz, M. Graven, A. Kuzniak, E. Simmt, & B. Xu (Eds.), *Invited lectures from the 13th International Congress on Mathematical Education (ICME13)* (pp. 679–698). Springer.

Vérillon, P., & Rabardel, P. (1995). Cognition and artifacts: A contribution to the study of thought in relation to instrumented activity. *European Journal of Psychology of Education, 10*(1), 77–101.

Voogt, J., Fisser, P., Pareja Roblin, N., Tondeur, J., & van Braak, J. (2013). Technological pedagogical content knowledge – A review of the literature. *Journal of Computer Assisted learning, 29*(2), 109–121.

Vygotsky, L. (1986). *Thought and language*. Cambridge, MA: MIT Press.

Zaslavsky, O., & Leikin, R. (2004). Professional development of mathematics teacher-educators: Growth through practice. *Journal of Mathematics Teacher Education, 7*, 5–32.

Zeichner, K. (2005). Becoming a teacher educator: A personal perspective. *Teaching and Teacher Education, 21*, 117–124.

APPENDIX

The following list of knowledge adaptations is presented to prospective mathematics teacher educators as a tool to be used in the task analysis:

- Partial recognition of ways of applying mathematical knowledge, (from recognizing variables and notations to recognizing formulas, conditions of applying them, etc.). For instance, to solve the task, one has to recognize the precise way for applying a theorem in a case slightly different from the case given in the textbook – may be the geometric figure is not the same, or the names of the points are different, and so on.
- Introduction of notations, points, or expressions as intermediaries.
- For instance, to solve the task, one has to give a name to a specific point so far without a name, or to extend a line segment into a straight line to make perceptible its intersection with another straight line that provides the possibility of using a theorem.
- Combination of several frames or concepts; changes, connections, interpretations of points of views, frameworks or registers.
- For instance, to solve the task, one has to use, in a geometric problem, an algebraic frame to calculate a length by the Pythagoras's theorem.
- Introduction of steps, organization of calculations or reasoning processes. This can range from the repeated use of the same theorem to reasoning *reductio ad absurdum*.
- For instance, for calculate a length by the Pythagoras's theorem, and solve the task, one has to introduce a first step, to verify that the triangle is rectangle, after having identified which may be the right angle (preliminary step).
- Use of previous questions in solving a problem.
- Making "strategic" choices, open or not (when only one will lead to the solution).

When the task to be analyzed involves a technology environment, some of the previous adaptations can be enriched, such as:

- Including a new frame that has to be combined to the others at stake.
- Identifying and interpreting of computers' feedbacks.
- Articulating knowledge involved in both environments at stake: technology and paper-and-pencil. Students must both adapt their paper-and-pencil knowledge to the use of technology and also adapt what they observe at the screen into paper-and-pencil knowledge.

Maha Abboud
LDAR
University of Cergy-Pontoise, UA, UPD, UPEC, URN

Aline Robert
LDAR
University of Cergy-Pontoise, UA, UPD, UPEC, URN

Janine Rogalski
LDAR
University of Paris Diderot, UA, UPC, UPEC, URN

OLIVE CHAPMAN, SIGNE KASTBERG,
ELIZABETH SUAZO-FLORES, DANA COX AND
JENNIFER WARD

7. MATHEMATICS TEACHER EDUCATORS' LEARNING THROUGH SELF-BASED METHODOLOGIES

Mathematics teacher educators can create spaces for self-awareness and self-understanding by conducting research exploring their lived histories and practices. In this chapter we discuss the creation of such spaces through the exploration of self-based methodologies with a focus on narrative inquiry, self-study, and auto-ethnography. We highlight the significance of these three self-based methodologies to show what they reveal about the professional self of mathematics teacher educators and the potential to reveal knowledge important to teacher education. Specifically, we discuss these methodologies from a theoretical perspective; in terms of related research literature of studies involving mathematics teacher educators; and based on their use in studies of the authors. Our studies collectively address how these methodologies have allowed us to empathetically and respectfully collaborate with students and teachers, while also giving us an opportunity to develop self-awareness of our identity, experiences, and bias. We, thus, draw on our experiences to highlight implications regarding the meaningfulness and usefulness of self-based methodologies in mathematics teacher educators' learning and research.

INTRODUCTION

The nature of mathematics teacher educators' professional knowledge and their learning to develop or enhance this knowledge are under-researched areas that need more attention given the growing emphases in many countries on teacher quality (Beswick & Goos, 2018). In this chapter, we focus on how mathematics teacher educators support their own learning to address the complexities in working with mathematics teachers. As teachers and researchers, mathematics teacher educators can learn by reflecting on their practice (Chapman, 2008a) from the perspective of being reflective practitioners (Schön, 1987) or from their research (Chapman, 2008b; Jaworski, 2001, 2008) through formal methodologies. However, this learning is often not acknowledged and articulated in published studies. Chapman (2008b) found that reports on studies in which mathematics teacher educators investigated instructional approaches they developed and implemented in their courses for

© KONINKLIJKE BRILL NV, LEIDEN, 2020 | DOI: 10.1163/9789004424210_008

prospective teachers were less about self-understanding to inform their practice and more about knowledge production to inform the field. Similarly, Jaworski (2008) pointed out,

> [for] teacher educators who have researched their own programmes of teaching teachers and reported the outcomes … [most] papers report outcomes for teachers from engagement in the programme and raise issues for teachers or for teacher education more generally. … Very few papers reflect critically … on what teacher educators themselves learn from engaging in teacher education, through reflecting on their own practice, and through research into the programmes they design and lead. And even fewer papers report on the learning of the teacher educator or on programmes designed to educate educators. (p. 3)

In contrast to such studies, we consider examples of mathematics teacher educators' research that have an explicit focus on self-understanding and methodologies (i.e., *self-based methodologies*) that support it. This is intended to help to build insight of how mathematics teacher educators are exploring their experiences to develop deep understanding of their practice in order to enhance it.

We use the term *self-based methodologies* to refer to approaches directed to the study of oneself; for example, one's own thinking, beliefs, actions, experiences. They focus both on the personal, in terms of improved self-understanding and enhanced understanding of the teaching and learning processes. These methodologies include: action research (McNiff, 2017), autobiographical research (Aleandri & Russo, 2015), narrative inquiry (Clandinin & Connelly, 2000), self-study (Feldman, 2003), and autoethnography (Ellis & Bochner, 2000). Our focus is on the last three that correspond to our research, which we intend to draw on to discuss how and what mathematics teacher educators learn from these approaches. Three of us (Chapman, Cox, and Suazo-Flores) have used narrative inquiry, one (Kastberg) has used self-study and one (Ward) autoethnography. As Cochran-Smith and Lytle (2004) argued, such approaches that place emphasis on self-understanding and self-exploration are essential to uncovering and enhancing the knowledge needed to understand, conceptualize and improve practice. On the other hand, Clandinin and Connelly (2000) argue that self-knowledge is, in the end, not important, but stress that, as a means, it is all-important.

Our goal, then, is to highlight the significance of these three self-based methodologies to show what they reveal about the professional self of mathematics teacher educators and the potential to reveal knowledge important to teacher education. To accomplish this, we first discuss the nature of the three self-based methodologies from a theoretical perspective followed by a literature review of mathematics teacher educators' work in which these self-based methodologies play a significant role in their learning. We include our work as extensions of the literature review for each self-based methodology and as contribution to the field through our stories of working with and learning from the self-based methodologies.

Thus, based on our work, we highlight implications regarding the meaningfulness and usefulness of these self-based methodologies in mathematics teacher educators' learning and research.

THEORETICAL PERSPECTIVES OF THREE SELF-BASED METHODOLOGIES

The three self-based methodologies, self-study, narrative inquiry, and autoethnography, are related in that they "privilege self in the research design, recognizing that addressing the self can contribute to our understanding of teaching and teacher education" (Hamilton, Smith, & Worthington, 2008, p. 17). Given our focus on them as approaches to mathematics teacher educators' learning, we consider the nature of each from the theoretical perspective of being particular ways of knowing and supporting learning. Thus, a description or an assessment of them from a research methodology perspective is beyond the scope of this chapter. Our discussion of each highlights characteristics that support their significance as a basis of framing mathematics teacher educators' learning.

Self-Study

Self-study as a methodology involves intentional, systematic inquiry of "our existential ways of being in the world" (Feldman, 2003, p. 27). Self-study methodology provides a way to "look at self in action, usually within educational contexts" (Hamilton et al., 2008, p. 17). Self-study researchers in education "focus on the nature and development of personal, practical knowledge through examining, in situ, their own learning, beliefs, practices, processes, contexts, and relationships" (Berry & Hamilton, 2017, para. 1). As a form of practitioner research (Borko, Liston, & Whitcomb, 2007), it involves studying self as teacher in context (LaBoskey, 2007), is self-initiated, and aims at improving one's practice (LaBoskey, 2007). It "functions as a means of better understanding the complex nature of teaching and learning and of stimulating educational change" (Berry & Hamilton, 2017, para. 1). Teacher educators, in particular, have embraced it as a viable means through which to investigate and learn from their own practice to better understand and shape teacher education (Bahr, Monroe, & Mantilla, 2018; Loughran 2005; Loughran, Hamilton, LaBoskey, & Russell, 2004) or to study themselves to understand the way they are teacher educators and to change their ways of being teacher educators (Feldman, 2003). Dinkelman (2003) argued that "self-study, as a form of reflection, ought to be an essential part of the activity of teacher educators" (p. 8). Self-study research "reveals us as researchers, as educators, and most importantly, as human beings" (Pinnegar & Russell, 1995, p. 6). It enables teacher educators to focus on their decisions and actions that construct who they are and accept responsibility for who they are.

Self-study as a research framework situates self as creator and investigator in the teaching environment, involves critical friends, and the documentation of and reflection about teaching practice (Lunenberg & Samaras, 2011; Samaras, 2011). It requires the teacher to examine and document their practice in order to analyze and understand their role in the teaching environment (Loughran et al., 2005; Samaras, 2011). It focuses on examining and making explicit beliefs and assumptions the teacher may have (Brandenburg, 2008; Lovin et al., 2012) in order to generate new knowledge that is sharable with the community. Dinkelman (2003) argued that "When teacher educators adopt self-study as an integral part of their own professional practice, the terrain of teacher education shifts" (p. 6).

Narrative Inquiry

Narrative is a major way in which people make sense of experiences, construct the self, and create and communicate meaning (Bruner, 2003). It is the way people organize their experiences into temporally meaningful episodes and permit past memories to be fully present in the moment toward shaping the future (Ellis, 2004). Narrative inquiry as a methodology is a way to study past experiences and construct meaning for future experiences; a way of thinking about and understanding experience (Clandinin & Connelly, 2000; Connelly & Clandinin, 1990, 2006), a way of finding out about one's self and the topic under investigation (Richardson, 1994), or a way of understanding one's own or others' actions (Chase, 2011).

Clandinin and Connelly (2000) explained that narrative inquiry researchers understand that what they know about themselves and their participants is a result of their interaction with them and the contexts around them. So, they must be attentive to the interaction of time, place, and social contexts that surround the participants' experiences; both personal and social issues by looking inward and outward and temporal issues by considering their influence with past and future. As Clandinin (2013) explained, "We think simultaneously backward and forward, inward and outward with attentiveness to place" (p. 41). The inward direction refers to internal conditions such as feelings and hopes, and outward refers to the environment.

Narrative inquiry was introduced to educational research through the work of Connelly and Clandinin. They explained, "The main claim for the use of narrative in educational research is that humans are storytelling organisms who, individually and socially, lead storied lives" (Connelly & Clandinin, 1990, p. 2). In education, narrative inquiry focuses on researching lived experiences of teachers, teacher educators, and students. It serves as both process and product of research and professional development (Connelly & Clandinin, 1990; Clandinin & Connelly, 2000; Lunenberg & Willemse, 2006). Thus, as Clandinin, Pushor and Orr (2007) noted, it is used by teachers and teacher educators interested in studying and improving their own practices. Epistemological dilemmas are understood narratively and teachers' narratives inform the understanding of professional knowledge landscapes (Clandinin & Connelly, 1995).

Autoethnography

In autoethnography, researchers seek to describe and systematically analyze personal experience in order to understand a culture or group in which they are members (Ellis, 2004; Ellis & Bochner, 2000; Ellis et al., 2008; Ellis, Adams, & Bochner, 2011). The researcher is the subject, and the researcher's interpretation of the experience is the data (Ellis & Bochner, 2000). Thus, the researchers' lives and experiences are the foci of the research (Reed-Danahay, 1997). They study themselves within a subculture and attempt to make meaning of all of the experiences in this setting. They draw on their own experiences written in the form of personal narratives to extend new knowledge (Sparkes, 2000). They study self as the main character with others as supporting actors in the lived experience (Chang 2008). Examining all aspects of a personalized experience allows the researchers greater opportunity to arrive at the core meaning of the experience.

There are various types and styles of autoethnography (Ellis & Bochner, 2000). For example, there is an analytic version in which personal stories undergo traditional content analysis and an evocative version in which the analysis is built into the story to let the story work on its own to build compassion and empathy for others (Atkinson & Delamont, 2006; Denzin, 2006; Ellis & Bochner, 2000).

According to Chang (2008), autoethnography has become a powerful source of research for practitioners as a way to give the researcher more insight about self and others. Thus, it offers teacher educators a means to understand themselves and enhance their practice and nurture an empathetic understanding of teachers.

Differences among the Three Methodologies

"These [self-based] methodologies privilege self in the research design, recognizing that addressing the self can contribute to our understanding of teaching and teacher education" (Hamilton, Smith & Worthington, 2008, p. 17). Thus, the boundary across the three methodologies could be blurred depending on how the researcher attends to the self or "I" and uses narratives or stories. However, as Hamilton et al. discuss, there are some clear differences from a methodological perspective that center on the focus of and the approach to the research design and how the questions are asked or the thinking underlying the research practices is revealed. A key difference is that narrative inquiry addresses the self in relation to others, autoethnography addresses the self within social or cultural contexts, and self-study addresses the self in action, usually within educational contexts.

In narrative inquiry, teacher educators learn through the study of experience as story by sharing and thinking about experiences. Such experiences usually involve teacher educators working with teachers to "share and learn from one another through exchanges about knowledge, skills, practices, and evolving understandings"

(Hamilton et al., 2008, p. 18). Thus, a variety of narratives or stories during the study are important (Clandinin & Connelly, 2006). It is also important to 'analyze' the stories in relation to the context or place, social interactions, and time surrounding the researcher-participants interactions (Clandinin & Connelly, 2000).

In autoethnography, teacher educators learn through writing about the personal and its relationship to culture. "A broad description of 'culture' would include evidence of shared patterns of thought, symbol, and action typical of a particular group. In an educational text, culture could be addressed as language, action, and/or interaction" (Hamilton et al., 2008, pp. 22–23). This focus on culture in relation to personal experience, as well as the way in which it is revealed in the text, differentiates autoethnography from narrative and self-study, which do not have to consider culture. Auto-ethnographers "situate themselves, contesting and resisting what they see using a multi-genre approach that can incorporate short stories, poetry, novels, photographs, journals, fragmented and layered writing" (Hamilton et al., 2008, p. 22).

In self-study, teacher educators learn through "whatever methods will provide the needed evidence and context for understanding their practice" (Hamilton & Pinnegar, 1998, p. 240). They engage in reflection "in a variety of ways including journaling, conversations with colleagues, graduate work, and thinking deeply about a teaching problem to search for solutions" (Hamilton et al., 2008, p. 24). The use of a critical friend distinguishes a self-study from the others.

Each of the three methodologies uses narratives or stories, but the way in which they are used varies in the context of the studies. For example, while narrative inquirers use a variety of stories to access experience during the study, auto-ethnographers write extensive personal narratives as they engage in their work, and self-study researchers write reflexive journals which could include stories to capture critical moments of their research. The difference is also related to the "position of the 'I'" as discussed by Hamilton et al. (2008):

> For narrative inquiry, the self in relation to others holds privilege in a storied, usually written, form. In auto-ethnography, it is the cultural I shaped by cultural contexts and complexities that takes the foreground. Where the other methodologies focus on relation or culture, self-study researchers focus on practice and improvement of practice, closely attending to self and others in and through their practice. (p. 25)

In keeping with the theme of this handbook, our focus is not on how or whether mathematics teacher educators have engaged in these methodologies from a strict theoretical research perspective, but whether engaging in them was useful or effective in supporting their learning as mathematics teacher educators. The examples we provide in the following sections of the chapter are intended to illustrate this.

MATHEMATICS TEACHER EDUCATORS' ENGAGEMENT IN SELF-BASED METHODOLOGIES

This section provides a review of related research literature involving studies conducted by mathematics teacher educators that used the three self-based methodologies of self-study, narrative inquiry, and autoethnography. It also includes descriptions of the experiences of four of the authors of this chapter who used these methodologies to support their learning as mathematics teacher educators.

Mathematics Teacher Educators' Engagement in Self-Study

Of the three self-based methodologies, self-study has received the most attention in studies involving mathematics teacher educators. There is a growing trend of mathematics teacher educators engaging in self-study as a means of exploring, understanding, and improving their own practice and the field of mathematics education. More recent studies continue to reflect this orientation to mathematics teacher educators' learning and growth with a focus on their thinking and practice in teaching mathematics education courses. The following examples of these studies highlight the nature of the mathematics teacher educators' learning resulting from the self-study methodology.

Schuck (2002), in her self-study of her practices in relation to the reform movement in mathematics teaching and learning, discovered that it was essential to be familiar with her students' beliefs as well as her own. She found that prospective primary school teachers often held beliefs about mathematics teaching and learning that constrained their access to rich and powerful ways of learning. She consequently revised her practice to help them challenge these beliefs. Further study of the new practices revealed some obstacles leading her to continue to strive to challenge her students' beliefs about mathematics and to engage them in developing their understandings of the content.

Goodell's (2006) self-study aimed at understanding how well the class activities she chose for her secondary school prospective teachers enabled them in developing their knowledge of, skills in, and dispositions towards teaching mathematics for understanding. Critically reflecting on her experiences in the classroom made her refine the practices in which she engaged her students. For example, her learning included how to foster reflection in her prospective teachers through the use of critical incidents and how to ensure that they were learning from their reflections. It also made her take an objective stance towards the prospective teachers' concerns and insisted that the incident reports were complete with analysis of the incident, as opposed to merely restating what happened.

Alderton (2008) engaged in self-study to understand the complex and context-based situations of her practice. Her goal was to improve her practice and to develop her knowledge and understanding of pedagogy in teacher education by living her values more fully in her practice. She became aware of a contradiction between her practice and beliefs that prompted her to question why she did not always live the values she professed. As a result, she started to transform the way she comprehended her practice regarding teaching and learning in teacher education and to examine and challenge some of her personal assumptions and beliefs about teaching and learning. She also developed more reflexive self-awareness which enabled her to problematize taken-for-granted assumptions relevant to her context of practice.

As a beginning mathematics teacher educator, Marin (2014) explored what she could learn about mathematics education and about herself as a teacher educator through self-study of her transition from teacher to teacher educator based on her experience teaching a course focused on inquiry. She described her experience and struggles in teaching the mathematics methods course for the first time. The self-study helped her to change both her teaching and herself. It helped her uncover her assumptions, challenge her beliefs, frame her practice, and understand who she was as a teacher educator. It enabled her to discover herself as a teacher educator and introduced her to the tools to continue to study, change, and grow. It also helped her to understand the prospective teachers' experiences in her methods course.

Hjalmarson (2017) conducted a self-study of her first-time teaching of an online course for mathematics specialists to support their development as mathematics coaches working with grades K-8 teachers to enhance mathematics teaching and learning. Her focus was on what guided her design decisions in facilitating learning in the online, project-based course for mathematics specialist students. Three themes about her role as course designer and teacher emerged from the self-study data analysis: supporting engagement and autonomy; creating authentic and practical learning experiences; and fostering collaboration and community.

Finally, unlike the previous examples that focused on the individual mathematics teacher educator, Bahr et al. (2018) engaged in a collaborative self-study of their practice that helped them refine their own teaching and learning as mathematics teacher educators. Their purpose was to create a comprehensive and cohesive set of outcomes to guide the teaching and learning of prospective and practising mathematics teachers with whom they worked and to guide their study of that teaching and learning. The self-study resulted in a refinement of their understandings and enabled them to focus their students more clearly and explicitly on a holistic picture that defined exemplary mathematics teaching and to help them schematize the complex work of teaching mathematics.

The preceding examples were intended to highlight the type of learning of mathematics teacher educators using self-study. We offer one more example based on the work of one of the authors of this chapter, Signe Kastberg, who has engaged in self-study as a methodology in her research and learning as an experienced mathematics teacher educator. Our intent with this example is to highlight the mathematics teacher educator's thinking, experiences, and learning with self-study over an extended period of her practice from her perspective. Thus, in what follows,

Kastberg provides a first-person account of her journey with self-study. This journey includes her knowing in relation to self with a focus on personal practical knowledge and improving practice. It also includes her knowing in relation to prospective teachers with a focus on having and conveying empathy and respect and receptivity to growing in relationship.

Signe Kastberg's journey with self-study. I share insights from self-studies I have conducted with colleagues over 10 years. I use characteristics of Kitchen's (2005a, 2005b) construct of relational teacher education to tell the story of my learning about mathematics teacher education through self-study. Kitchen's construct frames the role of relationship in my work. While Kitchen outlines seven characteristics of relational teacher education, I use two broader categories, knowing in relation to self and knowing in relation to prospective teachers to describe how awareness of each of these ways of knowing supported my learning to teach about teaching (Loughran, 2010). Initially I used self-study to gain control over my pedagogy and to become more intentional in my actions. Later, I came to know myself in relation to my teaching and to know prospective teachers as I worked in relation to them. My aim became to embrace and understand the process of teaching about teaching rather than to control it. I came to understand knowing in relation, not a mechanism to motivate or manipulate prospective teachers or myself, but rather as a process through which I can learn from the opportunities teaching affords. This story would not be possible without the support of my colleagues Alyson Lischka and Susan Hillman. These women have consistently supported me to improve my practice through shared inquiry, cradled by a sense of belonging to a community of mathematics teacher educators.

In the post-National Council of Teachers of Mathematics (2000) Standards era and pre-Common Core (National Governors Association Center for Best Practices, 2010), I had a certainty about how teaching mathematics should work. That certainty caused me to want to move from teaching mathematics content courses to teaching mathematics methods courses (pedagogy focused courses for prospective teachers). When a group of colleagues invited me to engage in a self-study of mathematics teacher educator beliefs, I was excited to learn about the methodology. The opportunity to collaborate with colleagues and discuss beliefs underlying our practice felt like a chance to gain insight about and control over my teacher actions. Initially, I did not know I was exploring my 'personal practical knowledge' (Clandinin, 1985). I felt certain about many things including how to teach mathematics to prospective teachers prior to their admission to a teacher education program. I am sure at some point we discussed the structure of the study, but I was more engaged with the collegial conversations about my pedagogy the study structure afforded (Lovin et al., 2012). Each conversation provoked awareness of my teaching actions and motivations for those actions. The study goal was to identify shared beliefs of a group of mathematics teacher educators. Beliefs we identified were the result of coming to know the mathematics teacher educator's individual values. My central value was fostering prospective teachers' autonomy as mathematicians, yet during conversations, I unearthed a "living contradiction" (Whitehead, 1989) between my value of autonomy and my practice of awarding points for particular sorts of classroom participation. For example, prospective teachers who shared during class time, received more points for their participation than those whose participation occurred online and outside of class. In my beliefs, the value of autonomy lived alongside a mathematics teacher educator's view of productive participation. My recognition of this contradiction led to further exploration of prospective teachers' views and approaches to participation in mathematics class.

In 2010 I began teaching mathematics methods courses at a different university. This shift in institutional context impacted my efforts to improve my practice. Work as a mathematics teacher educator teaching about teaching was fraught with complexities that I had not anticipated when I taught about learning mathematics. Prospective teachers seemed disinterested in exploring the mathematics and learning of mathematics that I loved. They considered themselves teacher-learners rather than mathematics-learners and made this clear in their evaluations of my teaching. One noted:

> I thought this was a METHODS course. I do not feel confident in knowing HOW TO TEACH math to students. This semester we were taught to investigate every student's thinking individually which is unrealistic when I have a classroom of 25 students. I thought this class was designed to teach me how to teach … not how to divide fractions, learn the Hindu Arabic Numeration System, and other math problems. We've had 3 courses learning math at Purdue … now we need to learn HOW to teach it. (Kastberg, 2012, p. 169)

This prospective teacher's reasoned critique pointed out what I knew to be my focus, namely mathematics learning. Prospective teachers felt they had learned mathematics and now, in the year before student teaching, they wanted to learn about mathematics teaching. I felt defeated and did not know how my knowledge of and experience with mathematics learning could be useful in supporting prospective teachers' learning to teach. At the same time, I was engaged in a self-study with critical friend Beatriz D'Ambrosio. The study (Kastberg, 2012) focused on describing learning mathematics through interactions with prospective teachers in mathematics methods courses using learning theory. Through the study I identified anticipated effects of activities I planned for the class and the actual effect. Differences between the anticipated and actual effects of activities I planned provoked new ideas about prospective teachers' mathematical concepts. Yet improving my practice resulted from wrestling with the question: How does my knowledge of mathematics help me teach about mathematics teaching and learning? It was through this question, that I began extending my attention from knowing in relation to self toward knowing in relation to prospective teachers. My awareness turned toward building relationships with and learning from prospective teachers rather than trying to provide instruction for them.

In another study with Alyson and Susan, who also had a goal of improving their practice, we focused on written responses prospective teachers gave to mathematics students (Kastberg, Lischka, & Hillman, 2018) in letter-exchanges. The structure of letter-exchanges was inconsistent with prospective teachers' views of supporting mathematics learners. Prospective teachers described challenges in posing and clarifying mathematical questions for mathematics learners. Their comments raised questions about the relevance of the letter-writing activities to prospective teachers. Prospective teachers viewed teaching as engaging learners face-to-face, rather than writing letters about mathematics. Initially, I defended the structure of the letter-exchange assignment, suggesting that written feedback was something every teacher needed to learn to provide and that written responses from the teacher were useful to families trying to support mathematics learners. Yet, I recognized that prospective teachers' critiques of the activity revealed challenges they were experiencing in the course. My consideration of prospective teachers' critiques made me aware of the challenges they faced and motivated me to construct activities more closely aligned with the work of teaching. I shifted my gaze toward how I could convey respect and empathy for prospective teachers and the challenges they faced in learning to teach. I moved toward knowing in relation to prospective teachers.

Alyson, Susan, and I had empathy and respect for prospective teachers, but struggled with how to convey these feelings. For us, conveying empathy involved taking up the dilemmas and challenges of prospective teachers (Noddings, 2010) even when planned class activities might need to be set aside to pursue prospective teachers' challenges. Gathering evidence of prospective teachers' experiences with written feedback and providing opportunities for prospective teachers to discuss their experience with written feedback began to convey our respect for those experiences. Prospective teachers discussions of their own feedback experiences developed my awareness of the need to explore our diversity of experiences with written feedback and other mathematics practices we had experienced.

Our exploration of prospective teachers' responses to mathematics learners raised a "living contradiction" (Whitehead, 1989) as we became aware that our written feedback to the prospective teachers fell short of expectations we had for their feedback practice. In particular, we expected them to use models of children's mathematics to construct written feedback, however we did not use models of their concepts of mathematics teaching and learning to craft our feedback. Instead, we found that in many cases we redirected them to attend to what we saw in children's mathematics. This realization provoked a new self-study of our written feedback to the prospective teachers (Kastberg, Lischka, & Hillman, 2016).

To learn from our written feedback, we characterized it using Hattie and Timperley's framework (2007) for effective feedback. Our characterization helped us identify areas for improvement such as attending to prospective teachers' mental processes used in constructing written feedback for mathematics learners. Yet we still struggled with the tensions involved in giving written feedback. Our insights into mathematics learners' thinking felt important to convey to prospective teachers, yet in telling them what we saw I felt we risked their autonomy as learners and teachers. We sought a different sort of relationship, one in which their ideas opened up possibilities for our own knowledge of mathematics teaching and learning. Our desire for such a relationship was evidence of our receptivity to growing in relation to prospective teachers.

In our most recent self-study (Kastberg, Lischka, & Hillman, 2018), Alyson, Susan and I turned toward our questioning practice. In this study I have focused on receptivity to growing in relationship with prospective teachers. In particular, I have worked toward the development of a perspective on relationship. My inquiry is driven by the question: What is my goal in constructing relationships with prospective teachers? This question highlights the shift from exploring the self, to exploring the self in relation to prospective teachers. I have learned that inquiry into self or other, necessarily always involves both. As teachers become receptive to growing in relation, relationships change. No longer are relationships tools to encourage development of prospective teachers' knowledge and of personhood, but instead relationships become a way to understand oneself and one's own practice. Growing in relationship moves beyond using the relationship to motivate or encourage prospective teachers to take risks and try new things; it becomes about seeing and knowing one's self. I use self-study methodology, not because I want to improve my practice by controlling the ways I implement particular routines or activities, not because I want to be intentional in my practice, but because engaging in such study helps me be conscious in the moment of teaching and act based on my experiences in ways that matter most to prospective teachers and to me. I have realized that my receptivity to growing in relationship to prospective teachers involves taking their models of mathematics teaching and learning not as barriers to my practice, but as emergent views.

Mathematics Teacher Educators Engagement in Narrative Inquiry

In contrast to the self-study methodology, there has been little attention in the research literature to the use of narrative inquiry as a methodology in mathematics teacher educators' learning or inquiry of themselves. There has been more attention to the use of narratives or stories as a tool for data collection and/or analysis in studies of the mathematics teacher. This includes the use of stories in the study of mathematics teacher identity (Drake, 2006; Drake, Spillane, & Hufferd-Ackles, 2001; Kaasila, 2007; Lutovac & Kaasila, 2014); learning and teaching of problem solving (Chapman, 2008c); motivation (Phelps, 2010); orientation toward mathematics (Kaasila, Hannula, Laine, & Pehkonen, 2008); knowledge of mathematics-for-teaching (Oslund, 2012); change in practice (Chapman & Heater, 2010); knowledge of curriculum and political contexts (de Freitas, 2004); and conversation on the teaching and learning of mathematics (Nardi, 2016). These studies addressed different aspects of the teachers' experiences through stories associated with the teachers' learning, personal lives, teaching, thinking, and change. For example: Drake (2006) used stories of key events in the lives of elementary school teachers as learners and mathematics teachers; de Freitas (2008) used narratives of prospective mathematics teachers' past experiences involving moments that were highly emotional for them; Oslund (2012) used experienced elementary school teachers' stories of new mathematics pedagogies; Lutovac and Kaasila (2014) used prospective elementary school teachers' stories they tell themselves or others about themselves as mathematics learners and teachers; Chapman and Heater (2010) used a high school mathematics teacher's narratives of shifts in her experience, thinking and practice; and Nardi (2016) used stories told by mathematicians engaged in conversation on the teaching and learning of mathematics.

While mathematics teacher educators learned from these studies, their reported goals of the studies were not self-exploration or self-understanding. Two studies in which the mathematics teacher educators used narrative inquiry to investigate themselves are Chauvot (2009) and Bailey (2008). Chauvot investigated her knowledge content, its structure, and her growth as a novice mathematics teacher educator-researcher from her doctoral program into her third year of a tenure-track faculty position at a large university. She explained that she used narrative inquiry as a process and product in identifying the knowledge she drew from to fulfill her role as a mathematics teacher educator-researcher. Her self-created narratives were analyzed to determine what knowledge she was seeking or used to inform decisions, to what she attributed gaining this knowledge, what categories this knowledge fell under, and how this knowledge was structured. Findings highlighted the different kinds of knowledge she needed to serve different roles as a mathematics teacher educator-researcher such as instructor of university courses and mentor of doctoral students.

Bailey (2008) explored her thinking about mathematics curriculum (how and why). She used narrative inquiry to examine her professional practice as a mathematics teacher educator with prospective primary school teachers in mathematics education. Through writing her stories and in later reflections on the stories, she discovered contradictions in her writing. She became aware of beliefs about mathematics and its learning that she did not know she held and that were contrary to what she espoused in the classroom. This resulted in new insights that led to subsequent changes in how she implemented curriculum. Resulting changes for her included using mathematical investigations as a teaching approach, accepting that this may involve periods of being stuck, using more questioning to support learning, and accepting that mathematics can be learned through collaboration.

In addition to these two examples to highlight what mathematics teacher educators can learn through narrative inquiry are the studies of two of the authors of this chapter, Dana Cox and Elizabeth Suazo-Flores, who have been using narrative inquiry as a methodology in their research and a way of learning as mathematics teacher educators. They have used it in different contexts that impacted their learning in different ways. Cox's work is with teacher leaders and prospective teachers while Suazo-Flores's work is with an experienced teacher. Both show how learning through narrative inquiry is based on relationships, that is, learning about themselves in relation to what they learn about their participants or students in working with them. For example, Cox learned about the importance of empathy and Suazo-Flores about caring relations in their research and teaching.

In what follows, Cox and Suazo-Flores provide first-person narrative accounts of their journeys to and with narrative inquiry. Cox's journey includes how her experience as a mathematics teacher educator led her to embrace narrative inquiry, which resulted in a shift in focus in her research and teaching in using it as a tool to explore her capacity for mathematical empathy and the capacity of narrative inquiry to inspire it in her students. In contrast, Suazo-Flores' journey includes how her experience as a mathematics teacher educator and PhD student led her to learn about and embrace narrative inquiry as a way of researching and learning about a mathematics teacher and herself. Through her PhD thesis, completed in 2017, she learned about her own and her participant's practical knowledge, the importance of the teacher's role in the classroom, and how to create caring relations in working with prospective teachers and other teachers. Thus, each of the following first-person accounts of their journeys as told by them highlights different aspects of engaging in narrative inquiry. Cox's account is presented first followed by that of Suazo-Flores.

Dana Cox's narrative inquiry journey. In my work with narrative inquiry, I adopted Clandinin and Connelly's (2000) perspective that it can be used to capture more personal and human dimensions of experience; to tell stories that balance a description of events as they occur but are then layered with reflection upon those experiences. While the original experiences are temporal and situated, their meaning expands through reflection. For me, narrative inquiry is also the means to acknowledge that I am a living contradiction (Whitehead, 1989); by virtue of being present and immersed in an experience, I am unable to articulate the contradictions between that lived experience and my deeply held values. It is only through reflection on lived experience that the nature of my contradiction emerges, and I am changed. To reflect again on that change is an iteration and, in this sense, the theory that I am able to create is living and ever-changing.

In my work with teachers, the act of writing a narrative is what helps me to become aware of my practice. I am choosing to share the story of how I embraced narrative inquiry as a methodology and how it has changed me in the form of a narrative. I begin with my origin story; the story of how I came to recognize narrative inquiry as a tool to both examine myself as a living contradiction and to share the result of that examination. While engaging with a Mathematics and Science Partnership project, my university colleagues and I planned a yearlong Leadership Academy that fitted under the 'train the trainers' model for professional development. Nine teacher leaders responded to our call and agreed to participate in our project, which adopted a leadership development stance akin to a game of telephone. We understood teacher leadership as an automatic by-product of providing long-term, high-quality professional development to teachers and giving them a platform to share what they had gained with others.

Midway through the year, the teacher leaders rejected a traditional leadership model based on expertise, individualism, and the transmission of knowledge. In this reflective time, we confronted a living contradiction (Whitehead, 1989) in our assumptions and stance. At all times, we believed that teachers deserved agency and voice and intended to empower them as mathematics education leaders. However, the game of telephone positioned teachers as message receivers and emerging experts (a designation that we would bestow) and denied teachers the very things we had intended the program to develop. This proved to be a critical moment wherein the project was democratically reinvented around a model of shared leadership (Schlechty, 2001) where teacher leaders were positioned as ambassadors of a culture of mathematical inquiry. This reinvention gave our teachers agency, which they used to modify more than just the leadership model. Our carefully designed curriculum and project rhythm became new as teachers brought personal inquiry into the center of our work together, which now allowed for multiple funds of knowledge (Moll, 1992). A consequence of this was that the story we felt able to tell was no longer about changes in participants, but about the process of breaking with norms as a project. We had proposed to develop our participants but found the collective story as well as our personal story far more compelling and legitimate.

With this shift in orientation of the project, we then sought out empathetic methodologies (D'Ambrosio & Cox, 2015) to help in telling the story of the successes and failures. Narrative Inquiry was one such methodology that helped construct a bridge between our lived experiences and our deeply held values. It offered new ways to tell important stories, to question, and to invite response. The main ideas that we wanted to convey were not test scores, but personal accounts of the vulnerabilities present when doing classroom research. Beatriz D'Ambrosio and I (2015) posited that methodologies that stem from a place of methodological belief (Elbow, 2008) might help the researcher understand more from the perspective of teachers and also understand themselves in relation. To truly use belief to scrutinize ideas that are different from our own, we must attempt to move beyond mere listening and restating and suspend doubt; we must believe in the idea's merit and see the truth within.

Viewing narrative inquiry as an empathetic methodology was consistent with us wanting to get away from the notion of representing teachers in our work and move toward teachers representing themselves. In this work, then, the only story that we felt comfortable sharing was that of our own self-awareness dawning as a result of the paradigmatic shifts in the professional development project. The only voice we felt comfortable using was our own, reflecting on our shared experiences and those narrative pieces our teachers had shared with us along the way. It was not our purpose to interpret what teachers told us, but to share the impact of the stories they told about our practice. My work with the teachers in my origin story helped me to discover the need for empathy in my research and brought me to understand the power of narrative inquiry to work with it.

I have used narrative as a tool to explore my capacity for mathematical empathy and the capacity of narrative inquiry to inspire it in others. I will illustrate this work with three examples. First, in an ongoing study, I have invited participants to write their own first-person narratives about a classroom activity centered on the act of defining mathematical terms. Along with co-researchers Jane Keiser and Suzanne Harper, I am using these narratives to uncover beliefs held by these prospective teachers about the fluidity of mathematical definitions and how collaborative writing as a medium might afford opportunities for mathematical empathy in the content classroom. Contrary to other methodologies used to examine mathematical knowledge for teaching, we are positioning prospective teachers as people from whom we can learn.

Second, I am a part of another study that positions students as teachers of problem solving. As a goal of this study, my co-researchers, Suzanne Harper and Todd Edwards, and I apply the methods of narrative inquiry to write narratives of how prospective secondary school mathematics teachers engage with mathematics and with technology (Cox & Harper, 2017). This has helped us to document authentic, articulated instances of both problem solving and problem posing in a geometric context (Cox, Harper, & Edwards, 2018). Here, narrative inquiry is used to capture more personal and human dimensions of experience.

Third, I was moved to search out other courses on my campus that focused on how to incorporate empathy into professional activity. During a course on Empathy in Design offered in our College of Creative Arts, I began to reimagine the work of doing classroom mathematics as mathematical design. I immediately began talking about these ideas with a former student, who was inspired to write her own narrative. This story focused on prototyping as a specific form of mathematical design thinking and was entirely situated in a context uniquely shared and shaped by us, the authors. For me this represents the collaborative power of narrative inquiry – neither of us could tell the story alone as it requires knowledge of the intentions and actions of both professor and student.

Elizabeth Suazo-Flores' narrative inquiry journey. My experience working as a mathematics teacher and professional development facilitator in Chile and the United States led me to understand research with teachers as collaboration. Exploring what research with teachers looks like and understanding of research with teachers as collaboration motivated my dissertation study. Narrative inquiry (Clandinin & Connelly, 2000) became the theoretical foundation that guided my dissertation work. It was the appropriate methodology for my study of an eighth-grade American mathematics teacher, Lisa, because it allowed me to talk about our existing relationship and explore our interactions as a construction that emerged from our ways of being. My intention was to study my interactions with Lisa while planning and implementing a lesson connected to the concept of area, embedded in the context of designing a miniature golf course, to learn about her personal practical knowledge (Elbaz, 1983). However, being someone who had valued traditional ways of knowing, my learning in this study also included learning about the narrative inquiry methodology and how to use it. Three central aspects of my learning in making the shift to using this process were being my authentic self (Rogers, 1961), data analysis, and reporting the findings of the study as a personal narrative.

Whenever I was with Lisa, I tried to be my authentic self. This for me meant to follow her energy, but also being honest with her about my concerns and motivations. This was rewarding, but at the same time difficult. It was rewarding because whenever I was with her, we were two mathematics teachers in the classroom. It was difficult because of Lisa's expectations of me as a researcher that were based on her understanding of research from a traditional perspective. I found that explaining to her that my goal was to document our lives in the school and she did not have to behave in any special way or hide anything because of my study was important to address this challenge. This helped to create a relationship between us in which Lisa understood that I was being my authentic self and she had her space to be herself as well. This relationship was important to access our authentic experiences for the narrative inquiry process (Suazo-Flores, 2016).

The second example of my learning to engage in narrative inquiry was the approach to data analysis. Data collection focused on both my experience in working with Lisa and Lisa's experience in planning and teaching the lesson on area. This included recording the many conversations that covered different topics with Lisa and in being our authentic selves and not setting boundaries to the conversations. It also included recording my thoughts in a journal and audio recording my thinking before and after leaving the field. Engaging in the data analysis was a learning process of being true to the narrative approach. A central aspect to the approach was to immerse myself in reading and re-reading the field texts and writing memos of my thoughts many times. I organized all the texts of the conversations and memos in a way that enabled me to look across the conversations and identified a plot (Polkinghorne, 1995) that later was validated by Lisa. This analysis process also led me to start trying to identify categories or themes in the field texts. I was also tempted to start dissecting my field texts. But I questioned myself about doing that. Dissecting field texts would make me disregard the characteristics of the participant's surroundings and ways of being, which would be in conflict with narrative inquiry. Therefore, I focused my attention on illustrating the teacher's personal practical knowledge and providing evidence of it in an interwoven narrative.

The third example of my learning was about how to report the findings as an interwoven narrative consistent with the narrative approach of Connelly and Clandinin (2000). The analysis resulted in too many stories to tell and I did not know how to write about them. With some support, I realized that I could build the narrative around my journey to, during and after the study. Writing about myself was easy because I felt confident describing how, at that time, I saw myself coming to propose my study. I wrote about myself as a learner and teacher back in Chile, to then later introducing my arrival to the American school where I met Lisa. Then, it was natural to introduce Lisa and our interactions planning and implementing the lesson. This process of writing the research text provoked an awakening experience (Connelly & Clandinin, 1994). Connelly and Clandinin (1995) used the concept of awakening to describe the experience of "becoming aware of the possibility of seeing oneself and the world in new ways" (p. 82).

In this study, in addition to learning about narrative inquiry, I learned not only about Lisa's personal practical knowledge, but also about my own personal practical knowledge, which brought me to learn about my teaching practices. For example, I learned that Lisa's motivations for working on lessons that involved real-world contexts could be traced back to her experiences as a child and in her first years of teaching mathematics. In her first 15 years teaching mathematics, she worked on mathematics lessons that involved field trips and working with colleagues for other disciplines. At the time of meeting her, these experiences were part of her past as a teacher. She was now using other ways to engage students in her mathematics lessons (i.e., using tiles, cubes, software, or playing board games). I learned that STEM lessons we started trying in her classroom resonated with her personal interest in engineering. Knowing about her personal experiences enabled me to support Lisa in her goal of planning a lesson with a real-world context, and introducing it as an engineering activity. Before learning about Lisa's personal experiences and interests, it would have been a challenge for me to empathize with her teaching priorities. Yet, knowing about her as a person contributed to my understanding of what brought her to make decisions and actions in her classroom.

Documenting our lives in the school planning and implementing a lesson brought me to learn about my personal practical knowledge. Lisa implemented this lesson in 2002 and remembered students enjoying it, which motivated her to try it again. We did adapt the lesson so that the concept of area was part of it. I later learned that I was the only one who was most interested in the mathematics part of the lesson. When I planned the study, I thought of providing students with an experience to work on the concept of area as a measurement concept. Lisa and I had implemented another lesson where we learned about students' difficulties with the concept of area when it is imbedded in a real-world context (Suazo-Flores, 2018). This experience awakened me and brought me to learn more about area as measurement, to then create spaces for students to explore it. In my personal experience, working with Lisa awaken me to question the value that society has imposed on instrumental or school knowledge (Fasheh, 2012). This allowed me to learn about myself, and my personal emphasis on school mathematics, but at the same time, it brought me to open myself to understand Lisa's ways of knowing in her classroom. The understandings I have gained from this study have become important bases for my work as a mathematics teacher educator of prospective teachers.

Mathematics Teacher Educators' Engagement in Autoethnography

Of the three self-based methodologies, autoethnography has received the least attention in mathematics teacher educators' learning. But it has been used by other teacher educators in studies of their practice. For example, Sanders, Parsons, Mwavita and Thomas (2015) used it to learn how they changed as literacy professional development leaders in a high-needs, culturally diverse, urban, school district in the United States. Taylor, Klein and Abrams (2014) used it to understand their role of supporting mentor teachers during their prospective teachers' field experience. Schneider and Parker (2013) used it to explore the impact of a study-abroad program for prospective teachers on their professional development and the implications for their work as teacher educators. Park (2014) used it to explore her experiences and gain deeper understanding of herself as a teacher educator in a TESOL (Teaching English to Speakers of Other Languages) program. These studies suggest that autoethnography could help mathematics teacher educators to gain deeper self-understanding and make meaningful changes to their practice. This is reflected in the work of one of the authors of this chapter, Jennifer Ward, who has engaged in autoethnography as a methodology in her research and learning as a classroom teacher and mathematics teacher educator.

Ward recently completed her Doctor in Education (EdD) while working as both a primary classroom teacher and a mathematics teacher educator of prospective teachers. The abstract to her thesis states:

> The purpose of this autoethnography was to explore the experiences, both successes and challenges, as I worked to teach mathematics using a social justice framework in a summer enrichment camp with four and five-year-old children. … Autoethnography was selected as a methodological approach in this study as I examined my own teaching experiences and journey engaging in teaching mathematics for social justice. Primary data sources include researcher reflective journal entries and videotaped lesson implementation while secondary sources include student work samples and artifacts. (Ward, 2017)

Ward gained insights into her experiences with teaching mathematics for social justice and questioned areas of her work related to power and control, perpetuating deficit views, relationship construction, and finding a balance between mathematics and social justice within the lessons (Ward, 2017). These insights became a basis of her knowledge and way of being in working with prospective teachers in her new tenured-track position at a university. In what follows, Ward provides further insights of her experience with this approach in her first-person account of her journey with autoethnography.

Jennifer Ward's journey with autoethnography. Plucked from the classroom to be involved in a hybrid role between my university and the local school district, I eagerly welcomed any endeavor to work with both prospective teachers and young children. Working with prospective teachers simultaneously in the field and coursework, fueled a need to stay connected with the lived experiences of teachers as they worked day to day with early learners. This drive continued throughout my studies, into the stages of my dissertation work, and now into my work as a mathematics teacher educator. As a classroom teacher, I relied on theory to begin to plan and guide my work, but I also placed emphasis on the situational context in which I existed. Many times, I became engrossed in the stories other teachers told of their classrooms and experiences. These stories captivated me and began the wheels in my head turning; thinking of how I might take these ideas and make sense of them, tweak them and use them within my own context. I also found that when engaged in coaching with both prospective teachers and practising teachers I could use the power of story or personal experience to develop connections, gain trust and push thinking forward for all parties involved.

This experience with story played a role in influencing my choice of research methodology to examine my teaching of mathematics for social justice in which both mathematics and social justice became foci for my lessons where children (ages four to five) could use mathematics to investigate life, power and societal issues (Gonzalez, 2009). I also needed a methodology that would place me as both the researcher and participant in this study. While other approaches could place me in both roles, I kept going back to the storied aspect of autoethnographic research; the intense focus on connecting with others engaged in similar work as well as telling an honest and transparent account of my teaching. Furthermore, doing all of this within a narrative where I could creatively interweave the events and how I saw them unfolding with the emotions stirring around in my head.

Engaging in autoethnography required that I write a narrative with authentic and open discussion about my experience (Dwyer & Buckle, 2009). The idea of presenting a truthful account; one that told *my* story, in *my* context as it made sense to *me,* resonated with my desire to connect with others engaging in this work, while presenting an account that did not seem to be perfected, but rather raw and honest. Having been fully immersed in the experience of being the participant and researcher, having a passion for the topic being explored and feeling connected with others doing similar work, I felt better prepared to craft a story that was authentic; transporting those who read the work into the classroom I had shared with my children that summer.

For me, this methodology equated to a form of therapy, as engaging in it to conduct research was an opportunity in which I was able to merge my role as a teacher and researcher addressing an emotional topic. Furthermore, by sharing such a candid piece of myself, I was able to elicit an emotional recall (Ellis, 2004). This emotional recall pushed me to reengage with the events surrounding the lessons I planned and implemented, seeing beyond the something as either going well or not. I began to see how the experience and myself were working in tandem, shaping each other in varying ways, rather than one solely impacting the other.

Throughout the continuous process of conducting and writing about my research, I felt as if I was constructing an understanding of myself that was ever changing. As I read, wrote, and weaved emotion and events together like a tapestry I felt as if I was forced, albeit willingly, to relive the memories from my study. Autoethnographic methods allow the researcher to critically examine their own experiences (Duncan,

2004; Spry, 2001), and because of this, I was able to critically examine my own narrative. For example, framing my thinking around who planned lessons, how they were planned and who decided when and how they were structured in the classroom allowed me to step back and re-enter my data to construct new thinking about how I had exhibited control over the children in a position of power. By immersing myself in the written story of the journey, I began to question my own stance of power within the classroom community. I witnessed firsthand a disconnect between my ideological beliefs on involving the children in co-constructing their mathematics learning experiences, and what actually occurred during my implementation, as I was the one planning lessons and became hyper focused on management issues during the initial lesson stages. In essence, I was contradicting the beliefs that had led me to do this work initially by not working to elicit and acknowledge the voices of my children. This made me aware of the idea of being a living contradiction; I used much of the language I did not believe in (deficit) and focused on management and control within the lesson, rather than children having power over the learning.

As I continuously reflected on these events, I was able to think more deeply and critically about my role as a mathematics teacher educator doing this work. This circled back to how I could support other teachers in their own journeys. For example, I could anticipate what they might encounter and share stories about the journey. This sharing could help me to coach by building empathetic relationships, recalling how I personally had felt in these situations and how I grappled with my decisions in the moment to grow and develop my own stance as a mathematics teacher educator. My reflection also led me to how I might leverage children's voices more in the mathematics lessons or support others in doing so; for example, sharing stories to challenge the idea that grades PreK-2 students cannot do so. Now, in my first year in a tenure track faculty role as mathematics teacher educator, I am able to use these experiences to work with prospective teachers during their work in PreK through grade 2 classrooms. Knowing that, in my experience, I resorted back to a focus on management and lost sight of the voice of my learners, I spend more time listening to my prospective teachers discuss their observations of, as well as their own teaching. I challenge them to examine children's perceptions of lessons and topics and to question the purpose of the prospective teachers' actions and those they see within the field. I have found that I am slower to make judgements about the beliefs of my prospective teachers based upon their actions in the field or things they say or write for class. Finally, the work of mathematics teacher educators is often collaborative in nature. I have found that being open and honest has helped me to connect with others and, similarly, I feel more at ease with other self-based methodologies that present the honest emotional side to their work.

IMPLICATIONS AND CONCLUSION

In this chapter, we highlighted three self-based methodologies as ways of knowing in mathematics teacher educators' learning with examples of the type of learning they support. These examples suggest that these methodologies are effective in helping mathematics teacher educators to learn about themselves, grow in their practice and contribute to the field of teacher education. While these self-based methodologies are similar in their focus on self and experience, they offer variations in the path to self-understanding. For example, mathematics teacher educators could engage in self-study to provoke awareness of elements of their knowledge and practices. They could engage in narrative inquiry to construct a bridge between their pedagogical experiences and deeply held values and to learn about themselves in relation to the perspectives of prospective teachers or practising teachers with whom they work. They could engage in autoethnography to critically examine their pedagogical experiences in relation to a particular cultural context or group. Our collective engagement in these self-based methodologies suggests the following implications and conclusions for mathematics teacher educators' learning through them and future support for their use in mathematics teacher educators' research.

Implications for Mathematics Teacher Educators' Learning

While there are possibly other implications for the use of self-based methodologies to support mathematics teacher educators' learning, the six we discuss are the ones that stood out for us based on our experiences with these methodologies. While they are all related in terms of supporting mathematics teacher educators' learning, self-understanding and growth, each is presented to draw attention to specific features mathematics teacher educators can attend to when engaging in the self-based methodologies.

Living awareness. These self-based methodologies allow mathematics teacher educators to view themselves as living awareness; awareness of and in action in relation to self and in relation to prospective teachers or practising teachers. This awareness involves a consciousness about practice that emerges from lived experiences of teaching and reflections on those lived experiences. The consciousness that emerges allows the mathematics teacher educators to focus their inquiry on particular elements of practice. For example, they could become mindful of how their choices of tasks and ways of interacting with prospective teachers impact them as learners and implement changes to address resulting complexities. Living awareness also suggests a stance or disposition that one needs to develop to continually revisit one's experience. For example, as Kastberg (co-author) noted:

> I use self-study methodology, not because I want to improve my practice by controlling the ways I implement particular routines or activities, not because I want to be intentional in my practice, but because engaging in such study helps me be conscious in the moment of teaching and act based on my experiences in ways that matter most to prospective teachers and to me.

Engaging in the self-based methodologies, then, allows mathematics teacher educators to experience living awareness, helps them in the development of it, and helps them to appreciate the importance of it to support or enhance their teaching. As Mason (1998) noted

> The key notions underlying real teaching are the structure of attention and the nature of awareness. (p. 244)
>
> … [T]o become an expert it is necessary to develop and articulate awareness of your awarenesses-in-action; to become a teacher in the full and most appropriate sense of that word, it is necessary to become aware of your awareness of those awarenesses-in-action. (p. 255)

Living contradictions. These self-based methodologies also enable mathematics teacher educators to examine themselves as living contradictions; contradictions between lived experiences and deeply held values regarding their practice or between anticipated and actual effect of their pedagogical decisions. For example, as in (co-author) Kastberg's case, a mathematics teacher educator could experience a living contradiction between her value of autonomy and her practice of awarding points for particular sorts of classroom participation; or as in (co-author) Ward's case, between her valuing giving children voice and her use of power and control in her classroom. Awareness of living contradictions in mathematics teacher educators' practice provides the opportunity for them to explore possibilities for the alignment of their actions and their ideas about bringing a coherence to their practice. They gain insights and, in some cases, a sense of control over their practice. Building coordination of their mental and physical actions during teaching in awareness allow them to "teach about teaching" (Loughran, 2004) from knowledge of learning about teaching as well as knowledge of mathematics learning.

Empathetic relations. These self-based methodologies are also considered empathetic methodologies (D'Ambrosio & Cox, 2015) based on their humanistic perspective of viewing self through empathetic lens. This includes acknowledgement of the other through listening and responding to achieve mutual understanding. Thus, through these self-based methodologies, mathematics teacher educators can understand more from the perspectives of the mathematics teachers in their studies and also understand themselves in relation to the teachers. For example, as Cox (co-author) found:

> My work with the teachers in my origin story helped me to discover the need for empathy in my research and brought me to understand the power of narrative inquiry to work with it. … I have used narrative as a tool to explore my capacity for mathematical empathy and the capacity of narrative inquiry to inspire it in others.

Engaging in these self-based methodologies can also help mathematics teacher educators to develop skills to support empathetic relations in their teaching. This includes empathetic listening skills; skills to scrutinize ideas that are different from their own and to attempt to move beyond mere listening and restating and suspend doubt (Harness, 2009); for example, to believe in an idea's merit and see the truth within when interacting with prospective teachers or practising teachers. This could significantly impact mathematics teacher educators' practice. For example, as a result of her autoethnographic experience, Ward (co-author) noted:

> I spend more time listening to my prospective teachers discuss their observations of, as well as their own teaching. … I have found that I am slower to make judgements about the beliefs of my prospective teachers based upon their actions in the field or things they say or write for class.

Similarly, as a result of her self-study experience, Kastberg (co-author) explained how she shifted her gaze toward how she could convey respect and empathy for her prospective teachers and the challenges they faced in learning to teach. She turned toward building relationships with and learning from prospective teachers rather than trying to provide instruction for them solely from her perspective.

Growing in relationships. The empathetic perspective of the self-based methodologies can also help mathematics teacher educators to understand how to support growth in their knowledge and teaching by moving from knowing in relation to self toward knowing in relation to their prospective teachers. As Kastberg (co-author) noted:

> I have realized that my receptivity to growing in relationship to prospective teachers involves taking their models of mathematics teaching and learning not as barriers to my practice, but as emergent views.

As teachers become receptive to growing in relation, relationships change. No longer are relationships tools to encourage development of prospective teachers' knowledge and of person-hood, but instead, relationships become a way to understand oneself and one's own practice. Growing in relationship moves beyond using the relationship to motivate or encourage prospective teachers to take risks and try new things; it becomes about seeing and knowing one's self.

Practical knowledge. Practical knowledge refers to the kind of knowledge teachers hold and use (Elbaz, 1983); knowledge of classroom situations and practical dilemmas they face in carrying out purposeful action in these settings (Carter, 1990). It is the knowledge that teachers themselves generate as a result of their experiences as teachers (Fenstermacher, 1994); that is, it originates in, and develops through, experiences in teaching. It is, therefore, personal, contextual, mainly tacit and guides teaching practice. As teachers of teachers, mathematics teacher educators hold practical knowledge, or personal practical knowledge, which underlies their classroom actions in supporting prospective teachers' learning. Since this knowledge tends to become taken-for-granted it also tends to become unexamined personal theories. The self-based methodologies provide a valuable means for mathematics teacher educators to think more deeply and critically about their practical knowledge and deepen their understanding of it and the opportunities and limitations it places on their students' learning. For example, stories of memorable or challenging mathematics teacher educators' experiences in working with mathematics teachers embody their practical knowledge. Exploring the plots or underlying themes of the stories reveals the nature of the practical knowledge (e.g., power and control in the classroom; deficit views) and its impact on their teaching.

Change in self and practice. The principal goal of the self-based methodologies in mathematics teacher educators' learning is self-understanding, however, not as an end in itself, but a means to develop new knowledge to enhance practice. In the section on mathematics teacher educators' engagement in self-based methodologies, we provided examples of changes mathematics teacher educators made to their practice. Of equal importance are the changes self-based methodologies support mathematics teacher educators to make to self. This is evident in our preceding discussion of living awareness, living contradictions, and empathetic relations, which require developing ways of being with skills that were previously less pronounced. The change in self is also directly related to change in practice. For example, as mathematics teacher educators learn to be more empathetic, the result is more meaningful relationships in the classroom to better support prospective teachers' learning and to model the type of interactions they can adopt with their students.

Future Support for Self-Based Methodologies

Despite their usefulness to mathematics teacher educators' learning, these self-based methodologies have not received much attention by mathematics teacher educators based on published studies. While self-study has received some attention, narrative inquiry and autoethnography are less considered as a means of mathematics teacher educators' self-understanding and growth. A possible reason for this is that mathematics teacher educators, who are also academics and researchers, need to get their research published and are concerned about acceptance of these methodologies or research reports on self in the mathematics education community. This could also be the reason why they do not address their learning in studies of their practice as pointed out by Chapman (2008) and Jaworski (2008). The two co-authors (Suazo-Flores and Ward) of this chapter who recently completed their doctorates expressed concerns about this. Regarding her use of narrative inquiry, Suazo-Flores explained,

> The challenge in engaging in the data analysis was the conflict of being true to the narrative process and feeling the need to work in a way to belong to the mathematics education community. Whatever way I decide to analyze

> the field texts, I would think to myself about how the mathematics education community would consider it, but also, how the narrative inquiry people would receive it. Therefore, I tried to satisfy both fields.

Regarding her use of autoethnography, Ward explained:

> As I began to journey into this work, I felt vulnerable to criticism I was sure would come from telling my story. In thinking about the construction and presentation of my study I feared I would not be taken seriously, as this methodology could be seen as vastly different from what had been accepted historically as research. These feelings welled up inside of me, as I worried how in the world I was I going to turn this work into something publishable, something presentable, something I could take on a job talk. My head spun wondering how I would be perceived from those outside my committee for doing this work.

For the field of mathematics teacher educator to move forward, there needs to be more openness to these self-based methodologies as a valid way of conducting research in which mathematics teacher educators can research themselves. On the other hand, mathematics teacher educators who use these methodologies have to make sure that they do so with close attention to theoretical guidelines that define them from a research perspective (e.g., Clandinin et al., 2007; Feldman, 2003).

CONCLUSION

While encouraging mathematics teacher educators to engage in self-based methodologies is important for self-understanding, of more importance is the implication for teacher education. In addition to supporting mathematics teacher educators' learning and improving practice, these methodologies can also result in deeper understandings about teacher education in general, thus, producing and advancing knowledge about teacher education. In general, these methodologies could result in significant learning about self and teacher education with implications for improvements of teacher education in general.

REFERENCES

Alderton, J. (2008). Exploring self-study to improve my practice as a mathematics teacher educator. *Studying Teacher Education, 4*(2), 95–104.

Aleandri, G., & Russo, V. (2015). Autobiographical questionnaire and semi-structured interview: Comparing two instruments for educational research in difficult contexts. *Procedia – Social and Behavioral Sciences, 197*, 514–524.

Atkinson, P., & Delamont, S. (2006). Rescuing narrative from qualitative research. *Narrative Inquiry, 16*, 164–172.

Bahr, D. L., Monroe, E. E., & Mantilla, J. (2018). Developing a framework of outcomes for mathematics teacher learning: three mathematics educators engage in collaborative self-study. *Teacher Education Quarterly, 45*(2), 113–134.

Bailey, J. (2008). Using narrative inquiry to explore mathematics curriculum. *Curriculum Matters, 4*, 147–163.

Berry, A., & Hamilton, M. L. (2017). Self-study of teacher education practices. In L. Meyer (Ed.), *Oxford bibliographies in education*. New York, NY: Oxford University Press.

Beswick, K., & Goos, M. (2018). Mathematics teacher educator knowledge: What do we know and where to from here? *Journal of Mathematics Teacher Education, 21*, 417–427.

Brandenburg, R. (2008). *Powerful pedagogy: Self-study of a teacher educator's practice*. Dordrecht: Springer.

Bruner, J. S. (2003). The narrative creation of self. In L. E. Angus & J. McLeod (Eds.), *The handbook of narrative and psychotherapy: Practice, theory and research* (pp. 3–14). Thousand Oaks, CA: Sage.

Carter, K. (1990). Teachers' knowledge and learning to teach. In W. R. Houston (Ed.), *Handbook of research on teacher education* (pp. 291–310). New York, NY: Macmillan.

Chang, H. (2008). *Autoethnography as method*. Walnut Creek, CA: Left Coast Press.

Chapman, O. (2008a). Educators reflecting on (researching) their own practice. In R. Even & D. L. Ball (Eds.), *The professional education and development of teachers of mathematics: The 15th ICMI study* (pp. 121–126). New York, NY: Springer.

Chapman, O. (2008b). Mathematics teacher educators' learning from research on their instructional practices: A cognitive perspective. In B. Jaworski & T. Wood (Eds.), *International handbook of mathematics teacher education: The mathematics teacher educator as a developing professional* (Vol. 4, pp. 110–129). Rotterdam, The Netherlands: Sense Publishers.

Chapman, O. (2008c). Narratives in mathematics teacher education. In D. Tirosh & T. Wood (Eds.), *International handbook of mathematics teacher education: Tools and processes in mathematics teacher education* (Vol. 4, pp. 15–38). Rotterdam, The Netherlands: Sense Publishers.

Chapman, O., & Heater, B. (2010). Understanding change through a high school mathematics teacher's journey to inquiry-based teaching. *Journal of Mathematics Teacher Education, 13*(6), 445–458.

Chase, S. E. (2011). Narrative inquiry. In N. K. Denzin & Y. S. Lincoln (Eds.), *The Sage handbook of qualitative research* (pp. 421–434). Thousand Oaks, CA: Sage Publications.

Chauvot, J. B. (2009). Grounding practice in scholarship, grounding scholarship in practice: Knowledge of a mathematics teacher educator–researcher. *Teaching and Teacher Education, 25*(2), 357–370.

Clandinin, J. (2013). *Engaging in narrative inquiry*. Walnut Creek, CA: Left Coast Press.

Clandinin, J. D. (1985). Personal practical knowledge: A study of teachers' classroom images. *Curriculum Inquiry, 15*(4), 361–385.

Clandinin, D. J., & Connelly, F. M. (1995). *Teachers' professional knowledge landscapes*. New York, NY: Teachers College Press.

Clandinin, D. J., & Connelly, F. M. (2000). *Narrative inquiry: Experience and story in qualitative research*. San Francisco, CA: Jossey-Bass.

Clandinin, D. J., Pushor, D., & Orr, A. M. (2007). Navigating sites for narrative inquiry. *Journal of Teacher Education, 58*(1), 21–35.

Cochran-Smith, M., & Lytle, S. (2004). Practitioner inquiry, knowledge and university culture. In J. J. Loughran, M. L. Hamilton, V. K. LaBoskey, & T. Russell (Eds.), *International handbook of self-study of teaching and teacher education practices* (pp. 601–650). Dordrecht: Kluwer Academic Publishers.

Connelly, F. M., & Clandinin, D. J. (1990). Stories of experience and narrative inquiry. *Educational Researcher, 19*(5), 2–14.

Connelly, F. M., & Clandinin, D. J. (1994). Telling teaching stories. *Teacher Education Quarterly, 21*(1), 145–158.

Connelly, F., & Clandinin, D. (1995). Narrative and education. *Teachers and Teaching, 1*(1), 73–85.

Connelly, F. M., & Clandinin, D. J. (2006). Narrative inquiry. In J. L. Green, G. Camilli, & P. Elmore (Eds.), *Handbook of complementary methods in education research* (3rd ed., pp. 477–487). Mahwah, NJ: Lawrence Erlbaum.

Cox, D., & Harper, S. (2017). Using narratives to articulate mathematical problem solving and posing in a technological environment. In E. Galindo & J. Newton (Eds.), *Proceedings of the 39th annual meeting of the North American Chapter of the International Group for the Psychology of Mathematics Education* (pp. 985–988). Indianapolis, IN: Hoosier Association of Mathematics Teacher Educators.

Cox, D. C., Harper, S. R., & Edwards M. T. (2018). Screencasting as a tool to capture moments of authentic creativity. In V. Freiman & J. Tassell (Eds.), *Creativity and technology in mathematics education* (pp. 33–57). New York, NY: Springer.

D'Ambrosio, B. S., & Cox, D. (2015). An examination of current methodologies in mathematics education through the lenses of purpose, participation, and privilege. *Perspectivas Da Educaçao Matemàtica, 8*(18), 687–708.

Denzin, N. (2006). Analytic autoethnography, or déjà vu all over again. *Journal of Contemporary Ethnography, 35*, 419–428.

Dinkelman, T. (2003). Self study in teacher education: A means and ends tool for promoting reflective teaching. *Journal of Teacher Education, 54*(1), 6–18.

Drake, C. (2006). Turning points: Using teachers' mathematics life stories to understand the implementation of mathematics education reform. *Journal of Mathematics Teacher Education, 9*(6), 579–608.

Drake, C., Spillane, J., & Hufferd-Ackles, K. (2001). Storied identities: Teacher learning and subject-matter context. *Journal of Curriculum Studies, 33*(1), 1–23.

Duncan, M. (2004). Autoethnography: Critical appreciation of an emerging art. *International Journal of Qualitative Methods, 3*(4), 28–39.

Dwyer, S. C., & Buckle, J. L. (2009). The space between: On being an insider-outsider in qualitative research. *International Journal of Qualitative Methods, 8*(1), 54–63.

Elbaz, F. (1983). *Teacher thinking: A study of practical knowledge*. Great Britain: Croom Helm.

Elbow, P. (2008). The believing game or methodological believing. *The Journal of the Assembly for Expanded Perspectives on Learning, 14*(1), 3.

Ellis, C. (2004). *The ethnographic I: A methodological novel about autoethnography.* Walnut Creek, CA: Rowman and Littlefield Publishers.

Ellis, C., Adams, T. E., & Bochner, A. P. (2011). Autoethnography: An overview. *Historical Social Research/Historische Sozialforschung, 36*(4) 273–290.

Ellis, C., & Bochner, A. P. (2000). Autoethnography, personal narrative, reflexivity: Researcher as subject. In N. K. Denzin, & Y. S. Lincoln (Eds.), *Handbook of qualitative research* (pp. 733–768). Thousand Oaks, CA: Sage.

Fasheh, M. J. (2012). The role of mathematics in the destruction of communities, and what we can do to reverse this process, including using mathematics. In O. Skovsmose & B. Greer (Eds.), *Opening the cage: Critique and politics of mathematics education* (pp. 93–105). Rotterdam, The Netherlands: Sense Publishers.

Feldman, A. (2003). Validity and quality in self-study. *Educational Researcher, 32*(3), 26–28.

Fenstermacher, G. D. (1994). The knower and the known: The nature of knowledge in research on teaching. *Review of Research in Education, 20*, 3–56.

Gonzalez, L. (2009). Teaching mathematics for social justice: Reflections on a community of practice for urban high school mathematics teachers. *Journal for Urban Mathematics Education, 2*(1), 22–51.

Goodell, J. E. (2006). Using critical incident reflections: A self-study as a mathematics teacher educator. *Journal of Mathematics Teacher Education, 9*(3), 221–248.

Hamilton, M. L., & Pinnegar, S. (1998). Conclusion: The value and the promise of self-study. In M. L. Hamilton (Ed.), *Reconceptualizing teaching practice: Self-study in teacher education* (pp. 235–246). London: Falmer Press.

Hamilton, M. L., Smith, L., & Worthington, K. (2008). Fitting the methodology with the research: An exploration of narrative, self-study and auto-ethnography. *Studying Teacher Education, 4*(1), 17–28.

Harkness, S. (2009). Social constructivism and the believing game: A mathematics teacher's practice and its implications. *Educational Studies in Mathematics, 70*, 243–258.

Hattie, J., & Timperley, H. (2007). The power of feedback. *Review of Educational Research, 77*, 81–112.

Hjalmarson, M. A. (2017). Learning to teach mathematics specialists in a synchronous online course: A self-study. *Journal of Mathematics Teacher Education, 20*, 3, 281–301.

Jaworski, B. (2001). Developing mathematics teaching: Teachers, teacher educators, and researchers as co-learners. In F.-L. Lin & T. Cooney (Eds.), *Making sense of mathematics teacher education* (pp. 295–320). Dordrecht: Kluwer Academic Publishers.

Jaworski, B. (2008). Mathematics teacher educator learning and development: An introduction. In B. Jaworski & T. Wood (Eds.), *The mathematics teacher educator as a developing professional: The mathematics teacher educator as a developing professional* (Vol. 4, pp. 1–13). Rotterdam, The Netherlands: Sense publishers

Kaasila, R. (2007). Using narrative inquiry for investigating the becoming of a mathematics teacher. *ZDM, International Journal of Mathematics Education, 39*, 205–213.

Kaasila, R., Hannula, M. S., Laine, A., & Pehkonen, E. (2008). Socio-emotional orientations and teacher change. *Educational Studies in Mathematics, 67*(2), 111–123.

Kastberg, S. (2012). Building an understanding of learning from moments in teaching. In L. B. Erickson, J. R. Young, & S. Pinnegar (Eds.), *Extending inquiry communities: Illuminating teacher education through self-study* (Proceedings of the Ninth International Conference on the Self-Study of Teacher Education Practices, pp. 167–170). Provo, UT: Brigham Young University.

Kastberg, S., Lischka, A., & Hillman, S. (2016). Exploring written feedback as a relational practice. In A. Ovens & D. Garbett (Eds.), *Enacting self-study as methodology for professional inquiry* (Proceedings of the Eleventh International Conference on the Self-Study of Teacher Education Practices, pp. 387–394). Auckland, NZ: The University of Auckland.

Kastberg, S., Lischka, A., & Hillman, S. (2018). Building questioning as a relational practice through self-study. In A. O. D. Garbett (Ed.), *Pushing boundaries and crossing boarders: Self-study as a means for knowing pedagogy* (Proceedings of the Twelfth International Conference on the Self-Study of Teacher Education Practices). Auckland, NZ: The University of Auckland.

Kastberg, S. E., Lischka, A. E., & Hillman, S. L. (2018). Characterizing mathematics teacher educators' written feedback to prospective teachers. *Journal of Mathematics Teacher Education*, 1–22 (online first). https://doi.org/10.1007/s10857-018-9414-6

Kitchen, J. (2005a). Conveying respect and empathy: Becoming a relational teacher educator. *Studying Teacher Education, 1*, 195–207.

Kitchen, J. (2005b). Looking backward, moving forward: Understanding my narrative as a teacher educator. *Studying Teacher Education, 1*, 17–30.

LaBoskey, V. K. (2007). The methodology of self-study and its theoretical underpinnings. In *International handbook of self-study of teaching and teacher education practices* (pp. 817–869). New York, NY: Springer.

Loughran, J. J. (2004). A history and context of self-study of teaching and teacher education practices. In J. J. Loughran, M. L. Hamilton, V. K. LaBoskey, & T. Russell (Eds.), *International handbook of self-study of teaching and teacher education practices* (pp. 7–39). Dordrecht: Springer.

Loughran, J. J. (2005). Researching teaching about teaching: Self-study of teacher education practices. *Studying Teacher Education, 1*(1), 5–16.

Loughran, J. J., Hamilton, M. L., LaBoskey, V. K., & Russell, T. (Eds.). (2004). *International handbook of self-study of teaching and teacher education practices*. Dordrecht: Springer.

Lovin, L. H., Sanchez, W. B., Leatham, K. R., Chauvot, J. B., Kastberg, S. E., & Norton, A. H. (2012). Examining beliefs and practices of self and others: Pivotal points for change and growth for mathematics teacher educators. *Studying Teacher Education, 8*(1), 51–68.

Lunenberg, M., & Samaras, A. P. (2011). Developing a pedagogy for teaching self-study research: Lessons learned across the Atlantic. *Teaching and Teacher Education, 27*(5), 841–850.

Lunenberg, M., & Willemse, M. (2006). Research and professional development of teacher educators. *European Journal of Teacher Education, 29*(1), 81–98.

Lutovac, S., & Kaasila, R. (2014). Pre-service teachers' future-oriented mathematical identity work. *Educational Studies in Mathematics, 85*(1), 129–142.

Marin, K. A. (2014). Becoming a teacher educator: A self-study of the use of inquiry in a mathematics methods course. *Studying Teacher Education, 10*(1), 20–35.

Mason, J. (1998). Enabling teachers to be real teachers: Necessary levels of awareness and structure of attention. *Journal of Mathematics Teacher Education, 1*(3), 243–267.

McNiff, J. (2017). *Action research: All you need to know*. London: Sage.

Moll, L. C. (1992). Bilingual classroom studies and community analysis: Some recent trends. *Educational researcher, 21*(2), 20–24.

Nardi, E. (2016). Where form and substance meet: Using the narrative approach of re-storying to generate research findings and community rapprochement in (university) mathematics education. *Educational Studies in Mathematics, 92*(3), 361–377.

National Council of Teachers of Mathematics. (2000). *Principles and standards for school mathematics*. Reston, VA: National Council of Teachers of Mathematics.

National Governors Association Center for Best Practices. (2010). *Common core state standards for mathematics*. Retrieved from http://www.corestandards.org/Math/Content

Noddings, N. (1984). *Caring, a feminine approach to ethics and moral education*. Berkeley, CA: University of California Press.

Oslund, J. A. (2012). Mathematics-for-teaching: What can be learned from the ethnopoetics of teachers' stories? *Educational Studies in Mathematics, 79*, 293–309.

Park, L. E. (2014). Shifting from reflective practices to reflexivity: An autoethnography of an L2 teacher educator. *English Teaching, 69*(1), 173–198.

Phelps, C. M. (2010). Factors that pre-service elementary teachers perceive as affecting their motivational profiles in mathematics. *Educational Studies in Mathematics, 75*(3), 293–309.

Polkinghorne, D. E. (1995). Narrative configuration in qualitative analysis. *International Journal of Qualitative Studies in Education, 8*(1), 5–23.

Reed-Danahay, D. (1997). *Auto/Ethnography: Rewriting the self and the social*. New York, NY: Berg.

Rogers, C. R. (1961). *On becoming a person: A therapist's view of psychotherapy*. New York, NY: Houghton Mifflin Harcourt.

Pinnegar, S., & Russell, T. (1995). Self-study and living educational. *Teacher Education Quarterly, 22*(3), 5–9.

Richardson, V. (1994). Conducting research on practice. *Educational Researcher, 23*(5), 5–10.

Samaras, A. P. (2011). *Self-study teacher research: Improving your practice through collaborative inquiry*. Thousand Oaks, CA: Sage.

Sanders, J. Y., Parsons, S. C., Mwavita, M., & Thomas, K. (2015). A collaborative autoethnography of literacy professional development work in a high-needs environment. *Studying Teacher Education, 11*(3), 228–245.

Schlechty, P. C. (2001). *Shaking up the schoolhouse: How to support and sustain educational innovation*. San Francisco, CA: Jossey-Bass.

Schneider, J., & Parker, A. (2013). Conversations in a pub: Positioning the critical friend as "peer relief" in the supervision of a teacher educator study abroad experience. *The Qualitative Report, 18*(32), 1–14.

Schön, D. A. (1987). *Educating the reflective practitioner*. Oxford: Jossey-Bass.

Schuck, S. (2002). Using self-study to challenge my teaching practice in mathematics education. *Reflective Practice, 3*(3), 327–337.

Sparkes, A. C. (2000). Autoethnography and narratives of self: Reflections on criteria in action. *Sociology of Sport Journal, 17*, 21–41.

Spry, T. (2001). Performing autoethnography: An embodied methodological praxis. *Qualitative Inquiry, 7*(6), 706–732.

Suazo-Flores, E. (2016). Working together: A caring relation between a teacher and a mathematics educator. *Purdue Journal of Service-Learning and Engagement, 3*, 34–37.

Suazo-Flores, E. (2018). Students' understanding of area: Combining practical and mathematical knowledge with a real world task. *International Journal for Research in Mathematics Education, 8*(1), 23-37.

Taylor, M., Klein, E. J., & Abrams, L. (2014). Tensions of reimagining our roles as teacher educators in a third space: Revisiting a co/autoethnography through a faculty lens. *Studying Teacher Education, 10*(1), 3–19.

Ward, J. (2017). *Early childhood mathematics through a social justice lens: An autoethnography* (Graduate Theses and Dissertations). University of South Florida, Tampa, FL. Retrieved from http://scholarcommons.usf.edu/etd/6975

Whitehead, J. (1989). Creating a living educational theory from questions of the kind, "How do I improve my practice?" *Cambridge Journal of Education, 19*(1), 41–52.

Olive Chapman
Werklund School of Education
University of Calgary

Signe Kastberg
College of Education
Purdue University

Elizabeth Suazo Flores
Department of Biological Sciences
Purdue University

Dana Cox
Department of Mathematics
Miami University

Jennifer Ward
Bagwell College of Education
Kennesaw State University

PART 3

MATHEMATICS TEACHER EDUCATORS LEARNING FROM PRACTICE

AINA APPOVA

8. CONCEPTUALIZATION AND ENACTMENT OF PEDAGOGICAL CONTENT KNOWLEDGE BY MATHEMATICS TEACHER EDUCATORS IN PROSPECTIVE TEACHERS' MATHEMATICS CONTENT COURSES

In this chapter, I discuss concerns regarding the adequacy of teacher education programs in preparing teachers for the classroom and the importance and nature of mathematics teacher educators' practices in courses for prospective teachers. I specifically focus on theoretical and empirical foundations that offer promising directions for moving the field forward in conceptualizing the practices of mathematics teacher educators (particularly in content courses) for providing opportunities for prospective teachers to develop pedagogical content knowledge. The chapter offers a theoretical framework to the field as a (beginning) conceptualization of mathematics teacher educators' practices. It includes empirical evidence from the findings of research with which I have been involved and literature in support of the proposed framework.

INTRODUCTION

There is a global debate on how best to recruit and prepare future teachers (Boyd et al., 2008; Levine, 2006). This debate is fuelled by educational reforms focused on ensuring high quality teacher education and development aligned with international standards, accountability, and improved student performance, as well as educational demands driven by economic and socio-political globalization (Cummings, 2003; Tatto, 2006). In fact, many nations across the world (e.g., Australia, Canada, Chile, Germany, Japan, Mexico and the United States) strive to reshape their educational systems to provide the necessary knowledge and skills for teachers needed to compete in the growing global economy (see Blomeke, 2006; Hooghart, 2006; LeTendre, 2002; Osborn, 2006; Shimahara, 2002; Tatto, 1999; Tatto, Schmelkes, Guevara, & Tapia, 2006; Weiss, Murphy-Graham, & Birkeland, 2005).

For example, *The Rainbow Plan* in Japan, also known as the Educational Reform Plan for the 21st Century, implemented by the Ministry of Education, Culture, Sports, Science, and Technology (MEXT, 2001), required a complete restructuring of the Japanese educational system from elementary school to university, to refocus

© KONINKLIJKE BRILL NV, LEIDEN, 2020 | DOI: 10.1163/9789004424210_009

teacher education and professional development on improving pupils' basic scholastic proficiency, aligning university programs to the international standards, training teachers as "real professionals" of education, and providing "private sector work experience for teachers" (Hooghart, 2006, p. 291). In contrast, the European ministers of education came together and agreed (by 2009) to unify the European university teacher education degrees by changing to a bachelor and master system, and converting to the European credit transfer system (ECTS) to make university student exchanges easier (see the European Ministers in charge of Higher Education, 1999, Bologna Declaration).

Many countries additionally continue to focus their efforts on improving teacher quality at the initial teacher preparation stages. For example, the Teacher Education Ministerial Advisory Group (TEMAG) in Australia, after examining a wide-range of evidence on initial teacher preparation (more than 170 submissions), reported that much of the challenge in their country was in the selection and "desirable balance between academic skills and personal characteristics" of teacher-candidates. TEMAG also reported that there are "mixed views" across various educational communities on "what teachers need to know, how they should teach, and how best to integrate theory and practice to have a measurable impact on student learning" (Craven et al., 2014, p. ix). TEMAG concluded that strengthening teacher quality begins with improving initial teacher preparation and that, "[n]ot all initial teacher education programs are equipping graduates with the content knowledge, evidence-based teaching strategies and skills they need to respond to different student learning needs" (Craven et al., 2014, p. vi).

Similarly, in the United States, there have been claims made that teacher education programs do not adequately prepare teachers for the classroom (Conference Board of Mathematical Sciences (CBMS), 2012; Grossman, 2008). A number of research studies have highlighted the variability in teacher preparation programs across different institutions and educational settings, particularly in regards to the number of required mathematics "content-based" courses designed specifically for prospective elementary school teachers (e.g., Darling-Hammond et al., 2000; Levine, 2006; Taylor & Ronau, 2006). As a result, CBMS proposed that all institutions preparing elementary school teachers must offer and require a minimum of nine (regular) semester credits[1] of mathematics "subject matter" or "content" courses (henceforward referred to as content courses) to help better prepare them for teaching mathematics in accordance with the vision that "teaching elementary [school] mathematics requires both a wide range of pedagogical skills and considerable mathematical knowledge" (CBMS, 2012, p. 55).

A wide range of pedagogical skills and considerable mathematical knowledge is often regarded as *pedagogical content knowledge*: a special type of teachers' knowledge that "represents the blending of content and pedagogy into an understanding of how particular topics, problems, or issues are organized, represented, and adapted to the diverse interests and abilities of learners, and presented for instruction" (Shulman, 1987, p. 8). Extensive research is available demonstrating that learning opportunities designed specifically around pedagogical content knowledge help to strengthen prospective teachers'[2] knowledge about the mathematics they will teach, as well as their ability to critically reflect on classroom instruction and improve students' learning (e.g., Baumert et al., 2010; Blömeke, Buchholtz, Suhl, & Kaiser, 2014; Burton, Daane, & Giesen, 2008; Carpenter, Fennema, Peterson, & Carey, 1988; Capraro, Capraro, Parker, Kulm, & Raulerson, 2005; Tirosh, Levenson, & Tabach, 2011; Vale, 2010).

However, pedagogical content knowledge-related opportunities have been primarily documented in mathematics methods courses, which are often taught in the education department and completed by prospective teachers the semester prior to student teaching (Greenberg & Walsh, 2008; Lutzer, Rodi, Kirkman, & Maxwell, 2007). Researchers argue that a methods course, toward the end of the program, is often not enough to provide the necessary learning experiences for prospective teachers to develop pedagogical content knowledge (Ambrose, 2004; Bass, 2005; Wideen, Mayer-Smith, & Moon, 1998). For example, Vacc and Bright (1999) found that even after a 2-year sequence of methods courses and student teaching experiences, specifically situated around pedagogical content knowledge, prospective teachers still were in the early stages of developing pedagogical content knowledge and struggling with content-specific pedagogies and addressing certain aspects of children's mathematical thinking. Most importantly, the authors reported that prospective teachers were "unable to use them [these skills] in their teaching" (Vacc & Bright, 1999, p. 107).

It has been argued that embedding pedagogical content knowledge-related learning opportunities into teacher education programs is critical, and that mathematics content courses might be ideal platforms for doing so, particularly since they are designed for prospective teachers to enhance their mathematical and pedagogical knowledge required for teaching (CBMS, 2012; also see Ambrose, 2004; Ball, Sleep, Boerst, & Bass, 2009; Greenberg & Walsh, 2008). However, in the United States, nearly all (90%) mathematics content courses are taught and developed by the mathematics department faculty and staff (Masingila, Olanoff, & Kwaka, 2012). These individuals often do not have formal training in mathematics education or preparing teachers, nor do they have experiences working with students and/or teaching mathematics to schoolchildren (Bass, 2005; Hodgson, 2001; Sztajn, Ball, & McMahon, 2006).

Masingila, Olanoff, and Kwaka (2012) reported that more than half of mathematics teacher educators who teach content courses feel unprepared and report lack of training, resources, and support at their institutions. Similarly, Sztajn, Ball, and McMahon (2006) noted that "trained as mathematicians or as teachers themselves, most teacher-developers lack knowledge about teachers as learners" (p. 151). Moreover, research is extremely limited on the nature of mathematics content courses, both nationally and internationally. We know very little about what these courses look like, especially across different institutions, programs, and contexts, including what content is taught and how it is taught to prospective teachers, as the practices of university faculty who teach content courses are not widely documented or disseminated (Bergsten & Grevholm, 2008; Even, 2008; Floden & Philipp, 2003; Hiebert, Morris, & Glass, 2003; McDuffie, Drake, & Herbel-Eisenmann, 2008). This lack of knowledge contributes not only to the absence of a shared vision and curriculum across mathematics teacher educators but also to variability across teacher education courses and programs (Ball et al., 2009; Hiebert & Morris, 2009; Zaslavsky, 2007).

In this chapter, I begin to address these gaps in the literature. I specifically focus on *theoretical* and *empirical* foundations that offer promising directions for moving the field forward in conceptualizing the practices of mathematics teacher educators (particularly in content courses) for providing opportunities for prospective teachers to develop pedagogical content knowledge. Thus, the structure of this chapter includes a theoretical framework that I offer to the field as a (beginning) conceptualization of mathematics teacher educators' practices. I additionally include empirical evidence from research findings with which I have been involved and literature in support of the proposed framework.

BACKGROUND

Mathematics Content Courses

National and international efforts are ongoing regarding the development of a thorough and comprehensive curriculum for mathematics content courses. For example, in Japan, the content of teacher preparation courses focuses more on "integrated studies," with an emphasis on cross-curricular thematic projects that teachers can later implement with their students to promote individuality, a sense of ethics, and "zest for living" (Hooghart, 2006, p. 291; MEXT, 2001). Some parts of the world (including the United States) have developed more unified approaches by integrating content and pedagogy into hybrid "content-methods" courses (European Agency for Development in Special Needs Education, 2011; Florian & Black-Hawkins, 2011; Hart, 2002; Ontario College of Teachers, 2017). Other countries promote a Science Technology Engineering and Mathematics (STEM)-based approach, blending mathematics courses with content from other sciences, to emphasize applied and contextual teaching of mathematics (Australian Curriculum and Assessment Reporting Authority [ACARA], 2014; OECD, 2005, 2010). Consequently, how, when, and to what extent "the content" is addressed in teacher preparation programs varies drastically.

For example, in the Canadian province of Ontario, prospective elementary teachers are not required to take any formal mathematics courses in their undergraduate degree (e.g., B.Ed.). However, the Ontario College of Teachers requires prospective teachers to complete a four-semester teacher education program beyond their postsecondary/undergraduate studies (e.g., Master of Teaching degree) to be qualified to teach. The majority of Master of Teaching mathematics course requirements are taught by the education department faculty (Ontario College of Teachers, see https://www.oct.ca).

In contrast, in the United States, most of the prospective elementary school teachers complete teacher education programs as undergraduates (e.g., Bachelor of Education) and are thereby qualified to teach. Most elementary teacher education programs, as mentioned earlier, require a minimum of nine semester-long credits of mathematics content courses, specifically designed for prospective grades K-8 teachers, to be completed in their first couple of years in the undergraduate teacher education program, in addition to at least one mathematics methods course (CBMS, 2012). However, in the United States, mathematics content courses are predominantly taught in the mathematics department, often by mathematicians who do not have experience in grades K-12 educational contexts. Also, most mathematics content courses do not include a "field" component as part of the course requirement (Greenberg & Walsh, 2008; Lutzer, Rodi, Kirkman, & Maxwell, 2007). The curriculum and combined syllabus for mathematics content courses typically includes topics that closely mirror the grades K-8 mathematics curriculum, placing heavier emphasis on topics in number and operations (95%), geometry (91%), measurement (88%), number theory (87%), probability (82%), algebra (80%), and statistics (77%) (Masingila et al., 2012).

Mathematics Teacher Educators

I broadly define mathematics teacher educators as "professionals who work with practicing and/or prospective teachers to develop and improve the teaching of mathematics" (Jaworski, 2008, p. 1). Furthermore, for the purpose of this chapter, I focus on the population of mathematics teacher educators who teach mathematics content courses to prospective teachers. However, since a variety of individuals (e.g., mathematicians, adjuncts, classroom teachers, graduate students) assume the role of mathematics teacher educators when teaching mathematics content courses (Greenberg & Walsh, 2008), I further refine this definition by adopting a prototype perspective proposed by Sternberg and Horvath (1995). In this perspective, I distinguish between "experience" and "expertise" of mathematics teacher educators, particularly because most mathematics department faculty are considered mathematics teacher educators, given their *experience* teaching prospective teachers, but may not have *expertise* in teacher preparation and certification or *experience* in teaching and conducting educational work with pupils and/or in school-based settings.

For example, in the research that I report here, in addition to their *experience* teaching mathematics content courses for prospective teachers, the participating mathematics teacher educators had also taught undergraduate mathematics courses for other majors (e.g., college algebra, precalculus, business calculus, statistics). Furthermore, the mathematics teacher educators' mathematics knowledge and competence was similar to their mathematics department colleagues and included a minimum of 24 semester credits of graduate-level mathematics coursework. However, unlike most of their mathematics department colleagues, the mathematics teacher educators' knowledge and insight additionally included *expertise* in grades K-12 education, curriculum, mathematics teaching to pupils, and teacher preparation and certification.

Thus, I adopt Sternberg and Horvath's (1995) conceptualization of *teaching expertise* by outlining three critical components: knowledge, efficiency and insight. The mathematics teacher educator that I discuss in this chapter, henceforth referred to as "*expert* mathematics teacher educators," include professionals with *expertise in teaching prospective teachers*, and who are similar to each other in quality and depth of knowledge, efficiency and insight in relation to teaching prospective teachers. They are different from novices and experienced colleagues in the mathematics departments. In other words, *expert* mathematics teacher educators have similar (to experienced colleagues) *lengths of experience* in teaching courses for prospective teachers, but they also have *experience* teaching in grades K-12 (students aged 5–18 years) school settings and *expertise* in teacher preparation and certification.

Studying Expert Mathematics Teacher Educators

Hiebert and Morris (2009) argue that improving mathematics teacher preparation will require an extensive research effort on the work of mathematics teacher educators with prospective teachers as well as a system to accumulate useable knowledge for the field. Thus, in studying *expert* mathematics teacher educators, I aimed to contribute to this knowledge in the field directly. Specifically, in this chapter, using empirical data and findings, I offer insights and classroom-based examples of commonly identified practices, in content courses, across a group of (ten) *expert* mathematics teacher educators who utilized these practices to provide opportunities for prospective teachers to develop pedagogical content knowledge.

The authors of *How People Learn: Brain, Mind, Experience, and School* (Bransford, Brown, & Cocking, 2000) suggest that, "the study of expertise shows what results in successful learning look like" (p. 31). Studying *expert* mathematics teacher educators offered avenues for contributing to the field the research evidence and directions for the mathematics teacher education community about the nature of (grades K-8) mathematics content courses taught by *expert* mathematics teacher educators, as well as what prospective teachers' learning may look like and the role of pedagogical content knowledge in these courses when taught by *expert* mathematics teacher educators.

THEORETICAL FRAMEWORK

When developing a theoretical framework for my research on expert mathematics teacher educators, I was faced with the dilemma of being at the intersection of extensive research available on mathematics (practising and prospective) teachers' pedagogical content knowledge and nearly non-existent literature on mathematics teacher educators' pedagogical content knowledge, especially the practices for teaching content courses. Furthermore, while examining the pedagogical content knowledge literature, I discovered discrepancies in theoretical interpretations of pedagogical content knowledge. For example, most pedagogical content knowledge studies seem to generally agree on the definition of pedagogical content knowledge and often use the original definition proposed by Shulman (1986; 1987), whereas theoretical models and classifications of pedagogical content knowledge components vary considerably in the field. Thus, when developing the framework, I utilized a *hybrid* approach that mirrored a process of considering the conceptualizations articulated in the literature (mainly teacher-based) and adopting grounded theory methods to identify additional (mathematics teacher educator-based) pedagogical content knowledge elements that emerged from my research and data. This process took several years and the resulting framework has gone through multiple iterations and revisions.[3] Nevertheless, I do not claim that this framework is a "finalized" product. Instead, I offer it as an emerging conceptualization of mathematics teacher educators' practices for developing prospective teachers' pedagogical content knowledge in the content courses, obtained through theoretical and empirical cross-analyses between my research findings and the pedagogical content knowledge models extant in the literature.

Pedagogical Content Knowledge: Three Recognized Components

Pedagogical content knowledge, originally coined by Shulman (1986), was identified as the "missing paradigm" in the field of unanswered questions that the research on teaching had largely overlooked. Shulman argued that "mere content knowledge was as useless pedagogically as content-free skill" and that a proper blend of the two (content and pedagogy) would require the field to refocus our teacher development efforts more toward "the content aspects of teaching" (Shulman, 1986, p. 8). He further clarified:

> [w]hat we miss are questions about the *content* of the lessons taught, the questions asked, and the explanations offered. … How do teachers decide what to teach, how to represent it, how to question students about it and how to deal with problems of misunderstanding? (Shulman 1986, p. 8)

In this quote, Shulman asks, "[h]ow do teachers decide what to teach …?" suggesting that *knowledge about curriculum* plays an important role in teachers' pedagogical content knowledge. Similarly, he was directly calling attention to the significance of teachers' pedagogical skills and *knowledge about instructional strategies* for teaching content-specific topics ("how to represent it [content]"), and teachers' *knowledge about students* and their learning, conceptions and misconceptions of these content-specific topics ("how to deal with problems of misunderstanding"). Hence, it is apparent from Shulman's delineations that these three knowledge components are at the heart of pedagogical content knowledge. Grossman (1990) built on the work of Shulman and proposed a fourth component of pedagogical content knowledge, namely *knowledge and beliefs about the reasons for teaching a subject*, which she defined as teachers' overarching conceptions of teaching a subject at different grade levels that "are reflected in teachers' goals for teaching particular subject matter" (Grossman, 1990, pp. 8–9).

These pedagogical content knowledge conceptualizations alone generated numerous research projects and offered various platforms for experts across many disciplines to explore different hypotheses related to pedagogical content knowledge. In the mathematics education literature, three of the pedagogical content knowledge components have been given much theoretical attention and have been empirically validated through research on mathematical knowledge for teaching (mathematics teacher educators). In their pedagogical content knowledge model, Hill, Ball and Schilling (2008) proposed that pedagogical content knowledge includes three components: knowledge of content and teaching, knowledge of content and students, and knowledge of curriculum (pp. 377–378; also see Ball, Thames & Phelps 2008; Hill, Rowan, & Ball, 2005). These three components closely mirror the pedagogical content knowledge components proposed by Shulman and Grossman (1990).

Hill et al. (2008) did not include the fourth component (proposed by Grossman, 1990) related to teachers' knowledge and beliefs about the reasons for teaching a subject. I argue that this component is critically needed in the (mathematics teacher educators-based) pedagogical content knowledge conceptualizations, particularly given my research findings on *expert* mathematics teacher educators' practices. I also argue that the "knowledge and beliefs" component is perhaps too limited (theoretically) and requires expansion to account for recent developments in the field (as well as my own findings) related to the *orientations toward teaching the subject* construct. Below, I provide justifications for these suggestions.

Pedagogical Content Knowledge: The "Missing" Component

A large number of studies are available in mathematics education literature on teachers' beliefs about teaching and learning mathematics (e.g., Swars Auslander, Smith, Smith, & Myers, 2019; Aguirre & Speer, 2000; Cohen, 1990; Ernest, 1989; Myers, Swars Auslander, Smith, Smith, & Fuentes, 2019; Maasz & Schlöglmann, 2009; Pajares, 1992; Thompson, 1992). Evidence from these studies strongly suggests that teachers' beliefs about mathematics teaching and learning have significant impact on their instructional practices (e.g., Cohen, 1990; Ernest, 1989; Kuhs & Ball, 1986; Thompson, 1992). A number of these studies also have clearly indicated that there may be a strong connection between teachers' beliefs and their *orientations toward teaching the subject*. For example, Kuhs and Ball (1986) argued that orientations toward teaching mathematics involve teachers' deep-seated beliefs and "ideal" images of mathematics teaching and learning. Similarly, Thompson, Philipp, Thompson and Boyd (1994) argued that teachers have different orientations toward mathematics and pedagogy and that these orientations have serious consequences for the teaching and learning that occurs in their classrooms. Yet, despite these suggestions, very little has been done in the field to further explore and operationalize the construct of orientations toward teaching mathematics.

Orientations toward teaching the subject, a construct initially introduced in the science education literature by Anderson and Smith (1987), is defined as "general patterns of thought and behavior" related to teaching and learning the subject (p. 99). Orientations toward teaching the subject are organized at the intersection between teachers' beliefs and purposes for teaching the subject and are manifested in teaching practices employed during instruction, ranging from "purely process or content to those that emphasize both and fit the national [United States] standards of being inquiry-based" (Magnusson, Krajcik, & Borko, 1999, p. 97; also Friedrichsen & Dana, 2003; Park Rogers et al., 2010; Musikul & Abell, 2009). Magnusson and colleagues (1999) articulated specific characteristics of science instruction that reveal the teacher's orientations toward teaching the subject based on the instructional strategies that she or he chooses to employ (e.g., hands-on explorations, investigations, discovery learning). However, it is not the use of a particular strategy, but the "purpose" behind employing that strategy that distinguishes a teacher's orientation toward teaching the subject (Magnusson et al., 1999, p. 97).

In many science education studies the *orientations toward teaching the subject* construct has been included as a pedagogical content knowledge component (see Abell & Bryan, 1997; Borko & Putnam, 1996; Friedrichsen et al., 2009; Friedrichsen, Van Driel, & Abell, 2010; Friedrichsen & Dana, 2003; Smith & Neale, 1989). Most of these studies, however, do not (per se) include Grossman's (1990) fourth component of pedagogical content knowledge (knowledge and beliefs about the reasons for teaching a subject). Instead, science education researchers argue that "orientations" are comprised of "teachers' knowledge and beliefs about the purposes and goals for teaching [a subject] at a particular grade level," including teachers' knowledge of grade-specific and grade-appropriate strategies and "overarching conceptions of teaching that subject" (Magnusson et al., 1999, p. 97). Many of these studies also distinguish *orientations toward teaching the subject* from *orientations toward the subject/discipline* (e.g., Abell & Smith, 1994), situated around teachers' knowledge and beliefs about *teaching the subject* rather than their beliefs about the *nature of the subject* as a discipline rooted in teachers' conceptions of the content "mirroring competing substantive structures" of the subject/discipline (Grossman et al., 1989, pp. 29–31; see also Feiman-Nemser & Buchmann, 1985; Grossman, 1987, 1991; Wilson & Wineburg, 1988).

Shoenfeld (2010) provided several heuristic examples as validations for his theoretical model for routine and non-routine, goal-oriented and knowledge-based teachers' behavior during teaching, including teachers' *orientations* toward teaching the subject. In his book, *How We Think*, Shoenfeld (2010) described that, during a lesson, teachers go through "acting in the moment" experiences that involve specific behaviors grounded in their *resources* (i.e., knowledge, including content knowledge, but also social and material resources available to them), pedagogical *goals* (classroom practices/actions based upon teachers' goals, which directly influence students' experiences), *orientations* towards teaching the subject (an abstraction of beliefs, including values and preferences), and *decision-making* (which can be modelled as a form of subjective cost-benefit analysis) (Shoenfeld, 2010; also see Shoenfeld, 2011a, 2011b, 2014, 2015). Specifically, teachers "act in the service of the goals they have established by selecting and implementing resources that will enable them to satisfy those goals," and their decision-making process in the moment of teaching "can be seen as the selection of goals consistent with the teachers' resources and orientations" (Shoenfeld, 2015, pp. 459–460).

Shoenfeld (2015) defines teachers' *orientations towards teaching the subject* as beliefs, values and preferences, as well as understandings and perceptions, related to the nature of mathematics, pedagogy, and students, on the basis of their experience. Similarly to the science education researchers, he suggests the use of the term *orientations*, instead of beliefs, to distinguish a broader construct that encompasses beliefs but also encompasses values, predilections, and insights, particularly because "beliefs alone cannot completely shape behavior: what one does is a function of what one decides are the most important things to do (the goals one sets, consistent with one's beliefs) and the resources that one has at one's disposal" (Shoenfeld, 2015, p. 459).

Pedagogical Content Knowledge: Proposed (Emergent) Framework

Consistent with recommendations in the literature and with the findings from the research study reported here, I argue for the inclusion of *orientations toward teaching the subject* as an additional pedagogical content knowledge component in the framework (shown in Figure 8.1).

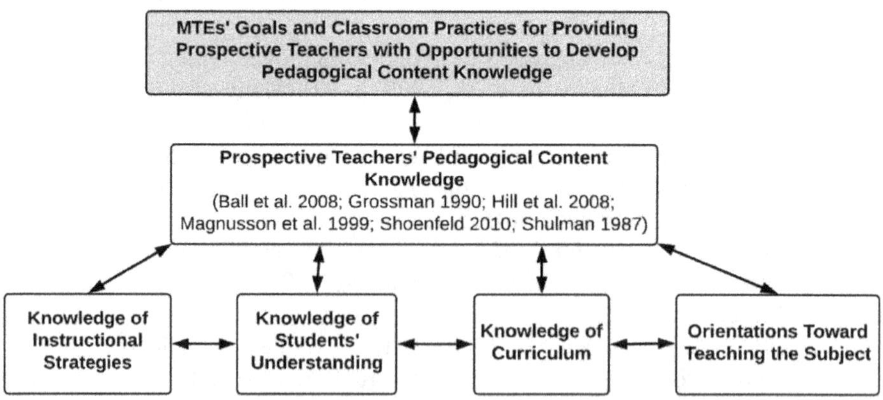

Figure 8.1. Model for mathematics teacher educators' (MTEs') goals and classroom practices for providing prospective teachers with opportunities to develop pedagogical content knowledge

In other words, I suggest that mathematics teacher educators' goals and classroom practices focused around providing opportunities for prospective teachers to develop pedagogical content knowledge (shaded in Figure 8.1) encompass four pedagogical content knowledge components: (1) knowledge of instructional strategies; (2) knowledge of students' understanding; (3) knowledge of curriculum; and (4) orientations toward teaching the subject. I define mathematics teacher educators' *goals* as what mathematics teacher educators want prospective teachers to learn in their mathematics content courses, particularly mathematics teacher educators' professional and personal intentions and purposes for prospective teachers' pedagogical content knowledge development. I define mathematics teacher educators' *classroom practices* as what mathematics teacher educators do during class (e.g., say, write, assign) to support their goals and intentions.

I define and interpret each pedagogical content knowledge component based on the definitions articulated in the literature (e.g., Ball et al., 2008; Grossman, 1990; Hill et al., 2008; Magnusson et al., 1999; Shoenfeld, 2010; also see Shoenfeld, 2011a, 2011b, 2014, 2015; Shulman, 1986) as they relate to the goals and classroom practices of mathematics teacher educators (in content courses) for providing opportunities for prospective teachers to develop and enhance their:

- *Knowledge of instructional strategies* and approaches for teaching mathematics responsibly and responsively, including knowledge about grade-level appropriate methods, activities, and manipulatives for teaching specific mathematical concepts;
- *Knowledge of students' understanding* and conceptions and misconceptions of particular mathematical topics, including teachers' knowledge about specific learning needs, approaches and strategies to be able to address the learning and understanding of specific mathematical concepts with students;
- *Knowledge of curriculum* and standards, including teachers' knowledge of curriculum goals, objectives, programs, and resources relevant to teaching mathematical content at specific grade levels and the horizontal/vertical curriculum structure of the subject;
- *Orientations toward teaching the subject*, which involve beliefs, purposes, values, preferences, understandings and perceptions about the nature of mathematics, pedagogy, and students based on their experience. Teachers' orientations toward teaching the subject define specific characteristics of their instruction and shape their teaching practices. However, although specific characteristics of instruction reveal teachers' orientations toward teaching the subject, based on the instructional strategies they choose to employ – it is the "purpose" behind employing that strategy that distinguishes a teacher's orientation toward teaching the subject.

In the next sections, I offer research accounts from my own findings and other studies reported in the field as empirical evidence and validations for the proposed definitions and components of this framework.

ABOUT THE STUDY

Research Goals and Perspectives

The findings reported here were from a phenomenographical study (Marton, 1981) which examined the work of ten *expert* mathematics teacher educators (six males; four females) from five different institutions in the Eastern portion of the United States. Phenomenography particularly helped to illuminate *qualitatively similar ways* (Bowden & Walsh, 2000) in which *expert* mathematics teacher educators perceived prospective teachers' learning and the goals they drew upon to provide opportunities for prospective teachers to develop pedagogical content knowledge, although the mathematics teacher educators were from different institutions (e.g., four-year colleges, master's degree granting, etc.) and represented a range of K-8 mathematics content courses (e.g., number and operations; geometry and measurement; algebra and numbers).

The objective of the project was to investigate the goals, purposes, intentions, and classroom practices articulated by *expert* mathematics teacher educators regarding their (grades K-8) mathematics content courses in general, and specifically in terms of providing opportunities for prospective teachers to develop pedagogical content knowledge (shaded in Figure 8.1). Mathematics teacher educators were asked to provide detailed reflections on and examples from their classroom practices (associated with their articulated goals), in particular their professional and personal intentions for prospective teachers' development and learning (including pedagogical content knowledge development) that may or may not have been included in their course syllabi and/or curriculum. At no point in the study (or data collection) was the term pedagogical content knowledge referenced or explicitly used with the mathematics teacher educators.

Expert Mathematics Teacher Educators and Content Courses

All the mathematics teacher educators in the study had an undergraduate degree in mathematics teaching (eight from mathematics education programs; two from mathematics programs) and were qualified to teach. The mathematics teacher educators were identified by: (a) having at least a master's degree in mathematics (three participants) or mathematics education (seven participants); (b) having at least fifteen (15) years of combined K-12 teaching experience and teaching mathematics content courses for prospective teachers at the university level (as a faculty member, not including graduate student teaching experience); and (c) being professionally active in mathematics teacher education by attending/presenting at local, state, and national meetings. All but one of the mathematics teacher educators also had a doctorate degree in mathematics education. Eight of the participants had more than 20 years of combined grades K-12 teaching experience and experience teaching prospective teachers at the university level, while the remaining two participants had sixteen (16) and eighteen (18) years of such experience. The average total years of teaching experience across the entire sample was 25.35 years and the median 21 years. The average total years of K-12 teaching experience was 9.1 years; the average for the total years teaching prospective teachers at the university level was 16.25 years.

The (ten) content courses that the mathematics teacher educators taught were all offered through mathematics departments, designed specifically for prospective teachers and completed by prospective teachers in the beginning years of their undergraduate programs. The topics addressed in the content courses included: (a) probability, statistics, geometry, and measurement (five courses), and (b) numbers, operations, and algebraic reasoning (five courses). All course syllabi described an emphasis on problem solving, modelling mathematics, recognizing connections among mathematical ideas, and prospective teachers as "doers" of mathematics (i.e., actively engaging in the mathematics rather than sitting passively in class). Six syllabi additionally contained references to an emphasis on goals and procedures addressed in current state and national standards for school mathematics. Providing the opportunity for prospective teachers to develop pedagogical content knowledge was not explicitly mentioned in any of the course syllabi.

Data Sources and Analyses[4]

Data were collected through two (1-hour) semi-structured individual interviews. The first (initial) interview was conducted at the beginning of the academic year (autumn semester). The second interview was conducted at the end of the academic year (spring semester).

During the initial interview, participants were asked about their educational background, prospective teachers' learning, and the goals/purposes for the K-8 mathematics content course they taught (particularly any goals that were not included in their syllabi). The mathematics teacher educators were also asked to reflect and provide specific examples of approaches they used to engage prospective teachers in order to address the identified goals and about how explicit they were with prospective teachers about these goals. The initial interviews were coded (to help prepare for the second interviews) by identifying common "goals/practices" codes across the entire sample of mathematics teacher educators. The second interview was used to follow-up with mathematics teacher educators on these commonly identified codes. All interviews were audio-recorded and transcribed verbatim.

Two types of data analysis were employed: a) open coding (Corbin & Strauss, 2008) to identify common "goal/practice" codes across the entire sample; and b) the constant comparison method (Boeije, 2002) to establish *qualitatively similar ways* that our mathematics teacher educators perceived, conceptualized and described commonly identified goals/practices. *Open coding* was conducted as discussed by Corbin and Strauss (2008) as "breaking data apart and delineating concepts to stand for blocks of raw data" (p. 195). *Open coding* particularly helped to identify specific common goals/practices (across the sample) that mirrored the pedagogical content knowledge conceptualizations from the literature. The *constant comparison* method involved several steps: (a) compiling individual mathematics teacher educators' interview responses by common "goal/practice" codes, (b) coding the compiled data, (c) developing "themes" and "theme tracking," and (d) drawing connections to/from the emergent codes, framework, and literature. The *constant comparison* method particularly helped to better understand each mathematics teacher educator's conceptualizations of the common pedagogical content knowledge-related goals and practices and *why* those goals and practices were important/critical to every mathematics teacher educator in this study.[5]

EMPIRICAL VALIDATIONS OF THE FRAMEWORK

In this section, I offer findings from this study and from the work of others in the field to provide empirical evidence in support of the pedagogical content knowledge definitions and components proposed in the framework. Although the study reported here focused on *expert* mathematics teacher educators who specifically taught content courses, other studies reported in the field included mathematics teacher educators who also taught methods and/or (hybrid) content-methods courses. In selecting these studies, I primarily focused on providing empirical evidence for conceptualizations of mathematics teacher educators' practices for developing prospective teachers' pedagogical content knowledge.

Overall, my study (of *expert* mathematics teacher educators) revealed more than 347 codes, which closely reflected the four pedagogical content knowledge components from the framework (see Table 8.1). These codes represent more

Table 8.1. Mathematics teacher educators' goals and practices related to providing opportunities to develop prospective teachers' pedagogical content knowledge (PCK)

Framework connections	*Mathematics Teacher Educators' goals and practices for developing prospective teachers' PCK*
PCK: knowledge of instructional strategies	Learning about manipulatives as instructional tools Learning about various models and representations Learning about student-friendly vocabulary Learning about/from worthwhile (K-8) lessons and tasks
PCK: knowledge of students' understanding	Learning about students' conceptions and ingenuities Learning about students' misconceptions and errors Learning about theories of student cognition and development
PCK: knowledge of curriculum	Learning about scope and sequence of K-8 curriculum Learning about mathematical connections beyond K-8 curriculum Learning about curriculum and policy documents
PCK: knowledge, beliefs and orientations toward teaching the subject	Learning about *Standards*-based mathematics teaching strategies Learning about meaningful mathematics learning practices

than 347 instances of the mathematics teacher educators mentioning a goal or a classroom example that they employed in their content courses, which were coded as pedagogical content knowledge-related. The codes were further refined and merged into "sub-themes," which are included as bullets (in Table 8.1) under each pedagogical content knowledge component. Every bullet in Table 8.1 reflects an instance/example mentioned at least once by every mathematics teacher educator in the study.

The goals and practices (in Table 8.1) were echoed in reflective descriptions of specific connections to grades K-8 school, curriculum, teaching, student learning, and beliefs and orientations about teaching mathematics that *expert* mathematics teacher educators provided for prospective teachers during content courses. Below, I elaborate on these findings in greater detail by using direct quotes from mathematics teacher educators as evidence. Please note, as with most interview-based research, some participants were more elaborate and descriptive than others. The direct quotes reported in this chapter were selected based on several criteria, including direct quotes from each mathematics teacher educator, and quotes that are more detailed and elaborate and yet representative of the entire sample. The direct quotes are not exhaustive of the examples provided by the mathematics teacher educators. All participants' names have been replaced with pseudonyms.

Mathematics Teacher Educators Provide Opportunities for Prospective Teachers to Develop Knowledge of Instructional Strategies

New findings are being reported in the literature regarding the knowledge and practices of mathematics teacher educators when working with prospective teachers, particularly in developing prospective teachers knowledge of content and pedagogy (e.g., Chick & Beswick, 2017; Even, 2008; Goodell, 2006; Superfine & Li, 2014). For example, in their pedagogical content knowledge framework, Chick and Beswick (2017) identified several practices that mathematics teacher educators employ to engage prospective teachers in developing knowledge about instructional strategies, including: using concrete materials to demonstrate a concept and describing or demonstrating ways to model or illustrate a concept, including specific representations, materials, and diagrams. The research findings from this study also included several similar accounts of these practices.

Findings from This Study

Mathematics teacher educators utilized four overarching practices to provide opportunities for prospective teachers to develop knowledge of instructional strategies for grades K-8 mathematics teaching, including prospective teachers' learning about the use of: (1) manipulatives as instructional tools, (2) various models and representations, (3) student-friendly vocabulary, and (4) worthwhile (grades K-8) lessons and tasks. Specifically, the *expert* mathematics teacher educators, in their content courses, reported using *physical and virtual manipulatives* to help prospective teachers become more familiar with them as instructional tools. The mathematics teacher educators shared that prospective teachers use manipulatives for their own explorations of mathematics and to "feel comfortable in seeing how to use those physical models when they're working with kids" (Ian). The mathematics teacher educators also suggested that it was important to demonstrate to prospective teachers the accessibility of manipulatives as well. For example, they mentioned the use of free virtual applets in-class and engaging prospective teachers in using physical manipulatives that can be created using supplies available at home or local stores, including "stack cubes, straws, and base-10 blocks. Stuff that they themselves might have access to in their own classroom and, if they didn't, it wouldn't be expensive to get" (Oliver).

The mathematics teacher educators also used (and required prospective teachers to use) *various diagrams, models, pictures*, and problem-solving techniques to promote their knowledge of multiple representations for specific mathematical ideas and concepts. For example, one mathematics teacher educator (Ella) shared that she introduced prospective teachers to different ways of calculating percentages: "[w] hen we do percent problems, we do what's called three types of models. Percent chart, percent diagram, and unit-percents." Other mathematics teacher educators mentioned different ways to model fractions, including "area models, and set models, and number line models, and how those are different and yet how they're the same, and why you [as a teacher] need more than one" (Ingrid). The mathematics teacher educators believed that prospective teachers' ability to represent mathematical concepts in more than one way deepens their mathematical knowledge and prepares them to become more effective teachers, particularly in being able to teach mathematics using flexible and meaningful ways. These findings are closely aligned with one of the five process standards of the *Principles and Standards for School Mathematics* (National Council of Teachers of Mathematics [NCTM], 2000), which explicitly states that effective teachers are able to *represent* mathematical ideas in a variety of ways, including "pictures, concrete materials, tables, graphs, number and letter symbols, spreadsheet displays, and so on" (p. 4).

The mathematics teacher educators also engaged prospective teachers in discussions about *using student-friendly vocabulary* and developing children's mathematical language, including modelling the use of specific phrases and expressions (as tactics) to emphasize the nuances associated with children's struggles with learning mathematical terminology. The mathematics teacher educators used phrases like "exchange" and "regrouping" (Oliver) when discussing subtraction with borrowing with prospective teachers, and explained whole-number subtraction using "appropriate" student-friendly mathematical language (and Base-10 place value mats): "[m]y purpose here is to get them [prospective teachers] to understand what they're doing [mathematically], so that they're not just blindly saying [to students] 'cross out this' and 'borrow' from here" (Trina).

The mathematics teacher educators additionally mentioned that explicitly discussing research-based findings related to *children's language difficulties* motivated prospective teachers to better understand the complexities of teaching mathematics to grades K-8 students and to better appreciate various pedagogical strategies that provide support for developing children's vocabulary and language. For example, when classifying polyhedral and non-polyhedral shapes, an mathematics teacher educators embedded a discussion about children and how "in kindergarten [prospective teachers] may not use that language" and instead they may need to anticipate children's language issues. For example, Adam noted, "I make them think about what is a kid going to call a vertex? So talking about the language kids use and trying to build on their language at appropriate grade levels." Overall, the *expert* mathematics teacher educators indicated that using student-friendly vocabulary with prospective teachers has a dual purpose: it helps prospective teachers reflect on the meaning of the mathematical terms and it helps prospective teachers to become more familiar with the language appropriate for communicating mathematical ideas to grades K-8 students.

The Conference Board of Mathematical Sciences (2012) strongly suggests that the content courses need to "make connections between the mathematics being studied [in the course] and mathematics prospective teachers will teach" (p 7). The *expert* mathematics teacher educators directly attended to this recommendation by adapting and using *worthwhile lessons and tasks from actual K-8 textbooks*, as exemplary teaching models, for prospective teachers to experience and observe firsthand the type of mathematical learning that is valued by the profession. The mathematics teacher educators emphasized that they purposely and carefully select these lessons/tasks focusing on rich, inquiry-driven, engaging, and active learning environments. The mathematics teacher educators also shared that they are explicit with prospective teachers about choosing these (grades K-8) lessons/tasks. For instance, Allen explained that "[t]his particular activity of making the shapes, drawing where they are, or filling in the shapes in an outline with the Tangram pieces, can be done in kindergarten." Overall, the *expert* mathematics teacher educators explained that utilizing worthwhile (grades K-8) mathematical lessons/tasks during content courses helped them to model specific teaching strategies that prospective teachers could utilize in their future classrooms.

Evidence from the Field

More than two decades of research exists on the use of mathematics manipulatives with prospective teachers, including the use of manipulatives and hands-on approaches for prospective teachers to solve problems and develop strategies to teach mathematic using concrete materials (e.g., Cakiroglu, 2000; Quinn, 1997; Vinson, 2001; Wenta, 2000) as well as field experiences for prospective teachers to try lessons involving manipulatives (e.g., Gresham, 2007; Swars, Daane, & Giesen, 2006). Most importantly, a number of studies show courses that embed the use of manipulatives to be effective in reducing prospective teachers' mathematics anxiety (e.g., Cakiroglu, 2000; Huinker & Madison, 1997; Swars, Daane, & Giesen, 2006; Vinson, 2001; Wenta, 2000).

Furthermore, numerous studies are available on addressing with prospective teachers the difficulties that children may experience when learning mathematics terminology, including pedagogical strategies that help to support and promote children's language development (e.g., Earp & Tanner, 1980; MacGregor, 1990; Rubenstein & Thompson, 2002; Thompson & Rubenstein, 2000; Thompson & Chappell, 2007; Whitin & Whitin, 2003). Language development is fundamental to students' problem solving, higher order thinking, and mathematical success (e.g., Riccomini, Sanders, & Jones, 2008; Seethaler, Fuchs, Star, & Bryant, 2011) and knowledge of mathematics vocabulary directly predicts pupils' mathematics performance (Riccomini, Smith, Hughes, & Fries, 2015; van der Walt, 2009; van der Walt, Maree, & Ellis, 2008).

Embedding and discussing worthwhile lessons and tasks from grades K-8 textbooks has been a widely documented teacher-development approach in the literature, for both prospective and practising teachers. Some studies suggest focusing on general teaching strategies, geared towards helping teachers to enhance and improve their overall practice; other studies focus on specific methods addressing a certain context, topic, lesson objective, and learning environment (e.g., Loucks-Horsley, Hewson, Love, & Stiles, 1998; Loucks-Horsley, Stiles, Mundry, Love, & Hewson, 2009; Sparks & Loucks-Horsley, 1989).

For example, Fung and Latulippe (2010) described an activity called "newspaper headlines," where prospective teachers were given quantitative data from the newspaper and asked whether the numbers were precise or if errors were reported. This activity provided an opportunity for prospective teachers to attend to estimation, precision, and the usefulness of real-world contexts when teaching and developing children's number sense. Similarly, Tirosh (2000) engaged prospective teachers in discussions and investigations of fractions-related activities (from K-8 textbooks), which helped to strengthen prospective teachers' knowledge of mathematics and offered exemplary lessons and strategies on teaching rational numbers to children. Mathematics teacher educators also report (in their content courses) using mathematically rich lessons from middle school textbooks and focusing the course around prospective teachers' learning from those particular tasks/lessons (see Lutz & Berglund, 2007).

Mathematics Teacher Educators Provide Opportunities for Prospective Teachers to Develop Knowledge of Students' Understanding

Knowledge about students is considered one of the most critical domains of pedagogical content knowledge (Ball, Thames & Phelps, 2008). Teachers must be able to anticipate what students are likely to think, do, and find confusing with mathematics content (Ball, Thames & Phelps, 2008). Research suggests that mathematics teacher educators engage in different practices mainly around prospective teachers development of knowledge about students. These involve two types of activities: direct and indirect interactions with students. Specifically, mathematics teacher educators provide opportunities for prospective teachers to visit local schools and work directly with "real-life" students. In contrast, indirect opportunities may involve prospective teachers analyzing students' thinking and learning by watching videos and examining authentic artifacts collected from students (e.g., work samples, written solutions).

Findings from This Study

The mathematics teacher educators utilized three overarching practices to provide opportunities for prospective teachers to develop their knowledge about students' understanding in their content courses, including: (1) learning about students' conceptions and ingenuities, (2) learning about students' misconceptions and errors, and (3) learning about theories of student cognition and development. The mathematics teacher educators used various artifacts to illustrate to the prospective teachers *students' conceptions and ingenuities* when solving problems and making sense of mathematics. They shared that they obtained these artifacts from various sources (e.g., professional literature, course textbook, online, local teachers). They did not simply present prospective teachers with students' artifacts – they developed/created tasks, prompts, and questions that required prospective teachers to reflect on specific aspects of students' thinking and learning. Ingrid's approach, for example, was to "have them [prospective teachers] watch videos out of class. I have some questions I want them to answer, or I have them write a response to a video."

The mathematics teacher educators also frequently recruited help from local schoolteachers to help arrange opportunities for prospective teachers to interact with students' mathematics and learning directly. For example, the prospective teachers had opportunities to visit a mathematics class. "Look at what these third graders are doing … they're doing lattice multiplication" (Ethan), or become pen pals with a local fifth-grade class "fifth graders are going to write an introduction letter with a math problem and send it to my students [prospective teachers]. Then [prospective teachers] will solve that math problem and send it back. Then [prospective teachers] will present them with a math problem" (Oliver).

The mathematics teacher educators provided prospective teachers with opportunities to learn about students' misconceptions and errors. They indicated that exposing prospective teachers to students' errors not only highlighted the range of possible students' misconceptions, but also helped to encourage prospective teachers to reflect upon their own mathematical mistakes and the knowledge needed to effectively address errors and misconceptions. Videos of children were used as powerful examples, "They [prospective teachers] will watch something and go, 'Huh? I'm going to have students like that?' … all of us at some point have misconceptions, how do you deal with that?" Another mathematics teacher educator (Ella) wanted prospective teachers to recognize that teaching builds upon both correct and incorrect mathematics: "I try to instil that. I talk about the value of making mistakes and analyzing mistakes, so we discuss their incorrect solutions."

Lastly, the expert mathematics teacher educators engaged prospective teachers in discussions about theories of child development and cognition (e.g., Van Hiele Levels of Geometric Thinking, Cognitively Guided Instruction, Bloom's Taxonomy). They used these discussions to provide prospective teachers with opportunities to apply theory in practice specifically in the context of mathematics teaching. One mathematics teacher educator described, "I show them examples from students who did the sorting. We go through and we talk about them relative to the Van Hiele levels" (Adam). Another mathematics teacher educator discussed applications of Bloom's Taxonomy in mathematics, "get into the language, focusing on properties, and not just 'here's what it [child's work] looks like' … building critical thinking and helping the child … getting into the higher levels in the Bloom's Taxonomy" (Ingrid). Overall, mathematics teacher educators suggested that these experiences particularly help prospective teachers to develop pedagogical and mathematical lenses for analyzing students' work.

Evidence from the Field

One of the most effective practices recommended in the literature for providing prospective teachers with opportunities to develop knowledge about students' understanding is engaging them in direct interactions with children using: individual or small-group interviews (Fernandes, 2012; Friel, 1998; Gee, 2006; Jenkins, 2010; Lannin & Chval, 2013; McDonough, Clark, & Clark, 2002; Spangler & Hallman-Shrasher, 2014); written assessments, prompts, or questions for students to respond to (e.g., Sjoberg, Slavit, & Coon, 2004; Stephens & Lamers, 2006); family "math nights" and after school activities (e.g., Bofferding, Kastberg, & Hoffman, 2016; Freiberg, 2004; Lachance, Benton, & Klein, 2007; Lachance, 2007); and pen-pal projects, which offer unique opportunities for prospective teachers to directly interact with students' mathematics without necessarily meeting them in-person (e.g., Crespo, 2003; Lampe & Uselman, 2008; Shokey & Snyder, 2007).

Working with children *directly* develops prospective teachers' awareness of the variety of problem-solving strategies and thinking that students develop and utilize when doing mathematics, and it develops prospective teachers' ability to adopt an *interpretative* rather than *evaluative* perspective when working with students and analyzing their solutions (Crespo, 2000; Mason, 2002). Lannin and Chval (2013) call these opportunities "powerful" as they provide prospective teachers with firsthand insights into students' learning and embed opportunities for prospective teachers to try out various instructional strategies to address students' misconceptions. Most importantly, these opportunities allow prospective teachers to recognize how difficult it is to gain insight into students' thinking, particularly the challenge of selecting a meaningful task or a question and to draw accurate conclusions about students' knowledge (Lannin & Chval, 2013).

Other studies report that *indirect interactions* with students are also effective for prospective teachers' pedagogical content knowledge development, including videos, transcripts, and case studies of students doing mathematics (e.g., Cognitively Guided Instruction (CGI), Integrating Mathematics and Pedagogy (IMAP), Annenberg Video Series, Teaching Channel, Show-Me Center), artifacts of students' written work and solutions (e.g., Cianca, 2013; Herbel-Eisenmann & Phillips, 2005; Jacobs, Lamb & Philipp, 2010) including samples of work illustrating students' errors and misconceptions (Borasi, 1994; Lim, 2014; Spangler & Hallman-Shrasher, 2014). Providing opportunities for prospective teachers to analyze students' work develops *professional noticing* skills (e.g., Amador, 2017; Carpenter, Fennema, Franke, Levi, & Empson, 1999; Jilk, 2016; McDuffie et al., 2014). *Professional noticing* provides critical lenses for prospective teachers to effectively analyze videos and students' work samples, by focusing prospective teachers' skills of making sense of the work by "attending to children's strategies, interpreting children's understandings, and deciding how to respond on the basis of children's understandings" (Jacobs, Lamb, & Philipp, 2010, p. 172; Philipp, 2008; Thomas et al., 2015; Sherin & van Es, 2003).

Mathematics Teacher Educators Provide Opportunities for Prospective Teachers to Develop Knowledge of Curriculum

In the past three decades, mathematics curricula have been a major focus of educational reforms in the United States and worldwide (Usiskin & Willmore, 2008). Curriculum theorists distinguish different categories of knowledge of curriculum. For example, knowledge of "intended" curriculum refers to the understanding of standards and curriculum frameworks outlined in the national, state, and local policies. Knowledge of "textbook" curriculum comprises teachers' experiences and expertise in teaching mathematics using various textbooks, resources, and materials, including the scope and sequence of specific (grade-level) topics included in those materials. The "intended" and "textbook" curricula provide pedagogical guidelines, but what transpires in the classroom ("taught" and "learned" curriculum) may look quite different (Gehrke, Knapp, & Sirotnik, 1992; Remillard, 2005). In the study reported here, the *expert* mathematics teacher educators primarily addressed the "intended" and "textbook" curriculum in their content courses. This is (likely) due to the fact that school teaching and field experiences are typically not required components of the content courses (in the United States).

Findings from This Study

The mathematics teacher educators utilized three overarching practices to provide opportunities for prospective teachers to develop knowledge of curriculum: (1) scope and sequence of grades K-8 curriculum, (2) mathematical connections beyond K-8 curriculum, and (3) curriculum and policy documents. Specifically, the *expert* mathematics teacher educators wanted prospective teachers to learn about the *scope and sequence of mathematical topics included in the grades K-8 curriculum*, including the grade levels where mathematical concepts are first introduced. The mathematics teacher educators explicitly connected the topics in their content courses to the topics in the grades K-8 curriculum to encourage prospective teachers' appreciation for studying these topics at greater depth, since they will be responsible for teaching them. For example, Ella said, "I tie in to the [State] Standards. I'll pull up 4th or 3rd grade standards and say, 'You know, you could be responsible for teaching this, so you need to have a deeper understanding of why things work'" (Ella). Overall, the mathematics teacher educators articulated that examining grades K-8 curriculum helps prospective teachers to better understand the development and evolvement of mathematical ideas over time (as grade-level progressions).

The mathematics teacher educators also shared that some topics in content courses provide opportunities for prospective teachers to explore *extensions and mathematical connections beyond grades K-8 curriculum*, however, many elementary school prospective teachers often challenged them by asking them to provide reasons for the need to learn mathematics beyond the grade-level that they would be qualified to teach. As a result, the mathematics teacher educators suggested that, rather than waiting for these inquiries to arise, they regularly (and explicitly) discussed with prospective teachers the scope and sequence connections and curriculum pathways, and how specific topics provide a foundation for students' mathematical success beyond grades K-8. For example, one mathematics teacher educator (Oliver) brought up extending and connecting the area model for integer multiplication to polynomials, noting that "the area model is a really effective approach that can be translated later into algebra and to the multiplication of polynomials."

The mathematics teacher educators suggested that content courses provide opportunities for prospective teachers to deepen their mathematical knowledge and develop content-specific pedagogy, and that curriculum plays a critical role in developing both. Thus, in the mathematics teacher educators' opinion, content courses should include mathematical extensions beyond prospective teachers' grade-level certifications. One mathematics teacher educator said, "Deeper understanding and richer connections between topics – that's always been one of my goals for the course. Prospective teachers at any level need to know more mathematics than the mathematics that they will be teaching" (Vance).

The mathematics teacher educators also indicated that they engaged prospective teachers in examining different *curriculum and policy documents* (e.g., *Adding It Up: Helping Children Learn Mathematics* (Kilpatrick, Swafford, & Findell, 2001); *Common Core State Standards* (National Governors Association, 2010); *Principles and Standards for School Mathematics* (NCTM, 2000)). They either did so by directly using these documents as course resources or by selecting a few chapters/key points for discussions and reflections. Their responses included, "the NCTM Process Standards and the Standards for Mathematical Practice of the Common Core. I love the new buzzword of 'sense making' and I am explicit with them [prospective teachers] about that" (Ethan). Overall, the mathematics teacher educators suggested that these documents offered prospective teachers a dual purpose for learning: becoming familiar with mathematics education standards and better understanding why their own learning in the content courses was structured in the ways that specifically mirrored these standards.

Evidence from the Field

Curriculum initiatives serve as catalysts for reform efforts to improve grades K-12 mathematics teaching and learning (Ball & Feiman-Nemser, 1988; Burstein, 1993). These efforts are aimed at providing mathematics teachers with quality curriculum, including scope and sequence recommendations to better support students' learning (Kuhs & Freeman, 1979; Weiss, Banilower, McMahon, & Smith, 2001). Curriculum initiatives have focused on mathematics instruction that moves students beyond surface-level rote and algorithmic procedural knowledge, and towards the development of a deeper conceptual knowledge and problem-solving skills (Schmidt & Houang, 2012). Mathematics teacher educators therefore have a responsibility to articulate and elucidate these curriculum efforts for prospective teachers. Several studies document mathematics teacher educators' practices to this end. Most include mathematics teacher educators embedding productive reflections, presentations, and discussions with prospective teachers about the nature of mathematics teaching, evolution of school curricula, and the development of *Standards* (Gurl, Fox, Dabovic, & Leavitt, 2016).

Gurl et al. (2016) reported that, in their courses, they facilitate prospective teachers' analyses of the Standards for Mathematical Practice (CCSSI, 2010) through "mini-lessons" that prospective teachers teach, as their peers role-play as school students. The authors suggested that prospective teachers' familiarity with the state and national curriculum standards is critical. Additionally, Chval, Lannin, Arbaugh, and Bowzer (2009) suggested using media, which can inspire discussions about the political nature and importance of the reform efforts in education. The authors, in their content-methods courses, show a video (*Prime Time Live with Diane Sawyer*, 1998), which includes interviews with prominent scholars about the results of the international studies for (mathematics and science) comparisons between the United States students and their foreign counterparts. Similarly, in the study reported here, several of the expert mathematics teacher educators mentioned discussions with prospective teachers about the political and public state of education. Ingrid, for instance, recounted one such discussion: "[l]ast year we talked about Common Core because it was in the news so much. You know, that one dad who sent back homework. My students [prospective teachers] brought it up, and I was glad that it came from them." Prospective teachers often brought to class a public opinion piece (from social media), asking the expert mathematics teacher educators to help make sense of the public pushback and criticism on the reform mathematics education movements.

Novice teachers use textbooks in a distinctly different way than more experienced mathematics teachers (Brown & Edelson, 2003; Christou, Eliophotou-Menon, & Philippou, 2004; Remillard & Bryans, 2004; Sherin & Drake, 2009). For example, Sherin and Drake (2009) found that before teaching each lesson, experienced teachers tended to evaluate and adapt their textbooks to their students' needs, whereas novice teachers mainly read the textbook noting the "details" rather than "big ideas." Similarly, Christou, Eliophotou-Menon and Philippou (2004) found that beginning teachers, when implementing new textbook curriculum, are primarily self- and task-oriented in contrast to more experienced teachers. These findings suggest that there is a need for prospective teachers to gain experience with curriculum resources in their teacher preparation programs, particularly understanding the role of textbooks for planning and teaching. Erb (1991) argued that "if teacher education is to contribute to breaking the inertia of curricular tradition, then programs must expose prospective teachers to the characteristics of curricular organization that are unique to the elementary and middle grades" (p. 25).

Mathematics Teacher Educators Provide Opportunities for Prospective Teachers to Develop Orientations toward Teaching the Subject

Prospective teachers need to experience effective models of mathematics teaching and learning in order to develop critical elements of practice valued by the profession (Ghousseini & Herbst, 2016; Grossman et al., 2009; Kuhs & Ball, 1986). Critical elements of practice are explicitly outlined in the *Standards* documents (NCTM, 2000, 2014). Extensive research exists showing that teachers' beliefs and orientations toward teaching mathematics are strong predictors of their classroom actions and behavior (Cooney, 1999; Ernest, 1989; Nespor, 1987; Pajares, 1992). We found that *expert* mathematics teacher educators, in their content courses, directly target prospective teachers' orientations and beliefs about critical elements of mathematical learning and teaching valued by the profession.

Findings from This Study

The *expert* mathematics teacher educators implemented two overarching practices to provide opportunities for prospective teachers to develop orientations toward teaching the subject (mathematics) by structuring the classroom learning experiences that deliberately confronted and expanded prospective teachers' views on: (1) *Standards*-based mathematics teaching strategies, and (2) meaningful mathematics learning practices. Specifically, the *expert* mathematics teacher educators challenged the prospective teachers' beliefs about teaching mathematics because, they explained, prospective teachers often have erroneous beliefs, based on their past experiences, about mathematics teaching being a procedural endeavour.

For example, Vance explained, “Many of them [prospective teachers] have a vision that teaching math is simply a matter of standing up in front of the class and telling the students what they need to do to solve a particular type of problem” (Vance). Therefore, the mathematics teacher educators wanted the prospective teachers to experience *what the standards-based teaching looks like in mathematics*. To do so, they modelled *standards*-based teaching practices by engaging the prospective teachers in group work, collaboration, and communication during class, “talking and sharing ideas and communicating … and not only have them [prospective teachers] talk with each other, but we do a lot of small group work where they work on problems and they share their reasoning” (Odessa). The mathematics teacher educators elaborated that working in small groups aids learning through the practice of “teaching each other” (Trina) and listening and challenging or critiquing each other’s reasoning.

The mathematics teacher educators were explicit with prospective teachers about their *standards*-based teaching practices. For example, one mathematics teacher educator (Oliver) revealed that *Principles to Actions* (NCTM, 2014) is a supplementary textbook for his content course, which helps him to better address the philosophical principles behind his *standards*-based teaching. The mathematics teacher educators also indicated that they provide prospective teachers with specific experiences that involve *meaningful mathematical learning* “to get them engaged in thinking about what it means to learn mathematics in a meaningful way” (Ian). They voiced concerns about prospective teachers being impatient during the problem-solving process, and that “they are unwilling to persevere in thinking through a mathematics problem. If they get frustrated they don’t ask follow-up questions” (Ian). As a result, the *expert* mathematics teacher educators deliberately chose more challenging “non-routine” mathematics problems and continually encouraged the prospective teachers to experience *perseverance* in problem-solving. They encouraged perseverance by guiding prospective teachers through productive struggle and continued cogitation, including giving specific advice on “re-entering the problem and thinking about it more deeply, being patient enough to recognize that all problems can’t be solved in less than one minute” (Vance). The mathematics teacher educator’s goal was for prospective teachers to recognize and appreciate that “there could be many different ways to get to the solution” (Ella) and to demonstrate that “solutions don’t come up right off, and it may go into the second day [of class]” (Trina). They called these “teachable moments” and shared that they make it a priority to regularly “assign problems that have a lot of thought provoking non-routine problems” (Ethan).

The data from this study also showed that the *expert* mathematics teacher educators deliberately embedded *cognitively dissonant* experiences into their content courses for the purpose of providing prospective teachers opportunities to continually endure mathematical struggles, puzzlements, and uncertainties, which help to confront prospective teachers’ habits, knowledge, and beliefs about mathematics learning. The mathematics teacher educators commonly described these opportunities as, "cognitive dissonance, where you shake them [prospective teachers] up a little bit from their 'security blanket' of knowing an algorithm" (Allen). One mathematics teacher educator (Odessa) stated, "I give them problems that push their limits of understanding of the math, having them struggle with it and explore with each other, share how they got through that struggle." During the *cognitive dissonance* process, the mathematics teacher educators required prospective teachers to work collaboratively, explain and share their thinking, verbalize solutions, learn to take insightful notes, and develop meaningful questions.

Evidence from the Field

The findings from the study reported here reflect the definitions of *orientations* outlined in the literature. Specifically, the mathematics teacher educators provided opportunities for the prospective teachers to observe and experience what *standards*-based teaching looks like in mathematics and emphasized "the national standards of being inquiry-based" (Magnusson, Krajcik, & Borko 1999, p. 97). These findings are particularly noteworthy because the *expert* mathematics teacher educators' descriptions of their teaching practices closely aligned with *constructivist* theories, in that they wanted prospective teachers to construct knowledge via specific experiences and to reflect on those experiences by engaging in collective inquiry and small-group interactions (e.g., Cobb, 1994; Ernest, 1994; Kroll & LaBoskey, 1996; Simon, 1995). They remarked on how, in their content courses, it was important for them (personally and professionally) to situate prospective teachers' learning around active learning, collaboration, group work, and communication.

In the literature, cases have been made for the need to examine teachers' *orientations toward teaching the subject*. For example, in studying teachers' work with children, Fennema, Carpenter, Franke, Jacobs, and Empson (1996) identified four levels of teachers' beliefs and orientations about children's mathematics learning: *Level A* involved teachers who believed that children learn best by being told how to do mathematics; *Level B* included teachers with conflicting beliefs who often questioned the notions that children need to be shown how to do mathematics; *Level C* was comprised of teachers who thought children learn mathematics best by solving problems and discussing their solutions; and *Level D* involved teachers who believed that children can solve problems without direct instruction and that mathematics teaching should be situated around children's abilities.

Additionally, teachers' instructional orientations were studied and classified by their beliefs and personal experiences with mathematics learning (Cooney, Shealy, & Arvold, 1998). The researchers found that teachers, who were identified as *naive idealists*, believed that learning entails absorbing what others believed to be true, without questioning it. In contrast, the *reflective connectionists* mainly engaged in reflections on the beliefs of others as compared to their own beliefs, and resolved conflicts via reflective thinking (Cooney et al., 1998). Cooney (1999) argued that the "inculcation of doubt and the posing of perplexing situations" (p. 173) were central to the shift from being a naive idealist to a reflective connectionist. The research findings from the study reported here directly reflect this suggestion, especially because the *expert* mathematics teacher educators deliberately incorporated *cognitive dissonance* into prospective teachers' learning experiences.

Furthermore, Shoenfeld (2015) suggested that research on teachers' *orientations* helps to describe and develop productive opportunities for teachers' learning. For example, he articulated that novice teachers often struggle with issues of "classroom management" and during this struggle they are unable to focus their attention on more subtle aspects of teaching that are prevalent in the practice of more seasoned and expert teachers (p. 243). Shoenfeld argued that, during a lesson, while "acting in the moment," teachers often *orient* themselves to different teaching situations and, on the basis of their beliefs and available resources, make decisions on how to pursue their pedagogical goals (Shoenfeld, 2015). Over time, as teachers develop teaching expertise, their lessons become "well practiced" domains – areas of professional practice in which "individuals have had enough time to develop a corpus of knowledge and routines that shape much of what they do" (Shoenfeld, 2015, p. 457; Shoenfeld, 2011). The findings from the study reported here reflect these recommendations, particularly showing that, *expert* mathematics teacher educators embed learning experiences for prospective teachers in content courses that reach beyond "learning content" and comprise a variety of learning opportunities that directly contribute to prospective teachers' beliefs and resources related to the nature of mathematics, teaching, and learning, that is, *orientations towards teaching the subject*.

IMPLICATIONS AND FUTURE DIRECTIONS

Cai et al. (2017) in an editorial panel stated that, "[w]e began our editorials in 2017 seeking answers to one complex but important question: [h]ow can we [the field] improve the impact of research on practice?" (p. 466). They suggested that we challenge the current divide between research and practice and adopt a system that emphasizes "learning opportunities as an integral element of research that has an impact on practice" (Cai et al., 2017, p. 466). My agenda in this chapter was to respond to this call by providing theoretical and empirical foundations for examining mathematics teacher educators' practices *as learning opportunities* that have an impact on the classroom instruction of university faculty who teach prospective teachers and enhance and develop their pedagogical content knowledge. In the sections that follow, I elaborate on these theoretical and empirical foundations and describe specific directions, based on the work reported here, for moving the field forward.

Theoretical Directions and Implications

I offered evidence depicting specific practices and learning opportunities that *expert* mathematics teacher educators provide to prospective teachers for developing their pedagogical content knowledge. I included a classification (Table 8.1) of specific pedagogical content knowledge-related practices, commonly articulated by *expert* mathematics teacher educators, offering the field new opportunities for studying mathematics teacher educators' practices and using this classification as a professional development tool for preparing novice mathematics teacher educators, including mathematics department faculty. This list of pedagogical content knowledge-related practices may not be exhaustive and may vary across different settings and mathematics teacher educator experiences. Nonetheless, it offers a foundation on which other scholars can build future research and upon which other mathematics teacher educators can build their practice related to teaching mathematics content courses.

Based on the research findings from the study reported here, *expert* mathematics teacher educators provide opportunities for prospective teachers to develop *representations of practice*, via prospective teachers' learning to use grades K-8 manipulatives as instructional tools, utilizing child-friendly language and various models/representations when explaining grades K-8 mathematics, and examining curriculum standards/documents to map the scope and sequence of mathematical topics included in the grades K-8 curriculum. Mathematics teacher educators also engaged prospective teachers in the actual grades K-8 mathematics lessons and arranged for them to visit local (grades K-8) classrooms, which developed prospective teachers' ways of "seeing and understanding elements of professional practice" (Ghousseini & Herbst, 2016, p. 80; also Feimen-Nemser, 2001; Moss, 2011; Sherin, 2001). Additionally, mathematics teacher educators incorporated online videos, student work, and professional literature readings to help prospective teachers examine grades K-8 students' mathematical thinking, cognition, and development and confront their own mathematical knowledge and understandings. These practices mirror the *approximations of practice* pedagogy, in the ways that mathematics teacher educators "engaged prospective teachers' mathematical knowledge with the aim to use it in light of particular content and particular students' understandings" (Ghousseini & Herbst, 2016, p. 100).

More research is needed to identify specific parallels and connections between research on *opportunities to learn* and the *pedagogical content knowledge-related expertise* that prospective teachers develop and experience in mathematics content courses, as these types of opportunities have been identified as critical for supporting prospective teachers' development of a profound knowledge about mathematics, knowledge about grades K-8 students, and of a beginning repertoire of instructional strategies for responsible and responsive ways to teach mathematics (Ball & Cohen, 1999; Ball & Wilson, 1996; Chapin, O'Connor, & Anderson, 2009; Chapin & O'Connor, 2012).

Furthermore, in this work, I provide new theoretical insights regarding mathematics teacher educators' knowledge and practices to help enrich the development of conceptual models and to strengthen the efforts of studying mathematics teacher educators' work and practices, including those engaged in the preparation of novice mathematics teacher educators. Research directions and implications from these conceptualisations are significant, particularly in their empirical groundings and validations reflecting the (four) pedagogical content knowledge components included in the proposed framework (Figure 8.1).

Specifically, I offer empirical evidence that *expert* mathematics teacher educators, in their content courses, engage prospective teachers in specific learning opportunities that support the development of pedagogical content knowledge, including knowledge about instructional strategies, knowledge about students' understanding, and knowledge about curriculum. I additionally provide evidence that *expert* mathematics teacher educators employing classroom practices that directly contribute to prospective teachers' learning and development of specific *orientations toward teaching the subject* (Magnusson et al., 1999; Shoenfeld, 2010). I encourage the field to explore the "orientations" construct further, investigating the nature of "orientations" and the similarities and differences between *orientations towards teaching the subject* and *orientations towards the subject/discipline*, particularly as it relates to the mathematics education of prospective teachers, their pedagogical content knowledge, and their development of professional practice. I offer the field an emerging framework as a foundation to test and build upon, including new directions for generating much needed research and usable knowledge in the field regarding mathematics teacher educators' practices in mathematics content courses.

Practitioner Directions and Implications

An, Kulm and Wu (2004) argued that enhancing prospective teachers' pedagogical content knowledge "should be the most important element in the domain of mathematics teachers' knowledge" (p. 146). It follows that a significant focus of mathematics teacher educators should be to provide opportunities for prospective teachers to develop pedagogical content knowledge, including in content courses specifically designed for prospective teachers. However, as a field, we know very little about these courses. The work reported here offers the field glimpses of the specific goals and teaching practices of *expert* mathematics teacher educators in the content courses, including their pedagogical content knowledge-specific classroom examples (commonly) identified and implemented in their content courses (Table 8.1).

Furthermore, it is important to study the classroom practices of university faculty who teach content courses, especially the role of mathematics teacher educators' pedagogical content knowledge in these courses. The findings from the study reported here offer convincing evidence suggesting that there is a (strong) presence of pedagogical content knowledge in mathematics content courses, when taught by *expert* mathematics teacher educators. I suggest that researchers consider qualitative methodologies and a situated perspective to investigate and depict pedagogical content knowledge "in action," especially to document what happens in the content courses and what matters to, and the orientations of, the mathematics teacher educators who teach these courses (Shoenfeld, 2010, 2015), and how their personal and professional goals impact prospective teachers' mathematical learning and professional development (Depaepe, Verschaffel, & Kelchtermans, 2013).

More research is needed providing insights on the teaching (and learning) that occurs in mathematics content courses. The findings revealed discrepancies between classroom practices of mathematics faculty/staff and the *expert* mathematics teacher educators when teaching mathematics content courses (documented in the literature). Specifically, current research shows that the majority of mathematics content courses are taught by the mathematics faculty/staff who mainly engage prospective teachers in lectures and occasionally in activity-based learning, with only a few institutions reporting the use of inquiry-based learning in their content courses (Masingila et al., 2012). Studies also document that "show-and-tell" continues to dominate college-level mathematics instruction, especially in lower level and service[6] courses, including grades K-8 mathematics content courses since prospective teachers complete these courses (as non-mathematics majors) in their first couple of years in undergraduate teacher education programs (Goldrick-Rab, 2007; Grubb, 1999; Stigler, Givvin, & Thompson, 2010). For example, Stigler et al. (2010) reported that in these courses, "students who failed to learn how to divide fractions in elementary school … are basically presented the same material in the same way yet again" (p. 4).

In contrast, I found that *expert* mathematics teacher educators' teaching practices are closely aligned with the constructivist and inquiry-based perspectives. The mathematics teacher educators in my study constantly encouraged prospective teachers' questioning and explanations, challenged prospective teachers with "cognitive dissonance," and structured their courses around prospective teachers constructing knowledge via working in small groups, hands-on learning, and contemplating non-routine mathematical tasks. These findings are significant given that "too often, the person assigned to teach mathematics to elementary teacher candidates is not professionally equipped to do so" (Greenberg & Walsh, 2008, p. 46) and that many current instructors do not teach content courses in ways that provide the type of mathematical support needed by prospective teachers (Lutzer et al., 2007; Masingila et al., 2012).

CONCLUSION

My work builds upon the research of scholars conceptualizing models of mathematics teacher educators' knowledge, expertise, and readiness to teach prospective teachers (e.g., Bass, 2005; Hodgson, 2001; Lutzer et al., 2007; Masingila et al., 2012; Superfine & Li, 2014; Sztajn et al., 2006; Zaslavsky, 2009). I invite the field to join the efforts in studying mathematics teacher educators, mathematics content courses and prospective teachers' learning in these courses, particularly paying closer attention to mathematics teacher educators' knowledge, classroom practices, and teaching philosophies. I encourage researchers to explore and document learning opportunities afforded to prospective teachers, specifically delineating opportunities that directly contribute to their pedagogical content knowledge development and orientations toward teaching the subject. I offer the field adaptations to current pedagogical content knowledge frameworks, based on the findings reported here, in an effort to develop more robust conceptual models that may help to better conceptualize the work and practices of mathematics teacher educators.

ACKNOWLEDGEMENT

The work reported here is part of a larger project and I would like to acknowledge the work of Cynthia Taylor (co-investigator) on this project. I would also like to thank Simon Goodchild and Kim Beswick for their feedback and suggestions, which helped to polish and strengthen this chapter.

NOTES

1 In the United States, a typical regular/full-length university semester is fifteen weeks long; one semester credit equals one hour of instruction per week or fifteen hours of instruction per semester.
2 Prospective teachers in this chapter are college students pursuing academic qualifications to teach mathematics to pupils aged 5–12 years.
3 For insights on the evolvement/development of this framework please see additional reports (e.g., Appova, 2018a,, 2018b; Appova & Taylor, 2017; Appova & Taylor, 2015, January; Appova & Taylor, 2014, April; Taylor & Appova, 2015).
4 For details on the methods and procedures from this study see Appova (2018a), Appova and Taylor (2017), Taylor and Appova (2015).
5 By "this study" I primarily refer to the work reported in Appova (2018a), Appova and Taylor (2017), and Taylor and Appova (2015).
6 Service mathematics courses are offered/taught in the mathematics department to serve students from different departments/disciplines or to serve students from a specific degree/program as a service to a department other than mathematics (e.g., engineering calculus, business mathematics).

REFERENCES

Abell, S. K., & Bryan, L. A. (1997). Reconceptualizing the elementary science methods course using a reflection orientation. *Journal of Science Teacher Education, 8*(3), 153–166.

Abell, S. K., & Smith, D. C. (1994). What is science? Preservice elementary teachers' conceptions of the nature of science. *International Journal of Science Education, 16*(4), 475–487.

Aguirre, J., & Speer, N. M. (1999). Examining the relationship between beliefs and goals in teacher practice. *The Journal of Mathematical Behavior, 18*(3), 327–356.

Akerlind, G. (2005). Variation and commonality in phenomenographic research methods. *Higher Education Research & Development, 24*(4), 321–334.

Amador, J. (2017). Preservice teachers' video simulations and subsequent noticing: A practice-based method to prepare mathematics teachers. *Research in Mathematics Education, 19*(3), 217–235.

Ambrose, R. (2004). Initiating change in prospective elementary school teachers' orientations to mathematics teaching by building on beliefs. *Journal of Mathematics Teacher Education, 7*(2), 91–119.

An, S., Kulm, G., & Wu, Z. (2004). The pedagogical content knowledge of middle school, mathematics teachers in China and the U.S. *Journal of Mathematics Teacher Education, 7*(2), 145–172.

Anderson, C. W., & Smith, E. L. (1985). Teaching science. In J. Koehler (Ed.), *The educator's handbook: A research perspective* (pp. 84–111). New York, NY: Longman.

Appova, A. (2018a). Developing prospective teachers' mathematics orientations in the content courses. In E. Bergqvist & M. Österholm (Eds.), *Proceedings of the 42nd Conference of the International Group for the Psychology of Mathematics Education* (Vol. 2, pp. 59–66). Umea, Sweden: PME.

Appova, A. (2018b, February). *Developing prospective teachers' "orientations" in content courses: Mathematics teacher educators' reflections on practice*. Paper presented Association of Mathematics Teacher Educators (AMTE), Houston, TX.

Appova, A., & Taylor, C. (2014, April). *Mathematics teacher educators' classroom practices: Mathematics content courses*. Paper presented at the American Educational Research Association (AERA), Philadelphia, PA.

Appova, A., & Taylor, C. (2015, January). *Instructional practices of experienced mathematics teacher educators teaching K-8 content courses*. Paper presented at the Association of Mathematics Teacher Educators (AMTE), Orlando, FL.

Appova, A., & Taylor, C. (2017). Expert mathematics teacher educators' purposes and practices for providing prospective teachers opportunities to develop pedagogical content knowledge in content courses. *Journal of Mathematics Teacher Education, 22*(2), 179–204.

Appova, A., & Taylor, C. E. (2020). Providing opportunities to develop prospective teachers' pedagogical content knowledge. *The Mathematics Enthusiast, 17*(2–3), 673–724.

Australian Curriculum and Assessment Reporting Authority [ACARA]. (2014). *Foundation to year 10 curriculum: Mathematics*. Retrieved from http://www.australiancurriculum.edu.au/mathematics/aims

Ball, D. L., & Feiman-Nemser, S. (1988). Using textbooks and teachers' guides: A dilemma for beginning teachers and teacher educators. *Curriculum Inquiry, 18*(4), 401–423.

Ball, D. L., & Cohen, D. K. (1999). Developing practice, developing practitioners: Toward a practice-based theory of professional education. In L. Darling-Hammond & G. Sykes (Eds.), *Teaching as the learning profession: Handbook of policy and practice,* (Vol. 1, pp. 3–22). San Francisco, CA: Jossey-Bass Education Series. Jossey-Bass.

Ball, D. L., Sleep, L., Boerst, T. A., & Bass, H. (2009). Combining the development of practice and the practice of development in teacher education. *The Elementary School Journal, 109*(5), 458–474.

Ball, D. L., Thames, M. H., & Phelps, G. (2008). Content knowledge for teaching: What makes it special? *Journal of Teacher Education, 59*(5), 389–407.

Ball, D. L., & Wilson, S. M. (1996). Integrity in teaching: Recognizing the fusion of the moral and intellectual. *American Educational Research Journal, 33*(1), 155–192.

Bass, H. (2005). Mathematics, mathematicians, and mathematics education. *Bulletin of the American Mathematical Society, 42*(4), 417–430.

Baumert, J., Kunter, M., Blum, W., Brunner, M., Voss, T., Jordan, A., Klusmann, U., Krauss, S., Neubrand, M., & Yi-Miau, T. (2010). Teachers' mathematical knowledge, cognitive activation in the classroom, and student progress. *American Educational Research Journal, 47*(1), 133–180.

Bergsten, C., & Grevholm, B. (2008). Knowledgeable teacher educators and linking practices. In B. Jaworski & T. Wood (Eds.), *The international handbook of mathematics teacher education: The mathematics teacher educator as a developing professional* (Vol. 4, pp. 223–246). Rotterdam, The Netherlands: Sense Publishers.

Blömeke, S. (2006). Globalization and educational reform in German teacher education. *International Journal of Educational Research, 45*(4–5), 315–324.

Blömeke, S., Buchholtz, N., Suhl, U., & Kaiser, G. (2014). Resolving the chicken-or-egg causality dilemma: The longitudinal interplay of teacher knowledge and teacher beliefs. *Teaching and Teacher Education, 37*, 130–139.

Boeije, H. (2002). A purposeful approach to the constant comparative method in the analysis of qualitative interviews. *Quality and Quantity, 36*(4), 391–409.

Bofferding, L., Kastberg, S., & Hoffman, A. (2016). Family mathematics nights: An opportunity to improve preservice teachers' understanding of parents' roles and expectations. *School Science and Mathematics, 116*(1), 17–28.

Borasi, R. (1994). Capitalizing on errors as "Springboards for inquiry": A teaching experiment. *Journal for Research in Mathematics Education, 25*(2), 166–208.

Borko, H., & Putnam, R. T. (1996). Learning to teach. In D. C. Berliner & R. C. Calfee (Eds.), *Handbook of educational psychology* (pp. 673–708). New York, NY: Simon & Schuster Macmillan.

Bowden, J. A., & Walsh, E. (2000). *Phenomenography.* Melbourne: RMIT University.

Boyd, D., Grossman, P. L., Hammerness, K., Lankford, R. H., Loeb, S., McDonald, M., et al. (2008). Surveying the landscape of teacher education in New York City: Constrained variation and the challenge of innovation. *Educational Evaluation and Policy Analysis, 30*(4), 319–343.

Bransford, J. D., Brown, A. L., & Cocking, R. R. (Eds.). (2000). *How people learn* (expanded ed.). Washington, DC: National Academy Press.

Brown, M., & Edelson, D. C. (2003). *Teaching as design: Can we better understand the ways in which teachers use materials so we can better design materials to support their changes in practice?* (LETUS design brief). Penn Graduate School of Education. Retrieved from http://www.gse.upenn.edu/~janiner/pdf/teaching_as_design-Final.pdf

Burstein, L. (1993). *Validating national curriculum indicators: A conceptual overview of the RAND/CREST NSF project.* Paper presented at the annual meeting of the American Educational Research Association, Atlanta, GA.

Burton, M., Daane, C. J., & Giesen, J. (2008). *Infusing mathematics content into a methods course: impacting content knowledge for teaching.* Institute of Education Sciences. Retrieved from https://files.eric.ed.gov/fulltext/EJ835496.pdf

Cai, J., Morris, A., Hwang, S., Hohensee, C., Robison, V., & Hiebert, J. (2017). Improving the impact of educational research. *Journal for Research in Mathematics Education, 48*(1), 2–6.

Cakiroglu, E. (2000). *Preservice elementary teachers' sense of efficacy in reform-oriented mathematics.* (Unpublished doctoral thesis). Indiana University, Bloomington, IN.

Capraro, R. M., Capraro, M. M., Parker, D., Kulm, G., & Raulerson, T. (2005). The mathematics content knowledge role in developing preservice teachers' pedagogical content knowledge. *Journal of Research in Childhood Education, 20*(2), 102–118.

Carpenter, T. P., Fennema, E., Franke, M. L., Levi, L., & Empson, S. B. (1999). *Children's mathematics: Cognitively guided instruction.* Portsmouth, NH: Heinemann.

Carpenter, T. P., Fennema, E., Peterson, P. L., & Carey, D. A. (1988). Teachers' pedagogical content knowledge of student's problem solving in elementary arithmetic. *Journal for Research in Mathematics Education, 19*, 385–401.

Chapin, S., & O'Connor, C. (2012). Project challenge: Using challenging curriculum and mathematical discourse to help all students learn. In C. Dudley-Marling & S. Michaels (Eds.), *High expectation curricula: Helping all students succeed with powerful learning* (pp. 113–127). New York, NY: Teachers College Press.

Chapin, S., O'Connor, C., & Anderson, N. (2009). *Classroom discussions: Using math talk to help students learn, grades K–6* (2nd ed.). Sausalito, CA: Math Solutions Publications.

Chick, H., & Beswick, K. (2017). Teaching teachers to teach Boris: A framework for mathematics teacher educators pedagogical content knowledge. *Journal of Mathematics Teacher Education, 21*(5), 475–499.

Christou, C., Eliophotou-Menon, M., & Phillippou, G. (2004). Teachers' concerns regarding the adoption of a new mathematics curriculum: An application of CBAM. *Educational Studies in Mathematics, 57*(2), 157–176.

Chval, K. B., Lannin, J. K., Arbaugh, F., & Bowzer, A. D. (2009). Videos and prospective teachers. *Teaching Children Mathematics, 16*(2), 98–105.

Cianca, S. A. (2013). Bird boxes build content area knowledge. *Mathematics Teaching in the Middle School, 19*(1), 22–29.

Cobb, P. (1994). Where is the mind? Constructivist and sociocultural perspectives on mathematical development. *Educational Researcher, 23*(7), 13–20.

Cohen, D. K. (1990). A revolution in one classroom: The case of Mrs. Oublier. *Educational Evaluation and Policy Analysis, 12*(3), 311–329.

Conference Board of the Mathematical Sciences. (2012). *The mathematical education of teachers, Part II.* Providence, RI & Washington, DC: American Mathematical Society and Mathematical Association of America.

Cooney, T. J. (1999). Conceptualizing teachers' ways of knowing. In P. Ernst (Ed.), *Forms of mathematical knowledge* (pp. 163–187). Dordrecht: Springer.

Cooney, T. J., Shealy, B. E., & Arvold, B. (1998). Conceptualizing belief structures of preservice secondary mathematics teachers. *Journal for Research in Mathematics Education, 29*(3), 306–333.

Corbin, J., & Strauss, A. (2008). *Basics of qualitative research: Techniques and procedures for developing grounded theory*. Thousand Oaks, CA: Sage Publications.

Craven, G., Beswick, K., Fleming, J., Fletcher, T., Green, M., Jensen, B., Leinonen, E., & Rickards, F. (2014). *Action now: Classroom ready teachers*. Australian Department of Education. Retrieved from https://docs.education.gov.au/system/files/doc/other/action_now_classroom_ready_teachers_accessible.pdf

Crespo, S. (2000). Seeing more than right and wrong answers: Prospective teachers' interpretations of students' mathematical work. *Journal of Mathematics Teacher Education, 3*(2), 155–181.

Crespo, S. (2003). Learning to pose mathematical problems: Exploring changes in preservice teachers' practices. *Educational Studies in Mathematics, 52*(3), 243–270.

Cummings, W. K. (2003). *The institutions of education*. London: Symposium Books.

Darling-Hammond, L., Macdonald, M. B., Snyder, J., Whitford, B. L., Rusco, G., & Fickel, L. (2000). *Studies of excellence in teacher education: Preparation at the graduate level*. New York, NY: AACTE Publications.

Depaepe, F., Verschaffel, L., & Kelchtermans, G. (2013). Pedagogical content knowledge: A systematic review of the way in which the concept has pervaded mathematics educational research. *Teaching and Teacher Education, 34*, 12–25.

Earp, N. W., & Tanner, F. W. (1980). Mathematics and language. *The Arithmetic Teacher, 28*(4), 32–34.

Ernest, P. (1989). The knowledge, beliefs and attitudes of the mathematics teacher: A model. *Journal of Education for Teaching, 15*(1), 13–33.

Ernest, P. (1994). Social constructivism and the psychology of mathematics education. In P. Ernst (Ed.), *Constructing mathematical knowledge: Epistemology and mathematical education* (pp. 62–71). London: The Falmer Press.

European Agency for Development in Special Needs Education. (2011). *Teacher education for inclusion across Europe: A synthesis of policy and practice in 25 countries*. Østre: Author.

European Ministers in charge of Higher Education. (1999). *The Bologna Declaration of 19 June 1999: Joint declaration of the European Ministers of Education*. European Association of Institutions in Higher Education. Retrieved from https://www.eurashe.eu/library/modernising-phe/Bologna_1999_Bologna-Declaration.pdf

Even, R. (2008). Facing the challenge of educating educators to work with practicing mathematics teachers. In B. Jaworski & T. Wood (Eds.), *The international handbook of mathematics teacher education: The mathematics teacher educator as a developing professional* (Vol. 4, pp. 57–73). Rotterdam, The Netherlands: Sense Publishers.

Feiman-Nemser, S. (2001). From preparation to practice: Designing a continuum to strengthen and sustain teaching. *Teachers College Record, 103*(6), 1013–1055.

Feiman-Nemser, S., & Buchman, M. (1985). Pitfalls of experience in teacher preparation. *Teachers College Record, 87*(1), 53–65.

Fennema, E., Carpenter, T. P., Franke, M. L., Levi, L., Jacobs, V. R., & Empson, S. B. (1996). A longitudinal study of learning to use children's thinking in mathematics instruction. *Journal for Research in Mathematics Education, 27*(4), 403–434.

Fernandes, A. (2012). Mathematics preservice teachers learning about English language learners through task-based interviews and noticing. *Mathematics Teacher Educator, 1*(1), 10–22.

Floden, R. E., & Philipp, R. A. (2003). Report of working group 7: Teacher preparation. In F. K. Lester & J. Ferrini-Mundy (Eds.), *Proceedings of the NCTM research catalyst conference* (pp. 171–176). Reston, VA: National Council of Teachers of Mathematics.

Florian, L., & Black-Hawkins, K. (2011). Exploring inclusive pedagogy. *British Educational Research Journal, 37*(5), 813–828.

Freiberg, M. (2004). Getting everyone involved in family math. *The Mathematics Educator, 14*(1), 35–41.

Friederichsen, P., van Driel, J. H., & Abell, S. K. (2010). Taking a closer look at science teaching orientations. *Science Education, 95*(2), 358–376.

Friedrichsen, P. J., Abell, S. K., Pareja, E. M., Brown, P. L., Lankford, D. M., & Volkmann, M. J. (2009). Does teaching experience matter? Examining biology teachers' prior knowledge for teaching in an alternative certification program. *Journal of Research in Science Teaching, 46*(4), 357–383.

Friedrichsen, P. M., & Dana, T. M. (2003). Using a card-sorting task to elicit and clarify science-teaching orientations. *Journal of Science Teacher Education, 14*(4), 291–309.

Friel, S. N. (1998). Reflections on practice: Exploring how one problem contributes to student learning. *Mathematics Teaching in the Middle School, 4*(2), 100–103.

Fung, M. G., & Latulippe, C. L. (2010). Computational estimation. *Teaching Children Mathematics, 17*(3), 170–176.

Gee, D. (2006). Math mates: Children guiding preservice teachers in learning about teaching. *Teaching Children Mathematics, 12*(8), 402–406.

Gehrke, N. J., Knapp, M. S., & Sirotnik, K. A. (1992). In search of the school curriculum. *Review of Research in Education, 18*(1), 51–110.

Ghousseini, H., & Herbst, P. (2016). Pedagogies of practice and opportunities to learn about classroom mathematics discussions. *Journal of Mathematics Teacher Education, 19*(1), 79–103.

Goldrick-Rab, S. (2007). *Promoting academic momentum at community colleges: Challenges and opportunities*. New York, NY: Columbia University, Teachers College, Community College Research Center.

Goodell, J. (2006). Using critical incident reflections: A self-study as a mathematics teacher educator. *Journal of Mathematics Teacher Education, 9*(3), 221–248.

Greenberg, J., & Walsh, K. (2008). *No common denominator: The preparation of elementary teachers in mathematics by America's education schools*. Washington, DC: National Council on Teacher Quality.

Gresham, G. (2007). A study of mathematics anxiety in pre-service teachers. *Early Childhood Education Journal, 35*(2), 181–188.

Grossman, P. L. (1987). *A tale of two teachers: The role of subject matter orientation in teaching* (Knowledge Growth in a Profession Publication Series). Stanford, CA: Stanford University, School of Education.

Grossman, P. L. (1990). *The making of a teacher: Teacher knowledge and teacher education*. New York, NY: Teachers College Press.

Grossman, P. L. (2008). Responding to our critics: From crisis to opportunity in research on teacher education. *Journal of Teacher Education, 59*(1), 10–23.

Grossman, P. L., Hammerness, K., & McDonald, M. (2009). Redefining teaching, re-imagining teacher education. *Teachers and Teaching, 15*(2), 273–289.

Grossman, P. L., Wilson, S. M., & Shulman, L. S. (1989). Teachers of substance: Subject matter knowledge for teaching. In M. Reynolds (Ed.), *Knowledge base for the beginning teacher* (pp. 23–36). Oxford: Pergammon Press.

Grubb, N. W., & Associates. (1999). *Honored but invisible: An inside look at teaching in community colleges*. New York, NY: Routledge.

Gurl, T. J., Fox, R., Dabovic, N., & Leavitt, A. E. (2016). Planning questions and persevering in the practices. *The Mathematics Teacher, 110*(1), 33–39.

Handal, B. (2003). Teachers' mathematical beliefs: A review. *The Mathematics Educator, 13*(2), 47–57.

Hart, L. C. (2002). Preservice teachers' beliefs and practice after participating in an integrated content/methods course. *School Science and Mathematics, 102*(1), 4–14.

Herbel-Eisenmann, B. A., & Phillips, E. D. (2005). Using student work to develop teachers' knowledge of algebra. *Mathematics Teaching in the Middle School, 11*(2), 62–66.

Hiebert, J., & Morris, A. K. (2009). Building a knowledge base for teacher education: An experience in K-8 mathematics teacher preparation. *The Elementary School Journal, 109*(5), 475–490.

Hiebert, J., Morris, A. K., & Glass, B. (2003). Learning to learn to teach: An "experiment" model for teaching and teacher preparation in mathematics. *Journal of Mathematics Teacher Education, 6*(3), 201–222.

Hill, H., Ball, D. L., & Schilling, S. G. (2008). Unpacking pedagogical content knowledge: Conceptualizing and measuring teachers' topic-specific knowledge of students. *Journal for Research in Mathematics Education, 39*(4), 372–400.

Hill, H. C., Rowan, B., & Ball, D. L. (2005). Effects of teachers' mathematical knowledge for teaching on student achievement. *American Educational Research Journal, 42*(2), 371–406.

Hodgson, B. (2001). The mathematical education of school teachers: Role and responsibilities of university mathematicians. In D. Holton (Ed.), *The teaching and learning of mathematics at university level* (pp. 501–518). Dordrecht: Kluwer Academic Publishers.

Hooghart, A. M. (2006). Educational reform in Japan and its influence on teachers' work. *International Journal of Educational Research, 45*(4–5), 290–301.

Howard, P., Perry, B., & Tracey, D. (1997). *Mathematics and manipulatives: Comparing primary and secondary mathematics teachers' views.* European Association of Institutions in Higher Education. Retrieved from https://files.eric.ed.gov/fulltext/ED461502.pdf

Huinker, D., & Madison, S. K. (1997). Preparing efficacious elementary teachers in science and mathematics: The influence of methods courses. *Journal of Science Teacher Education, 8*(2), 107–126.

Jacobs, V. R., Lamb, L. L. C., & Philipp, R. A. (2010). Professional noticing of children's mathematical thinking. *Journal of Research in Mathematics Education, 41*(2), 169–202.

Jaworski, B. (2008). Mathematics teacher educator learning and development: An introduction. In B. Jaworski & T. Wood (Eds.), *The international handbook of mathematics teacher education: The mathematics teacher educator as a developing professional* (Vol. 4, pp. 1–13). Rotterdam, The Netherlands: Sense Publishers.

Jenkins, O. (2010). Developing teachers' knowledge of students as learners of mathematics through structured interviews. *Journal of Mathematics Teacher Education, 13*(2), 141–154.

Jilk, L. M. (2016). Supporting teacher noticing of students' mathematical strengths. *Mathematics Teacher Educator, 4*(2), 188–199.

Kilpatrick, J., Swafford, J., & Findell, B. (Eds.). (2001). *Adding it up: Helping children learn mathematics.* Washington, DC: National Academy Press.

Kroll, L. R., & Laboskey, V. K. (1996). Practicing what we preach: Constructivism in a teacher education program. *Action in Teacher Education, 18*(2), 63–72.

Kuhs, T. M., & Ball, D. L. (1986). *Approaches to teaching mathematics: Mapping the domains of knowledge, skills, and dispositions.* East Lansing, MI: Michigan State University, Center on Teacher Education.

Lachance, A. M. (2007). Family math nights: Collaborative celebrations of mathematical learning. *Teaching Children Mathematics, 13*(8), 404–408.

Lachance, A. M., Benton, C. J., & Klein, B. S. (2007). The school-based activities model: A promising alternative to professional development schools. *Teacher Education Quarterly, 34*(3), 95–111.

Lampe, K. A., & Uselman, L. (2008). Pen pals: Practicing problem solving. *Mathematics Teaching in the Middle School, 14*(4), 196–201.

Lannin, J. K., & Chval, K. B. (2013). Challenge beginning teacher beliefs. *Teaching Children Mathematics, 19*(8), 508–515.

LeTendre, G. K. (2002). Setting national standards: Educational reform, social change, and political conflict. In G. DeCoker (Ed.), *National standards and school reform in Japan and the United States,* (pp. 19–32). New York, NY: Teachers College Press.

Levine, A. (2006). *Educating school teachers.* New York, NY: The Education Schools Project.

Lim, K. H. (2014). Error-eliciting problems: Fostering understanding and thinking. *Mathematics Teaching in the Middle School, 20*(2), 106–114.

Loucks-Horsley, S., Hewson, P. W., Love, N., & Stiles, K. E. (1998). *Designing professional development for teachers of science and mathematics.* Thousand Oaks, CA: Corwin Press.

Loucks-Horsley, S., Stiles, K. E., Mundry, S., Love, N., & Hewson, P. W. (2009). *Designing professional development for teachers of science and mathematics.* Thousand Oaks, CA: Corwin Press.

Lutz, P. M., & Berglund, J. (2007). Using NSF-Funded middle school materials in a university mathematics content course. *Teaching Children Mathematics, 13*(6), 342–347.

Lutzer, D. J., Rodi, S. B., Kirkman, E. E., & Maxwell, J. W. (2007). *Statistical abstract of undergraduate programs in the mathematical sciences in the United States.* Washington, DC: American Mathematical Society.

Maasz, J., & Schlöglmann, W. (2009). *Beliefs and attitudes in mathematics education: New research results.* Rotterdam, The Netherlands: Sense Publishers.

MacGregor, M. (1990). Writing in natural language helps students construct algebraic equations. *Mathematics Education Research Journal, 2*(2), 1–10.

Magnusson, S., Krajcik, J., & Borko, H. (1999). Nature, sources, and development of pedagogical content knowledge for science teaching. In J. Gess-Newsome & N. G. Lederman (Eds.), *Examining pedagogical content knowledge* (pp. 95–132). Dordrecht: Kluwer Academic Publishers.

Marton, F. (1981). Phenomenography – Describing conceptions of the world around us. *Instructional Science, 10*(2), 177–200.

Marton, F. (1986). Phenomenography – A research approach to investigating different understandings of reality. *Journal of Thought, 21*(3), 28–49.

Marton, F., & Booth, S. A. (1997). *Learning and awareness.* New York, NY: Psychology Press.

Masingila, J. O., Olanoff, D. E., & Kwaka, D. K. (2012). Who teaches mathematics content courses for prospective elementary teachers in the United States? Results of a national survey. *Journal of Mathematics Teacher Education, 15*(5), 347–358.

Mason, J. (2002). *Researching your own practice: The discipline of noticing.* London: Routledge Falmer.

McDonough, A., Clarke, B., & Clarke, D. M. (2002). Understanding, assessing and developing children's mathematical thinking: The power of a one-to-one interview for preservice teachers in providing insights into appropriate pedagogical practices. *International Journal of Educational Research, 37*(2), 211–226.

McDuffie, A. R., Drake, C., & Herbel-Eisenmann, B. A. (2008). The elementary mathematics methods course. In B. Jaworski & T. Wood (Eds.), *The international handbook of mathematics teacher education: The mathematics teacher educator as a developing professional* (Vol. 4, pp. 247–264). Rotterdam, The Netherlands: Sense Publishers.

McDuffie, A. R., Foote, M. Q., Bolson, C., Turner, E. E., Aguirre, J. M., Bartell, T. G., … Land, T. (2014). Using video analysis to support prospective K-8 teachers' noticing of students' multiple mathematical knowledge bases. *Journal of Mathematics Teacher Education, 17*(3), 245–270.

Ministry of Education, Culture, Sports, Science, and Technology (MEXT). (2001). *Education reform plan for the 21st century: Rainbow plan: Seven priority strategies.* Retrieved from http://www.mext.go.jp/b_menu/hakusho/html/hpac200101/index.html

Moss, P. A. (2011). Analyzing the teaching of professional practice. *Teachers College Record, 113*(12), 2878–2896.

Murphy, E. (1997). *Constructivism: From philosophy to practice.* European Association of Institutions in Higher Education. Retrieved from https://files.eric.ed.gov/fulltext/ED444966.pdf

Musikul, K., & Abell, S. K. (2009). *Professional development for elementary teachers of science in Thailand: A holistic examination.* Paper presented at the annual international meeting of the National Association for Research in Science Teaching, Garden Grove, CA.

Myers, K. D., Swars Auslander, S., Smith, S. Z., Smith, M. E., & Fuentes, D. S. (2019). Developing the pedagogical capabilities of elementary mathematics specialists during a K-5 mathematics endorsement program. *Journal of Teacher Education.* https://doi.org/10.1177/0022487119854437

National Governors Association Center for Best Practices, Council of Chief State School Officers. (2010). *Common Core State Standards Initiative [CCSSI mathematics].* National Governors Association Center for Best Practices, Council of Chief State School Officers, Washington DC.

National Council of Teachers of Mathematics. (1989). *Curriculum and evaluation standards for school mathematics.* Reston, VA: National Council of Teachers of Mathematics.

National Council of Teachers of Mathematics. (1991). *Professional standards for teaching mathematics.* Reston, VA: National Council of Teachers of Mathematics.

National Council of Teachers of Mathematics. (1995). *Assessment standards for school mathematics.* Reston, VA: National Council of Teachers of Mathematics.

National Council of Teachers of Mathematics. (2000). *Principles and standards for school mathematics*. Reston, VA: National Council of Teachers of Mathematics.

National Council of Teachers of Mathematics. (2014). *Principles to actions: Ensuring mathematical success for all*. Reston, VA: National Council of Teachers of Mathematics.

Nesbitt Vacc, N. N., & Bright, G. W. (1999). Elementary preservice teachers' changing beliefs and instructional use of children's mathematical thinking. *Journal for Research in Mathematics Education*, 89–110.

Nespor, J. (1987). The role of beliefs in the practice of teaching. *Journal of Curriculum Studies, 19*(4), 317–328.

OECD. (2005). *Teachers matter: Attracting, developing and retaining effective teachers*. Paris: OECD Publications.

OECD. (2010). *Educating teachers for diversity: Meeting the challenge*. Paris: OECD Publications.

Ontario College of Teachers. (2017). *Annual report*. Retrieved from https://reports2017.oct.ca/en

Osborn, M. (2006). Changing the context of teachers' work and professional development: A European perspective. *International Journal of Educational Research, 45*(4–5), 242–253.

Pajares, M. F. (1992). Teachers' beliefs and educational research: Cleaning up a messy construct. *Review of Educational Research, 62*(3), 307–332.

Park Rogers, M. A., Abell, S. K., Marra, R. M., Arbaugh, F., Hutchins, K. L., & Cole, J. S. (2010). Orientations to science teacher professional development: An exploratory study. *Journal of Science Teacher Education, 21*(3), 309–328.

Philipp, R. A. (2008). Motivating prospective elementary school teachers to learn mathematics by focusing upon children's mathematical thinking. *Issues in Teacher Education, 17*(2), 7.

Quinn, R. J. (1997). Effects of mathematics methods courses on the mathematical attitudes and content knowledge of preservice teachers. *The Journal of Educational Research, 91*(2), 108–114.

Remillard, J. T., & Bryans, M. B. (2004). Teachers' orientations toward mathematics curriculum materials: Implications for teacher education. *Journal for Research in Mathematics Education, 35*(5), 352–388.

Remillard, J. T. (2005). Examining key concepts in research on teachers' use of mathematics curricula. *Review of Educational Research, 75*(2), 211–246.

Riccomini, P. J., Sanders, S., & Jones, J. (2008). The key to enhancing students' mathematical vocabulary knowledge. *Journal on School Educational Technology, 4*(1), 1–7.

Riccomini, P. J., Smith, G. W., Hughes, E. M., & Fries, K. M. (2015). The language of mathematics: The importance of teaching and learning mathematical vocabulary. *Reading & Writing Quarterly, 31*(3), 235–252.

Rubenstein, R. N., & Thompson, D. R. (2002). Understanding and support children's mathematical vocabulary development. *Teaching Children Mathematics, 9*(2), 107.

Schmidt, W. H., & Houang, R. T. (2012). Curricular coherence and the common core state standards for mathematics. *Educational Researcher, 41*(8), 294–308.

Schoenfeld, A. H. (2014). What makes for powerful classrooms, and how can we support teachers in creating them? A story of research and practice, productively intertwined. *Educational Researcher, 43*(8), 404–412.

Seethaler, P. M., Fuchs, L. S., Star, J. R., & Bryant, J. (2011). The cognitive predictors of computational skill with whole versus rational numbers: An exploratory study. *Learning and Individual Differences, 21*(5), 536–542.

Sherin, M. G. (2001). Developing a professional vision of classroom events: Teaching elementary school mathematics. In *Beyond classical pedagogy: Teaching elementary school mathematics* (pp. 75–93). Hillsdale, NJ: Erlbaum.

Sherin, M. G., & Drake, C. (2009). Curriculum strategy framework: Investigating patterns in teachers' use of a reform-based elementary mathematics curriculum. *Journal of Curriculum Studies, 41*, 467–500.

Sherin, M. G., & van Es, E. A. (2003). A new lens on teaching: Learning to notice. *Mathematics Teaching in Middle School, 9*(2), 92–95.

Shimahara, N. K. (2002). *Teaching in Japan: A cultural perspective*. New York, NY: Routledge Falmer.

Shoenfeld, A. H. (2010). *How we think: A theory of goal-oriented decision making and its educational applications*. New York, NY: Routledge.

Shoenfeld, A. H. (2011a). Reflections on teacher expertise. In Y. Li & G. Kaiser (Eds.), *Expertise in mathematics instruction* (pp. 327–341). Boston, MA: Springer Science+Business Media.

Shoenfeld, A. H. (2011b). Toward professional development for teachers grounded in a theory of decision making. *ZDM Mathematics Education, 43*(4), 457–469.

Shoenfeld, A. H. (2015). How we think: A theory of human decision-making, with a focus on teaching. In S.J. Cho (Ed.), *The proceedings of the 12th international congress on mathematical education.* doi:10.1007/978-3-319-12688-3_16

Shokey, T. L., & Snyder, K. (2007). Engaging preservice teachers and elementary-age children in transformational geometry: Tessellating t-shirts. *Teaching Children Mathematics, 14*(2), 83–87.

Shulman, L. S. (1986). Those who understand: Knowledge growth in teaching. *Educational Researcher, 15*(2), 4–14.

Shulman, L. S. (1987). Knowledge and teaching: Foundations of the new reform. *Harvard Educational Review, 57*(1), 1–22.

Simon, M. A. (1995). Reconstructing mathematics pedagogy from a constructivist perspective. *Journal for Research in Mathematics Education, 26*(2), 114–145.

Sjoberg, C. A., Slavit, D., & Coon, T. (2004). Take time for action: Improving writing prompts to improve student reflection. *Mathematics Teaching in the Middle School, 9*(9), 490–493.

Smith, D. C., & Neale, D. C. (1989). The construction of subject matter knowledge in primary science teaching. *Teaching and Teacher Education, 5*(1), 1–20.

Spangler, D., & Hallman-Thrasher, A. (2014). Using task dialogues to enhance preservice teachers' abilities to orchestrate discourse. *Mathematics Teacher Educator, 3*(1), 58–75.

Sparks, D., & Loucks-Horsley, S. (1989). Five models of staff development. *Journal of Staff Development, 10*(4), 40–57.

Stephens, A. C., & Lamers, C. L. (2006). Assessment design: Helping preservice teachers focus on student thinking. *Teaching Children Mathematics, 13*(2), 118–123.

Sternberg, R. J., & Horvath, J. A. (1995). A prototype view of expert teaching. *Educational Researcher, 24*(6), 9–17.

Stigler, J. W., Givvin, K. B., & Thompson, B. J. (2010). What community college developmental mathematics students understand about mathematics. *MathAMATYC Educator, 1*(3), 4–16.

Superfine, A. C., & Li, W. (2014). Exploring the mathematical knowledge needed for teaching teachers. *Journal of Teacher Education, 65*(4), 303–314.

Swars, S. L., Daane, C. J., & Giesen, J. (2006). Mathematics anxiety and mathematics teacher efficacy: What is the relationship in elementary preservice teachers? *School Science and Mathematics, 106*(7), 306–315.

Swars Auslander, S., Smith, S. Z., Smith, M. E., & Myers, K. (2019). A case study of elementary teacher candidates' preparation for a high stakes teacher performance assessment. *Journal of Mathematics Teacher Education.* https://doi.org/10.1007/s10857-018-09422-z

Sztajn, P., Ball, D. L., & McMahon, T. A. (2006). Designing learning opportunities for mathematics teacher developers. In K. Lynch-Davis & R. L. Rider (Eds.), *The work of mathematics teacher educators: Continuing the conversation* (pp. 149–162). San Diego, CA: Association of Mathematics Teacher Educators.

Tatto, M. T. (1999). Education reform and state power in Mexico: The paradoxes of decentralization. *Comparative Education Review, 43*(3), 251–282.

Tatto, M. T., Schmelkes, S., Guevara, M. D. R., & Tapia, M. (2006). Implementing reform amidst resistance: The regulation of teacher education and work in Mexico. *International Journal of Educational Research, 45*(4–5), 267–278.

Tatto, M. T. (2006). Education reform and the global regulation of teachers' education, development and work: A cross-cultural analysis. *International Journal of Educational Research, 45*(4–5), 231–241.

Taylor, C., & Appova, A. (2015). Mathematics teacher educators' purposes for K-8 content courses. In K. Beswick, T. Muir, & J. Fielding-Wells (Eds.), *Proceedings of 39th Psychology of Mathematics Education Conference* (Vol. 4, pp. 241–248). Hobart, Tasmania, Australia: PME.

Taylor, P. M., & Ronau, R. (2006). Syllabus study: A structured look at mathematics methods courses. *AMTE Connections, 16*(1), 12–15.

Thomas, J. N., Eisenhardt, S., Fisher, M. H., Schack, E. O., Tassell, J., & Yoder, M. (2015). Professional noticing: Developing responsive mathematics teaching. *Teaching Children Mathematics, 21*(5), 294–303.

Thompson, D. R., & Chappell, M. F. (2007). Communication and representation as elements in mathematical literacy. *Reading & Writing Quarterly, 23*(2), 179–196.

Thompson, D. R., & Rubenstein, R. N. (2000). Learning mathematics vocabulary: Potential pitfalls and instructional strategies. *The Mathematics Teacher, 93*(7), 568–574.

Thompson, A. G. (1992). Teachers' beliefs and conceptions: A synthesis of the research. In D. A. Grouws (Ed.), *Handbook of research on mathematics teaching and learning: A project of the National Council of Teachers of Mathematics* (pp. 127–146). New York, NY: Macmillan Publishing Co, Inc.

Thompson, P. W., Philipp, R., & Boyd, B. (1994). Calculational and conceptual orientations in teaching mathematics. In A. Coxford (Ed.), *Professional development for teachers of mathematics. Yearbook of the NCTM* (pp. 79–92). Reston, VA: National Council of Teachers of Mathematics.

Tirosh, D. (2000). Enhancing prospective teachers' knowledge of children's conceptions: The case of division of fractions. *Journal for Research in Mathematics Education, 31*(1), 5–25.

Tirosh, D., Tsamir, P., Levenson, E., & Tabach, M. (2011). From preschool teachers' professional development to children's knowledge: comparing sets. *Journal of Mathematics Teacher Education, 14*, 113–131.

Usiskin, Z., & Willmore, E. (Eds.). (2008). *Mathematics curriculum in Pacific Rim Countries China, Japan, Korea, and Singapore: Proceedings of a Conference*. Charlotte, NC: Information Age Publishing.

Vacc, N., & Bright, G. W. (1999). Elementary preservice teachers' changing beliefs and instructional use of children's mathematical thinking. *Journal for Research in Mathematics Education, 30*(1), 89–110.

Vale, C. (2010). Supporting "out-of-field" teachers of secondary mathematics. *Australian Mathematics Teacher, 66*(1), 17–24.

van der Walt, M. (2009). Study orientation and basic vocabulary in mathematics in primary school. *South African Journal of Science and Technology, 28*, 378–392.

van der Walt, M., Maree, K., & Ellis, S. (2008). A mathematics vocabulary questionnaire for use in the intermediate phase. *South African Journal of Education, 28*(4), 489–504.

Vinson, B. M. (2001). A comparison of preservice teachers' mathematics anxiety before and after a methods class emphasizing manipulatives. *Early Childhood Education Journal, 29*(2), 89–94.

Weiss, C. H., Murphy-Graham, E., & Birkeland, S. (2005). An alternate route to policy influence: How evaluations affect DARE. *American Journal of Evaluation, 26*(1), 12–30.

Weiss, I. R., Banilower, E. R., McMahon, K. C., & Smith, P. S. (2001). *Report of the 2000 national survey of science and mathematics education.* Chapel Hill, NC: Horizon Research.

Wenta, R. G. (2000). *Efficacy of preservice elementary mathematics teachers* (Unpublished doctoral thesis). Indiana University, Bloomington, IN.

Whitin, D. J., & Whitin, P. (2003). Talk counts: Discussing graphs with young children. *Teaching Children Mathematics, 10*(3), 142–150.

Wideen, M., Mayer-Smith, J., & Moon, B. (1998). A critical analysis of the research on learning to teach: Making the case for an ecological perspective on inquiry. *Review of Educational Research, 68*(2), 130–178.

Wilson, S. M., & Wineburg, S. S. (1988). Peering at history through different lenses: The role of disciplinary perspectives in teaching history. *Teachers College Record, 89*(4), 525–539.

Zaslavsky, O. (2007). Mathematics-related tasks, teacher education, and teacher educators. *Journal of Mathematics Teacher Education, 10*(4), 433–440.

Zaslavsky, O. (2009). Mathematics educators' knowledge and development. In R. Even & D. L. Ball (Eds.), *The professional education and development of teachers of mathematics* (pp. 105–111). New York, NY: Springer.

Aina Appova
College of Education and Human Ecology
The Ohio State University

YINGKANG WU, YILING YAO AND JINFA CAI

9. LEARNING TO BE MATHEMATICS TEACHER EDUCATORS

From Professional Practice to Personal Development

This chapter capitalizes on the strengths of recent research on effective teacher professional development to address how mathematics teacher educators learn from their practice to function as teacher educators and how their practice can be supported to promote their development. Specifically, we examine the development of mathematics teacher educators with a focus on practice as a means for changing their knowledge and beliefs. The chapter aims to provide new insights on mathematics teacher educators' development through their practice by systematically reviewing and analysing literature in the area. Based on the literature, we identified five features of teacher professional development: a focus on content, building on student learning and thinking, closely aligning with (classroom teaching) practice, building a learning community, and professional development that is ongoing. In the context of (mathematics) teacher educators' development through practice, these five features were expanded in terms of both their meaning and scope and used as a framework for our selection, review, and analysis of literature related to mathematics teacher educators' development through their practice. The chapter also discusses methodological issues of research in the area and points out future directions of research related to mathematics teacher educators' development through practice.

INTRODUCTION

Mathematics teachers are generally recognized to be key to students' opportunities to learn mathematics (Even & Ball, 2009; Even & Krainer, 2014). Thus, mathematics teachers and their work have been studied extensively, from small-scale studies to large-scale investigations that have examined mathematics classrooms around the world (e.g., Clarke, Keitel, & Shimizu, 2006). Teacher learning has also been the subject of sustained international studies (Darling-Hammond & Cobb, 1995; Tatto et al., 2012). Correspondingly, mathematics teacher educators are key to teachers' opportunities to learn to teach mathematics. However, despite the significant resources that have been devoted to studying teaching and teacher learning, studies of teacher educators in general and mathematics teacher educators in particular have only just recently emerged and begun evolving (Adler, Ball, Krainer, Lin, & Novotna,

© KONINKLIJKE BRILL NV, LEIDEN, 2020 | DOI: 10.1163/9789004424210_010

2005; Even & Ball, 2009; Jaworski & Wood, 2008; Tzur, 2001; Wang, Spalding, Odell, Klecka, & Lin, 2010). The education of mathematics teacher educators is critical for professional development of mathematics teachers (Even & Krainer, 2014). However, there is little research about the development of mathematics teacher educators and about effective ways to support educators to initiate, guide, and facilitate teacher learning (Even, 2008).

Recent years have seen an increased interest in research on mathematics teacher educators' learning (Krainer, Chapman, & Zaslavsky, 2014). Most opportunities for mathematics teacher educators to learn are not offered in formal courses (Even & Krainer, 2014). Instead, most of them become mathematics teacher educators through reflections from their own professional practices. Zaslavsky (2008) proposed a model that describes the roles of teacher educators as facilitators and designers of mathematics tasks to foster teacher learning and at the same time captures the dynamic nature of mathematics teacher educators' learning from reflecting on their own teaching practice with teachers. Self-study is considered as a form of reflection on practice and is defined as intentional and systematic inquiry of one's own practice. It is conducted by individual teacher educators as well as groups working collaboratively to understand problems of practice more deeply, and it is argued to serve a dual purpose by functioning as a means to promote reflective teaching and as a substantive end of teacher education (Dinkelman, 2003).

The accumulation of research studies in recent years on teacher educators in general and mathematics teacher educators in particular creates an opportunity to step back and examine what is known about teacher educators' learning. In this chapter, we review research studies published from 2001 to 2018 to synthesize what we know about teacher educators' learning from their professional practice. More specifically, this chapter aims to address the following two research questions: (1) How do mathematics teacher educators learn to function as teacher educators from their professional practices? and (2) how can mathematics teacher educators' practices be supported to promote their development? Answers to these two questions will provide significant insights on mathematics teacher educators' development through their practices by systematically reviewing and analysing relevant literature in the area.

CONCEPTUAL FRAMEWORK: PROFESSIONAL PRACTICE OF TEACHER EDUCATORS

Teaching is one of the main responsibilities of teacher educators. For example, the Association of Teacher Educators (2008) established a broad set of nine standards for teacher educators that includes teaching, cultural competence, scholarship, professional development, program development, collaboration, public advocacy, contributing to improving the teacher education profession, and vision. The primacy of teaching in those standards reflects a deep-seated commitment to teaching in the profession. Troyer (1986) noted that teacher educators were strongly focused on teaching, with a typical teaching load being three to four courses per academic period. Nearly 20 years later, Koster, Brekelmens, Korthagen, and Wubbels (2005) reported that teacher educators identified a host of teaching-related tasks as among the most important tasks that they performed in their work. A focus on teaching has clearly continued to dominate the activity of many teacher educators. Hence, we have situated our analysis of the literature related to (mathematics) teacher educators within their teaching-related professional practice, which is defined broadly by five elements shown in the left diagram in Figure 9.1.

Teaching responsibilities related to the preparation of prospective teachers and professional development (PD in Figure 9.1) for practising teachers are at the center of the diagram, indicating that these two elements are most prominent and at the core position among all the identified professional practices. Although both elements involve working with teachers, the design and implementation of the teaching activities with respect to prospective teachers and practising teachers are obviously different. These two elements are, therefore, separated and regarded as independent but related. The other three elements – school teaching, research, and teacher educators' education and professional development – are regarded as supporting elements of the central ones.

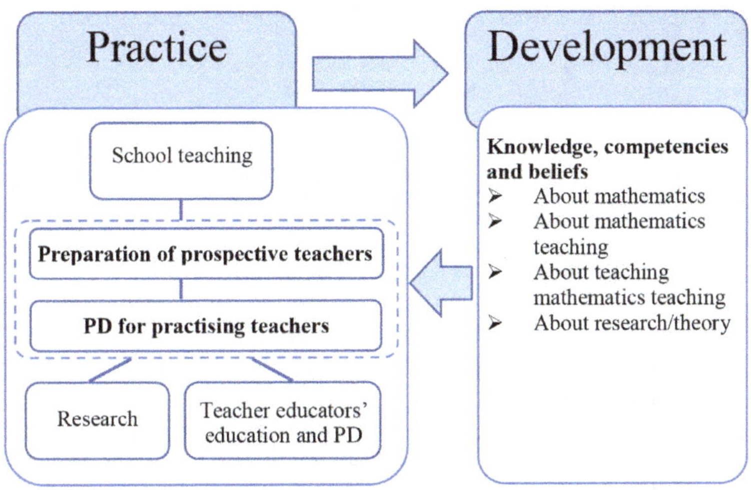

Figure 9.1. Conceptual framework

School-teaching experience is an indicator of teacher knowledge structure (Borko & Livingston, 1989) and is closely related to the teaching responsibilities of teacher educators, regardless of whether they are preparing prospective teachers or guiding practising teachers, because teacher knowledge is taken as an indispensable component of teacher educator knowledge as suggested by Perks and Prestage (2008). It has also been reported that prior school-teaching experience might have an effect on the ways that mathematics teacher educators perceive and deal with challenges in their teaching of mathematics method courses (Wu, Hwang, & Cai, 2017). Research articles analyzed for this chapter are confined to those related to the teaching responsibilities of teacher educators without considering mathematics education research in a general sense. Results from research focused on the teaching practice of teacher educators could bring insights and validated practices to support and improve the teaching work of mathematics teacher educators. The practice of teacher educators' education and professional development refers to any formal courses or informal activities that aim to develop teacher educators' teaching-related knowledge and competencies, cultivate productive beliefs, and improve their teaching practice. In fact, the five elements of professional practice are interconnected and can be mutually supportive of each other.

We view practice and development as a spiral process in mathematics teacher educators' learning to become mathematics teacher educators. As indicated by the arrows in Figure 9.1, knowledge, competencies, and beliefs can serve as the foundation to support teacher educators' practice, and practice can serve as a means and lead to changes in their knowledge, competencies, and beliefs, whereas changes in knowledge, competencies, and beliefs can further lead to changes in practice. In this chapter, however, our discussion will focus on practice as a means to promote mathematics teacher educators' learning and development.

METHODOLOGY

In this section, we report our methods for identifying, summarizing, and coding research articles related to (mathematics) teacher educators and for synthesizing how mathematics teacher educators learn and develop from their practices.

Identifying Research Articles

Step 1: Search. We conducted searches in Springer Link database[1] and China Academic Library and Information System (CALIS) Foreign Journal Web[2] using the key terms "mathematics teacher educator" and "teacher educator," respectively. Approved by the State Council of China, CALIS has devoted itself to the construction and sharing of information resources for higher education in China (Xiao & Chen, 2005). CALIS Foreign Journal Web includes a collection of more than 14,000 journals and represents an optimal approach to find non-Chinese journal papers. We specified nine academic journals as listed in Table 9.1 when conducting search in CALIS foreign journal web. These nine English-language journals consist of seven journals on mathematics education, all of which have been recognized as very high quality (Toerner & Arzarello, 2012), and the top two journals on general teacher education. Using this search, we obtained 33 (using "mathematics teacher educator"

Table 9.1. The nine research journals

Number	*Title*
1	Educational Studies in Mathematics
2	Journal for Research in Mathematics Education
3	Mathematics Thinking and Learning
4	ZDM – Mathematics Education
5	Journal of Mathematics Teacher Education
6	Journal of Mathematical Behavior
7	International Journal of Science and Mathematics Education
8	Teaching & Teacher Education
9	Journal of Teacher Education

as the key term) and 228 (using "teacher educator" as the key term) articles spanning 1974 to 2018.

Step 2: Selection. The search via Springer Link database resulted in articles from a variety of journals and books. To ensure that they met high standards of academic quality and could be reviewed in a timely manner, we narrowed the selection of articles to those published in 2001 and later within the nine journals shown in Table 9.1. This selection process resulted in a total of 86 articles from the original 261 obtained from both database searches. We reviewed the 86 articles and excluded four review articles, nine editorials, eight discussion articles with no supporting data, and two articles irrelevant to the practices and development of (mathematics) teacher educators, leaving a remainder of 63 research articles that included data.

Step 3: Supplement. We conducted the database searches again, using the key terms "teaching researcher" and "teaching research specialist," to include as many research articles related to teacher educators as possible. Teaching researcher or teaching research specialist is a unique type of teacher educator in China who provides practical guidance mainly to practising teachers (Huang et al., 2017; Paine, Fang, & Jiang, 2015). We followed the same steps as described above and obtained two additional research articles related to mathematics teaching researchers, both published in *ZDM – Mathematics Education.* Thus, the total number of articles included for review in this chapter is 65. Details of the 65 articles are provided in the Appendix.

Summarizing and Coding the Research Articles

Two of the authors read, summarized, and coded the 65 articles. Each article was summarized and coded in a spreadsheet with respect to key features, including publication year, author, journal, key words, participants, research method, the professional practice involved, description of the practice, aim, and main findings of the study. In addition, we took notes on each article, recording the theoretical orientation and a summary of any major themes suggested by the study. Participants were coded according to their number, country, and specialized subject discipline – that is, whether the area of teacher educator(s) studied was mathematics, another discipline, or general education. Research methods were coded as qualitative, quantitative, or mixed. If the article was not related to any of the five elements of practice shown in Figure 9.1, it was coded as Others.

By referring to the coded data and the recorded notes, the same two authors checked each other's codes and finalized the codes for the 65 articles. Inconsistencies were resolved through discussion and justification until consensus was reached. The aim of this coding process was to acquire essential information about each article.

Analysing, Aggregating, and Deriving Themes

We combined the 65 articles according to the five elements of practice and reviewed the coded data and the notes carefully for the purpose of aggregating and deriving themes to capture the main results in each element. Subthemes were identified where necessary, and we returned to the research articles for any needed clarification. Similar to the coding process, two authors worked together to derive the themes in each element of practice and any inconsistencies were again resolved through discussion and justification until consensus reached.

For example, 38 articles among the 65 are related to the preparation of prospective teachers' practice. The emergent codes for these 38 articles were reviewed and grouped, resulting in five themes that represent the reported study results. These five themes are as follows: (1) *nature of the practice*; (2) *complexity of the practice*; (3) *knowledge, competencies, and beliefs required for being a teacher educator*; (4) *goals and strategies in teaching prospective teachers*; and (5) *collaboration within and across communities*. We present the details of the identified themes for the other four elements of practice in the next section.

RESULTS

In this section, we first report the overall characteristics of the 65 articles and then report the results for each practice.

Overall Characteristics of the Articles

Table 9.2 summarizes the methodological approach, teacher educators' specialized subject area, and numbers of teacher educators studied in the 65 reviewed articles. Most of the studies (87.7%) used qualitative methods (de Freitas, Lerman, & Parks, 2016) for data collection and analysis, including case study, self-study, action research, interviews, observation, video and audio transcripts, discourse analysis, the phenomenographical approach, and the narrative approach. Only three (4.6%) and five (7.7%) studies adopted quantitative and mixed methods, respectively. Moreover, the sample sizes of the studies were small, with nearly 70% of the studies having less than or equal to 20 teacher educators participating in the study and 20% of the studies having only one teacher educator as the subject. This is consistent with the observation that most of the studies used qualitative methods in which thick and intensive description of the cases was required. In addition, 28 (43.1%) of the 65 studies involved mathematics teacher educators, with 24 (36.9%) articles related to teacher educators in other subjects such as English, science, history, and literacy and 13 (20%) articles about teacher educators in general. Considering that the actual number of articles about mathematics teacher educators was quite small, all of the 65 articles were retained for analysis to provide a comprehensive review. Moreover, we aimed to make comparisons between research studies on mathematics teacher educators and those on other teacher educators to identify any existing differences, so as to offer insights into the particularities of mathematics teacher educators' learning from practice.

Table 9.2. Number of articles with respect to methodological approaches used, teacher educators' specialized subject area, and number of teacher educators studied

Methodological approach		
Qualitative	Quantitative	Mixed
57	3	5

Teacher educators' specialized subject area		
Mathematics	Other subject	General
28	24	13

Number of teacher educators studied						
1	2–5	6–10	11–20	21–100	>100	No information
13	19	7	6	13	5	2

Table 9.3 shows the number of articles within each of the five elements of teacher educators' teaching-related practices as defined in Figure 9.1. Among the 65 articles, five articles were identified as not directly related to teacher educators' teaching-related practices. These five articles addressed boundary-crossing experiences (Trent, 2013), teacher educators' research engagement (Borg & Alshumaimeri, 2012), views about teacher competency (Pantić & Wubbels, 2010), the demand and supply of United States teacher educators (Twombly, Wolf-Wendel, Williams, & Green, 2006), and teacher educators' identity formation (Bullough, 2005).

Table 9.3. Number of articles in each practice category

Element of practice	*Math teacher educator*	*Other teacher educator*	*Overall*
Teaching-related practices			
Preparation of prospective teachers	16	22	38
Professional development for practising teachers	10	3	13
School teaching	3	3	6
Research	10	9	19
Teacher educators' education and professional development	1	5	6
Others	0	5	5

Note: Individual articles could be assigned to multiple practice categories.

The other 60 articles were categorized according to the five elements of teaching-related professional practice. It should be noted that individual articles could be assigned to multiple practice categories. Thirty-eight articles were assigned to the category of preparation of prospective teachers practice, indicating that research on teacher educators' teaching work in preparing prospective teachers is the most prominent. In contrast with this, only six articles addressed school teaching practice and teacher educators' education and professional development, respectively, indicating that these two elements of practice have not attracted much attention in research related to teacher educators in general and mathematics teacher educators in particular.

Preparation of Prospective Teachers

The practice category containing the most articles was the preparation of prospective teachers category, indicating that this practice is the most emphasized among the five elements of practice. Among the 38 articles in this category, 16 are about mathematics teacher educators and the other 22 about other teacher educators. This practice seems to generally involve university-based teacher educators. We derived five themes to synthesize the research results of these 38 studies, as shown in Table 9.4.

Theme 1: Nature of the practice. The five studies in this theme sketch the contours of teacher educators' work in preparing prospective teachers. As suggested by Ellis, McNicholl, Blake, and McNally (2014), teaching and maintaining the good relationships with schools necessary for teaching have intensified to become a defining characteristic of teacher educators' work. Relationship maintenance involves building, sustaining, and repairing the networks of personal relationships

Table 9.4. Number of articles with results related to themes of preparation of prospective teachers practice

Themes	*Number of articles*		
	Math teacher educator	*Other teacher educator*	*Overall*
1. Nature of the practice	1	4	5
2. Complexity of the practice	5	4	9
3. Knowledge, competencies, and beliefs required for being a teacher educator	5	10	15
4. Goals and strategies in teaching prospective teachers	10	7	17
5. Collaboration within and across communities	2	2	4

Note: Individual articles could be assigned to multiple themes.

which allow teacher education programs and school partnerships to function smoothly. Williams (2014) reported that university-based teacher educators are actually working in the third space between universities and schools in the provision of teacher preparation programs and need to negotiate complex and sometimes difficult professional relationships with schools. Similarly, Wu et al. (2017) reported that mathematics teacher educators from Chinese universities were also involved in dealing with logistical issues related to supervising student teaching, such as mediating the relationship between student teachers and their cooperating teachers, and negotiating with schools to provide student teachers with more opportunities to deliver mathematics lessons. Therefore, the uniqueness of the preparation of prospective teachers practice of teacher educators relies in its teaching-dominant nature and the increasing demands that they work with schools compared to other academic staff working at universities.

However, there is apparently a shortage of structures that could help teacher educators to develop their practice in this regard. Berry and Van Driel (2013) found that teacher educators' personal background and individual career paths played out quite differently in their teaching of prospective teachers, and they explained that this was related to the absence of an induction program or planned professional learning for teacher educators. Their view is confirmed by Goodwin et al. (2014), who reported that there were no codified or coherent courses integrated into the preparation of doctoral students, especially those intent on a professional life as a teacher educator.

Theme 2: Complexity of the practice. The nine articles in this theme characterize the complexity of the preparation of prospective teachers practice in two ways. First, teacher educators must cope with a variety of external challenges in their teaching work. Murray and Male (2005) reported that one of the challenges encountered in their transition from schoolteachers to university-based teacher educators is in developing a pedagogy for higher education teaching work. As stated in the standards for teacher educators (Association of Teacher Educators, 2008), teacher educators need to apply for culturally relevant pedagogy in teacher education to promote social justice. Specifically, Galman, Pica-Smith, and Rosenberger (2010) reported that teacher educators need to pay attention to antiracist pedagogy in their own teaching practice for the purpose of transforming prospective teachers' beliefs and developing practice with antiracist pedagogy. Meanwhile, Willemse, Lunenberg, and Korthagen (2005) explored the preparation of prospective teachers for moral education in the Netherlands and found that the process of preparing prospective teachers for moral education remained largely implicit and that the practices of teacher educators were hardly directed by any systematic or critical analysis of the relations between goals, objectives, teaching and learning methods, and outcomes when implementing the moral aspects embedded in their teacher education curriculum. Tillema and Kremer-Hayon (2002) investigated how Dutch and Israeli teacher educators were committed to promoting self-regulated learning in their prospective teachers and noted several professional dilemmas, such as the dilemma of theory-centered teaching versus practice-centered teaching and dilemmas of teacher delivery versus student initiation, which complicated their teaching work.

Besides the challenge of developing a pedagogy for higher education teacher work, teacher educators also experience pressures related to ensuring the quality of teacher education programs. Based on TEDS-M data, Hsieh et al. (2011) examined the views of prospective teachers and teacher educators toward mathematics teacher education quality, reporting that in all of the countries involved, prospective teachers were less approving of the courses or content arrangement of teacher education programs than were teacher educators, thus perhaps lowering educators' motivation to improve the arrangement.

Second, teacher educators need to draw on and transition between internal resources to deal with the nested structure of teaching triads (Jaworski, 1992), with the teaching triad of mathematics teachers embedded as the content component of the teaching triad of teacher educators when working with prospective teachers (Zaslavsky & Leikin, 2004). Leikin, Zazkis, and Meller (2018) extended this mathematics teacher educator teaching triad by incorporating mathematics challenges for teachers as another challenging content piece besides teachers' existing didactical challenges. This suggests that mathematics teacher educators need to make use of mathematics knowledge and mathematics pedagogical knowledge to function as mathematics teachers as well as didactical knowledge for teaching mathematics teaching to function as mathematics teacher educators. This difference was also noticed in the study by Wu et al. (2017), who pointed out that the role of a mathematics teacher educator is more demanding than the role of a mathematics teacher. Moreover, Chick and Beswick (2018) conducted a self-study to examine one of the authors' own teaching practices with prospective teachers, finding cases that demonstrated that mathematics teacher educators' work may require them to simultaneously and flexibly utilize both mathematics teachers' pedagogical content knowledge and mathematics teacher educators' pedagogical content knowledge while they teach.

Theme 3: Knowledge, competencies, and beliefs required for being a teacher educator.

- Knowledge and competencies

At present there is no consensus on the knowledge and competencies required to be a teacher educator. Using case studies, John (2002) revealed teacher educators' knowledge to be characterized by a number of dimensions, including intentionality, practicality, subject specificity, and ethicality. Smith (2005) surveyed 40 novice teachers and 18 teacher educators to investigate their view of professional knowledge for teacher educators, finding that subject-matter knowledge, pedagogical and didactic knowledge, and knowledge of interpersonal communication were considered to be very important. Based on personal experience as a school teacher and a university-based teacher educator, Zeichner (2005) suggested that teacher educators need to have knowledge about conducting self-study of their practices and about conceptual and empirical literature on teacher education. Using a Delphi study approach, Koster et al. (2005) identified four areas of competencies regarded as necessary to teacher educators: content competencies, communicative and reflective competencies, organizational competencies, and pedagogical competencies. These findings all provide theoretical implications for the knowledge base for teacher educators.

Goodwin et al. (2014) surveyed 293 teacher educators to identify the knowledge essential to teacher educating. Using Cochran-Smith and Lytle's (1999) theory of "relationships of knowledge and practice," they found that there was a lack of knowledge-for-teacher educating practice and that knowledge-in-teacher educating practice had become a somewhat haphazard and dysfunctional process. This implies that the field of teacher education has paid little attention to what teacher educators should know and be able to do, although discussions on the conception of knowledge and competencies for teacher educators have appeared gradually.

Efforts have been made to explore the specific knowledge required for being a mathematics teacher educator by examining their teaching practices. Zazkis and Zazkis (2011) used interview data with five mathematics teacher educators to exemplify how mathematics teacher educators' mathematical knowledge contributed to their teaching, including in their task design, their choice of problem-solving activities, and their responses to students' questions. Leikin et al. (2018) noted that mathematics teacher educators must cope with mathematical challenges when teaching prospective teachers, including mathematical aspects such as mathematical contents and concepts, particular problem-solving strategies and proof techniques; and meta-mathematical aspects such as the rigor of the mathematical language, the essence of proofs, the meaning of theorems and definitions, the development of logical thinking, and abstraction. Castro Superfine and Li (2014) explored the knowledge demands of mathematics school teacher educators as they teach mathematics content to prospective elementary teachers and identified three forms of knowledge that are different from knowledge that teachers use in their work. These forms of knowledge include knowledge of connecting student errors to instructional moves, connecting algorithms to the school curriculum, and connecting research to mathematics content learning.

In addition to the aforementioned research studies exploring mathematics teacher educators' mathematical knowledge requirements, studies exist that discuss mathematics teacher educators' pedagogical content knowledge and research knowledge. Chick and Beswick (2018) presented a framework for the pedagogical content knowledge required for mathematics teacher educators as they work to develop prospective teachers' pedagogical content knowledge for teaching mathematics. This framework builds on the existing research on pedagogical content knowledge and categorizes aspects of work of teacher education. Chauvot (2009) investigated the knowledge content, structure, and growth of a novice mathematics teacher educator–researcher using a self-study approach. By examining two contexts – teaching university courses and mentoring doctoral students – she drew a knowledge map to illustrate the concept of mathematics teacher educator-researcher knowledge. Specifically, she proposed and emphasized the knowledge of conducting, interpreting, and writing about research as a unifying component of mathematics teacher educator-researcher knowledge.

- Beliefs

Among the five articles coded to beliefs, three are related to identity. These articles stressed a need to develop an identity as a teacher educator and researcher (Doecke, 2004; Murray & Male, 2005; Williams, 2014). Williams (2014) identified working in the third space (between universities and schools) and reflecting on the implications of this work on their pedagogy and identity as teacher educators as a substantial factor which could enable teachers to make the transition to being teacher educators.

Based on qualitative data from 12 experienced Flemish educators, Vanassche and Kelchtermans (2014) revealed three types of teacher–educator positioning, each of which constituted "a coherent pattern of normative beliefs about good teaching and teacher education, the preferred relation with prospective teachers, and valuable approaches and strategies to enact these assumptions in practice" (p. 125). They based their work on the assumption that teacher educators' beliefs and practices are consistent. However, Ariza, Pozo, and Toscano (2002) investigated the conceptions of 28 Spanish teacher educators on the principles, contents, methods, and evaluation of ongoing teacher education and reported major contradictions between their conceptions and practices. They explained that this might be because the teacher educators experienced difficulty in transforming their conceptions on teacher education into coherent, procedural enactments.

Theme 4: Goals and strategies in teaching prospective teachers.

- Goals

Four articles were related to the goals that teacher educators possess in teaching teacher-education courses. All of them expressed similar results, which were that teacher educators established goals to focus on conceptual understanding of the content and to develop pedagogical content knowledge to engage school students' learning (Appova & Taylor, 2019; Berry & Van Driel, 2013; Leikin et al., 2018; Li & Castro Superfine, 2018). For example, Li and Castro Superfine identified four goals of mathematics content courses for prospective elementary school teachers from the perspective of six mathematics teacher educators: (1) help prospective teachers develop conceptual understanding of the content that they will teach, (2) expand the mathematics content that is needed for teaching, (3) ensure that prospective teachers have a positive experience with mathematics, and (4) create collaborative and safe learning environments.

- Strategies

Various instructional strategies were reported in the 17 articles coded to this theme. We grouped them into three categories. The first category is modelling good teaching. In fact, modelling effective instruction to meet the needs of diverse learners is one of the teaching standards in the document *Standards for Teacher Educators* (Association of Teacher Educators, 2008). Lunenberg, Korthagen, and Swennen (2007) defined modelling by teacher educators as "the practice of intentionally displaying certain teaching behavior with the aim of promoting student teachers' professional learning" (p. 589), and they found that teacher educators lacked the knowledge and skills needed to use modelling in a productive way. However, Kaufman (2009) reported perceived academic and affective benefits for prospective teachers when adopting modelling practices in his teaching. Loughran and Berry (2005) illustrated how a teacher educator began to conceptualize the practice of explicitly modelling through the collaborative self-study approach. Li and Castro Superfine (2018) reported mathematics teacher educators' use of modelling to connect prospective teachers' mathematics learning to teaching practice.

The second category of strategies is implementing worthwhile tasks in teacher education classrooms. To deepen prospective teachers' conceptual understanding of mathematics, mathematics teacher educators focused on mathematical tasks, including multiple ways of solving problems, and the use of hands-on tasks, and tried to maintain the cognitive demand of high-level tasks (Li & Castro Superfine, 2018; Masingila, Olanoff, & Kimani, 2017). Using various artifacts of learning and teaching to develop prospective teachers' pedagogical content knowledge is another type of task in teacher education practice. The artifacts could be videos of children doing mathematics or solving mathematics problems in classroom settings or interview settings (Appova & Taylor, 2019; Li & Castro Superfine, 2018); anecdotal stories about what children would think about important mathematical concepts, especially their misconceptions or cognitive challenges (Li & Castro Superfine, 2018); or multimedia cases of school teaching practice (Doerr & Thompson, 2004; Wu et al., 2017), which provide a site for analysis, reflection, and reasoning about the practices.

The third category of strategies is providing prospective teachers with opportunities to experience enactment and reflection processes, which are fundamental to teachers' growth (Clarke & Hollingsworth, 2002). Having prospective teachers practice components of teaching was the most common strategy in teacher education (e.g., Berry & Van Driel, 2013; Brock, Moore, & Parks, 2007; Kinach, 2002; Li & Castro Superfine, 2018; Masingila et al., 2017). For example, Peled and Hershkovitz (2004) asked prospective teachers to solve a standard application problem and compare their solutions; then, in a class discussion setting, they were encouraged to evaluate a variety of children's answers to the same problem from interviews. Group discussion, in the case of the standard application problem, supported teachers to better understand why the solution was relevant and how it was realized in the specific situation; it also functioned as a model for teachers to create a similar environment for their own students to develop deeper mathematical understanding.

The development of prospective teachers' reflective thinking is a goal of many teacher education programs (Hatton & Smith, 1995). At the same time, reflection is a critical means to support teachers' learning, as shown by many studies' findings (e.g., Brock et al. 2007; Lunenberg & Korthagen, 2003). Based on a self-study, Tzur (2001) suggested that reflecting on the activity–effect relationship was a mechanism that promoted his own growth. A variety of artifacts of reflection such as journal writing and discussion were reported in studies of teacher education practice. For example Gelfuso (2017) examined the intentional language use of a teacher educator as she engaged in reflective conversations with prospective teachers, and Goodell (2006) asked prospective mathematics teachers to discuss, analyze, and write critical incidents encountered in their practicum and found that this activity provided a structured framework for developing prospective teachers' as well as his own reflective-practice skills.

Theme 5: Collaboration within and across communities. Four articles were coded to this theme, with two about collaboration within a community and the other two about collaboration across communities. In the study by Masingila et al. (2017), two novice mathematics teacher educators and an expert mathematics teacher educator worked together and reflected on their teaching in a community of practice while helping prospective mathematics teachers to develop their mathematical knowledge for teaching. Gallagher, Griffin, Parker, Kitchen, and Figg (2011) examined the professional development of teacher educators through the establishment of a self-study community of practice. Both studies reported that the community of practice provides a safe space for teacher educators to discuss, share, and more deeply comprehend relevant issues in their teacher education practice.

Bleiler (2015) investigated the collaboration between a mathematician and a mathematics teacher educator in their team teaching of a mathematics content and methods course for prospective secondary school mathematics teachers. She found that the collaboration and participation in the practice of "the other" enabled the mathematician and the mathematics teacher educator to increase the awareness of their own practice and the practices characterizing their respective communities. Williams (2014) examined how university-based teacher educators managed to work with mentor or cooperating teachers in schools and suggested that collaborations in the third space between universities and schools provided an opportunity for all participants to work together to gain new knowledge and understandings of teaching and learning and to develop boundary practices to enhance the learning of teacher educators, school teachers, student teachers, and school students. These findings illustrate the potential of collaborations within and across communities as a form of professional development for teacher educators.

Professional Development for Practising Teachers

The professional development for practising teachers practice category includes 13 articles, 10 of which address mathematics teacher educators and the other three of which address other teacher educators. We further coded them to three themes, as shown in Table 9.5.

Table 9.5. Number of articles with results related to themes of professional development for practising teachers practice

Themes	*Number of articles*		
	Math teacher educator	*Other teacher educator*	*Overall*
1. Various perspectives in designing and organizing professional development activities	8	2	10
2. Strategies in working with practising teachers	8	3	11
3. Developing an identity of teacher educator	1	1	2

Note: Individual articles could be assigned to multiple themes.

Theme 1: Various perspectives in designing and organizing professional development activities. Ten articles were coded to this theme. Among them, eight articles are about mathematics teacher educators and the other two are about teacher educators in general. Various perspectives in designing and organizing professional development activities were observed, including using theory to inform teaching practice, providing tasks and resources to teachers, and working in a community of practice.

Lin, Yang, Hsu, and Chen (2018) investigated mathematics teacher educator-researchers' perspectives on the use of theory in facilitating teacher growth based on the framework of the theory-centered scholarship triangle. Qualitative analysis revealed three distinct types of perspectives: a perspective on research, a perspective on practice, and a perspective on the connection between research and practice. The results showed that mathematics teacher educators-researchers tended to use theory to solve teachers' pedagogical problems and emphasized the importance of context in the facilitation of teacher learning. Gu and Gu (2016) examined the characteristics of mathematics teaching researchers' mentoring of practising teachers in the context of Chinese lesson study.

Furthermore, several papers reported that teacher educators provided or designed tasks and resources for teachers. Based on the activity theory, Chen, Lin, and Yang (2018) designed and revised the tools for a mathematics teacher educators-researcher's planned and implemented design-based professional development workshops for practising mathematics teachers. Yang, Hsu, Lin, Chen, and Cheng (2015) explored the educative power of an experienced mathematics teacher educator-researcher who provided his insights and strategies on design-based teacher professional development programs. In this program, a mathematics teacher educator-researcher mainly guided teachers to experience the process of the hypothetical learning cycle and to develop mathematics teachers' professional practice on communication, reasoning, and connection. Similarly, Peled and Hershkovitz (2004) examined how mathematics teacher educators made teachers working on a proportional reasoning problem aware of the nature of students' alternative solutions, which may be correct or incorrect.

Lastly, five papers described action research and communities of practice as important and effective professional development techniques for practising teachers (Hopkins & Spillane, 2014; Huang, Su, & Xu, 2014; Ponte, Ax, Beijaard, & Wubbels, 2004; Sakonidis & Potari, 2014; Zaslavsky & Leikin, 2004). We review the specific practices described in these papers in theme 2.

Theme 2: Strategies for working with practising teachers. Eleven articles were coded to this theme. We further categorized them along four perspectives according to their types of education practice: implementing design-based tasks in teacher education, conducting action research or related teaching research in teacher education, collaborating with teachers or teacher educators, and using online resources.

The first category in this theme is implementing worthwhile design-based tasks in teacher education. There are several ways for mathematics teacher educators to improve teachers' mathematical power which pertains to the ability to draw on knowledge that is required to improve the mathematical and pedagogical power of teachers (Zaslavsky, Chapman, & Leikin, 2003) and develop teachers' task design and implementation as well as their ability to understand students' mathematical thinking. For example, mathematics teacher educators focused on designing a hypothetical learning cycle; providing a workshop for diagnostic conjecturing tasks with learner-centered activities; and using alternative viewpoints to analyze students' answers, regardless of whether they were correct or incorrect to mathematics problems (Chen et al., 2018; Peled & Hershkovitz, 2004; Yang et al., 2015).

The second category is conducting action research or related teaching research in teacher education. Ponte et al. (2004) reported three phases in the action research for teachers' professional knowledge, including teacher educators encouraging the teachers to develop their own interpretations, solutions, and additions to the content; translating the teachers' questions about content into questions about how action research can be used to tackle these issues; and communicating with teachers about their concrete experiences. Hopkins and Spillane (2014) explored teacher educators in the school as a site for teacher education. They examined how beginning primary school teachers' learning about instruction was supported by school-based teacher educators via advice- and information-seeking and providing interactions. They found that a supportive school culture and interaction among school staff members of the same school subject in the formal organizational structures inside school were important in developing novice teachers' instructional ability.

The third category emphasizes that collaboration is an important strategy for practising teachers' professional development. Zaslavsky and Leikin (2004) provided teachers with opportunities to collaborate with mathematics teacher educators on specific mathematical tasks during professional development workshops and characterized the mathematics teacher educators' professional growth along the four dimensions of action, reflection, autonomy, and networking. They identified reflection, self-analysis, and collaboration with teachers as benefits for teacher educators' professional development. Similarly, Sakonidis and Potari (2014) proposed that mathematics teacher educators collaborate with mathematics teachers in the context of a community of inquiry and methods courses. Furthermore, Huang et al. (2014) examined the co-learning of practising mathematics teachers and mathematics teaching researchers through parallel lesson study in China. They reported that mathematics teacher researchers developed their professional competencies by carrying out teaching research activities, mentoring teachers, and deepening their own understanding of teaching.

Lastly, Gueudet, Sacristán, Soury-Lavergne, and Trouche (2012) considered the interaction between mathematics teacher educators and the resources that they use for their online teaching work. They discussed two teaching paths by two educator teams – individualization and inquiry with dynamic geometry – as resources for teacher educators. The individualization path involves a "scenario grid," which is associated with two other grids: one for the observation of a lesson and one for the report of the lesson. In the "inquiry" path, the lesson description is inserted into a more general "resource template," which includes a student sheet, some post-test reports, and some examples of students' productions.

In summary, few studies addressed teacher educators' knowledge of professional development for practising teachers, instead emphasizing the teaching strategies that teacher educators should apply to practising teachers' professional development.

Theme 3: Developing an identity as teacher educator. Two articles reported teacher educators' identity transformation during their professional development process. Tzur (2001) reflected on his own experiences of becoming a mathematics teacher educator, noticing that integrating his teaching research and practice was a very important step in the process of becoming a mathematics teacher educator. However, he also faced challenges in his growth process regarding the lack of significant change in his work with practising teachers. In addition, Day and Leitch (2001) described the role of emotion in practising teachers' and teacher educators' lives. They identified the importance of raising awareness of "commitment," "caring," "courage," "compromise," and "fragmentation of personal time" in their teaching.

School Teaching

In many countries, school-based teacher educators play a key role in teacher education (European Commission/EACEA/Eurydice, 2015). School-based teacher educators' goal is to model the teacher role through their own teaching and provide an example to teachers for implementing teaching practices (Jaspers, Meijer, Prins, & Wubbels, 2014; Lunenberg, 2010). They may also develop their knowledge competencies about learning and teaching. Six studies addressed the school teaching practice and were categorized into two themes as shown in Table 9.6.

Table 9.6. Number of articles with results related to themes of school teaching practice

Themes	*Number of articles*		
	Math teacher educator	*Other teacher educator*	*Overall*
1. Own school-teaching experience	3	1	4
2. Views related to school teaching	0	2	2

Note: Individual articles could be assigned to multiple themes.

Theme 1: Own school-teaching experience. Understanding how teacher educators notice their own professional teaching and use it in their practice is important for knowing how to support the development of their professional noticing as they work with teachers (Amador, 2016). Amador examined novice teacher educators' teaching experiences and how they modelled students' mathematical thinking in their teaching process. School-teaching experience provides mathematics teacher educators with opportunities to notice students' mathematical thinking before teaching others to do so as well as opportunities to focus on understanding how teachers engage with mathematics. Amador found that novice teacher educators lacked in-depth interpretive analysis about student thinking and rarely made connections between students' thinking and the broader principles of teaching and learning. The incorporation of teaching experiments and model building provided a context in which the novice teacher educators could assume a self-reflective stance to consider their own professional noticing. Meanwhile, McGraw, Lynch, Koc, Budak, and Brown (2007) employed multimedia tools to conduct a mathematics teaching project with prospective and practising mathematics teachers, mathematicians, and a mathematics teacher educator. In this activity, mathematics teacher educators served in a unique role as discussion facilitators as well as participants and all groups examined viewed academic and practical knowledge as important for teaching.

Tzur (2001) and Zeichner (2005) reflected on their own experiences, as they transitioned from mathematics learners to mathematics teacher educators and from teachers to teacher educators in the school context, and revealed several other factors besides participating in specific teaching activities that teacher educators should use or notice in their teaching practice. Based on their school-teaching experiences, they considered the differences between teaching school students and teaching novice teachers as well as the differences in working with different types of novice teachers. These two studies suggest that reflection and inquiry or critique of practices are important techniques that teacher educators should practice in their teaching.

Theme 2: Views related to school teaching. Uibu, Salo, Ugaste, and Rasku-Puttonen (2017) and Caspersen (2013) investigated teacher educators' beliefs and valuing of knowledge and teaching practice in the school context. They found that all of the prospective teachers, novice teachers, experienced teachers, and teacher educators paid attention to students' cognitive development and some talked about academic knowledge and practical skills, such as relevant content knowledge, children's development, and mastering several modes of teaching. However, they still held different views about different teaching responsibilities; for example, teacher educators were seen as paying more attention to students' development of social skills including reflexive skills, independence, and social competence than prospective teachers, and, compared with teacher educators, teachers in schools were seen as placing a greater emphasis on practical skills, and so on. Despite its importance for teacher educators, very few studies addressed teacher educators' school-teaching experiences or practice. Thus, research and practice related to teacher educators' professional development should pay closer attention to teacher educators' school-teaching practice.

Research

Nineteen articles addressed research practice and were grouped into three themes as shown in Table 9.7. Ten of the articles referred to mathematics teacher educators'

Table 9.7. Number of articles with results related to themes of research practice

Themes	*Number of articles*		
	Math teacher educator	*other teacher educator*	*Overall*
1. Self-study on knowledge, competencies, and beliefs required for being a teacher educator	6	4	10
2. Self-study on teaching strategies	3	2	5
3. Survey study on views related to teacher educators' teaching	1	3	4

Note: Individual articles could be assigned to multiple themes.

research practice and the other nine discussed teacher educators' research practice in general.

Theme 1: Self-study of the knowledge, competencies, and beliefs required for being a teacher educator. Ten studies focused on teacher educators' expert knowledge, which mainly includes mathematics content knowledge, mathematics pedagogical knowledge, and knowledge for teaching teachers. For example, the authors of three studies reflected on their own growth from a mathematics student or teacher to becoming a mathematics teacher educator, finding that different identities require different sets of knowledge. They generally believed that the mathematics content knowledge and mathematics pedagogical content knowledge that mathematics teachers ought to have also constitute necessary knowledge for mathematics teacher educators (Chauvot, 2009; Chick & Beswick, 2018; Zeichner, 2005).

In addition to having knowledge comparable to that of mathematics teachers, mathematics teacher educators also need knowledge that is specific to teaching others to teach mathematics, such as an awareness of the perspectives that underlie their teacher education practice, the design of tasks for teachers that focus on students' learning, and connecting research with the practice of teaching (Doecke, 2004; Tzur, 2001; Yang et al., 2015; Zeichner, 2005). Moreover, Masingila et al. (2017) reported on the mathematical knowledge for teaching teachers that mathematics teacher educators use and develop when they work together and reflect on their teaching when helping prospective primary school teachers generate their own mathematical knowledge.

Collaboration awareness and competence are also very important to teacher educators' development. Two studies addressed this topic from different viewpoints. Sakonidis and Potari (2014) described mathematics teacher educators collaborating with both experienced and novice teachers in two contexts: a community of inquiry into mathematics teaching and its development, and a research methods course aspiring to initiate participating teachers into research practice through inquiry. The results showed that mathematics teacher educators could learn deeply from mathematics teachers on identity transformation and understanding teaching and learning. Furthermore, mathematics teacher educators were able to provide support that was more effective and relevant to the teachers' needs once the relationship between research and teaching practices was better realized. Collaborating with teacher educators in a community of practice was described in a study by Gallagher et al. (2011) in which they outlined the professional development of pre-tenure teacher educators through the establishment of a self-study group in a community of practice. In this community, authentic conversations among teacher educators would be generated, and they would have resonance and a sense of belonging to the community of practice.

Lastly, teacher educators' professional identities and beliefs related to education and professional development were addressed by three self-studies (Doecke, 2004; Zeichner, 2005; Galman et al., 2010). Zeichner (2005) briefly examined various aspects of his transition from a classroom teacher to a cooperating teacher and suggested that teacher educators should pay consideration to their role in conducting teachers' professional development activities. Similarly, Doecke (2004) and Galman et al. (2010) reflected on their own roles within a changing policy and curriculum landscape as well as antiracist pedagogy which challenges their professional identity and practice.

Theme 2: Self-study of teaching strategies. Researching teacher educators' own teaching strategies is another way to facilitate their professional development. We coded five studies to this theme. Based on teacher educators' teaching methods, we identified two major categories. The first category involves teacher educators researching their teaching and reflecting on their teaching methods and content to analyze whether their teaching could contribute to teacher education. For example, Kinach (2002) used Liping Ma's teaching method, which focuses on teaching for understanding of the conceptual knowledge of school mathematics, to teach prospective secondary school teachers in their methods course. The author concluded that prospective secondary school teachers can deepen their relational understanding of mathematics within a method course by focusing on instructional explanations. Goodell (2006) asked prospective mathematics teachers to reflect on and discuss critical incidents that they encountered in their field experiences and she herself engaged in self-reflection on this teaching approach of critical-incident discussion. Both of these authors found that these two teaching methods can not only contribute to prospective teachers' development but can also benefit mathematics teacher educators' development.

The second category of teacher educators' self-study of their own teaching methods involves explicit modelling which can be used for teacher education. Loughran and Berry (2005) see this method as a new way for teacher educators to learn about teaching prospective teachers, and Kaufman (2009) revealed five benefits of academic and affective components based on classroom discourse by examining the effect of his modelling processes. Similarly, Peled and Hershkovitz (2004) used three research phases to help teachers analyze incorrect answers obtained from solving standard problems, with the goal being that teachers would know how to evaluate children's answers to the same problem during their teaching.

Theme 3: Survey study on views related to teacher educators' teaching work. Four studies examined teacher educators' views related to their own teaching. Two of the four studies investigated teacher educators' and teachers' views about the kinds of competencies and beliefs that teacher educators should possess. First, the studies reported that teacher educators should possess content competencies, communicative and reflective competencies, pedagogical competencies, and so on (Koster et al., 2005; Smith, 2005). Second, two studies found that teacher educators' beliefs and emotions related to students, teachers, the subject, and education are critical for their professional development (Day & Leitch, 2001; Felbrich, Müller, & Blömeke, 2008). For teachers' professional development, Koster et al. (2005) conducted a Delphi study to identify several tasks that warrant consideration by teacher educators, including working on their own development as well as that of colleagues, providing a teacher education program, and taking part in policy development and the development of teacher education.

Teacher Educators' Education and Professional Development

Six articles fell into the category of teacher educators' education and professional development practice as shown in Table 9.8. These six studies provided more detailed description of the mathematics teacher educators' and other teacher educators' education modes, allowing us to identify two themes: preparation programs for teacher educators and learning within communities. It is worth noting that there is just one study focused on mathematics teacher educators' education and professional development.

Table 9.8. Number of articles with results related to themes of teacher educators' education and professional development practice

Themes	*Number of articles*		
	Math teacher educator	*other teacher educator*	*Overall*
1. Preparation program for teacher educators	1	0	1
2. Learning within communities	0	5	5

Note: Individual articles could be assigned to multiple themes.

Theme 1: Preparation program for teacher educators. Someone who will become a teacher educator or develop their professional ability should also be a learner. One study referred to a preparation program for mathematics teacher educators

(Psycharis & Kalogeria, 2018). The Greek educational system is highly centralized with a single curriculum that teachers must follow strictly, and practising teachers have limited opportunities for development because of the lack of long-term teacher education structures. Thus, Psycharis and Kalogeria investigated how practicum processes facilitated trainee teacher educators' transition to the professional level of teacher educators through an emphasis on the trainee teacher educators' didactical design. In this program, trainee teacher educators were considered as active partners in the integration of technology in teaching practice and experienced practicum activities, including the sequential processes of observation–reflection–design–implementation-reflection based on their documentation work. Meanwhile, trainee teacher educators could select teacher education support courses to observe and implement their designs as instructor, explainer, and facilitator. The authors found that the teacher educators' growth appeared interrelated with their practice and reflection.

Theme 2: Learning within communities. Five of the studies described teacher educators' learning within different types of communities.

- Learning within teaching-practice communities

Three articles discussed the findings of a series of studies focusing on teacher educators' professional development in a professional development community based on thinking education (Hadar & Brody, 2010; Brody & Hadar, 2011; Hadar & Brody, 2016). The professional development community for teacher educators was applied to the university context and was also a learning community and a community of practice. Such communities can serve as opportunities for organizational improvement, professional development, innovation, and enhancement of practice (McLaughlin & Talbert, 2001) as well as reducing traditional teacher isolation. This series of studies organized several professional development communities for teacher educators in which they provided activities including talking about student learning. In particular, it consisted of three studies: building a professional development community among teacher educators, identifying teacher educators' personal professional trajectories in a professional development community, and verifying the function of talking about student learning in community.

In the first study, Hadar and Brody (2010) provided activities for teacher educators from different disciplines and encouraged them to talk about student thinking. The results showed that this kind of professional development community could improve teacher educators' teaching skills and could help reduce their work isolation. Especially helpful for promoting teacher educators' collaboration and professional learning was having teacher educators from different disciplines talk about their students' thinking.

The second study (Brody & Hadar, 2011) identified a four-stage dynamic model of personal professional trajectories that represent teacher educators' general development process: anticipation and curiosity, withdrawal, awareness, and change. Like the findings from the first study, the authors also emphasized teacher educators' willingness and motivation to change as a result of learning from the professional development community in this study.

Finally, Hadar and Brody (2016) studied teacher educators' professional growth by documenting how they talked about student learning. They investigated the characteristics of talk about student learning among teacher educators in a professional learning community and functions of such talk for teacher educators' professional learning. The characteristics identified were managing understanding of student learning, advisory talk with each other (teacher educators' colleagues could offer alternative pedagogical solutions and suggest methods of evaluating the effectiveness of these alternatives), and meta-analytic talk (including connecting experiences from a variety of sources and relating to a theoretical framework). Correspondingly, three main functions were identified, including developing an inquiry stance toward practice, awareness of the connection between teaching and learning, and awareness of their own learning process and roles.

These three studies provide useful information about the benefits of a professional development community for teacher educators' training, the process of teacher educators' professional development growth, and ways of talking about student learning to promote teacher educators' professional development practice.

- Learning to identity transformation within communities

In the field of research on teacher educators' practice, several studies exist that focus on teacher educators' identity and agency transformation. One important reason for this could be that teacher educators generally begin as teachers or university students. For example, Hökkä, Vähäsantanen, and Mahlakaarto (2017) investigated teacher educators' collective professional agency and identity within an identity coaching program. The program sought to (a) support participants' professional identity work, (b) help participants to clarify their work roles amid changes, (c) strengthen participants' professional agency in their work communities, and (d) increase well-being at work (Kalliola & Mahlakaarto, 2011). To achieve these goals, the coaching program included four main thematic components of identity: personal identity (e.g., a person's own life story, personal strengths, and developmental areas), professional identity (e.g., a person's professional history, competencies, and fundamental values), relationship identity (a person's social relations, roles, and relationships within the work communities), and organizational identity (a person's position, opportunities, and ways to influence the organization as well as the sense of belonging and commitment to the organization). This study revealed a transformative pathway for the collective professional identity and agency of a teacher educators' group, which is helpful for understanding and facilitating teacher educators' identity education.

- Learning within inquiry communities

Aside from specific programs for teacher educators' education, another area of research focuses on curriculum provision for teacher educators' professional development. Cochran-Smith (2003) analyzed four teacher educator communities in different contexts which allowed them to inquire collaboratively about assumptions and values, professional knowledge and practice, the contexts of schools as well as higher education, and their own as well as their students' learning. The major lesson learned from across these different contexts is that inquiry as a stance is a valid, valuable, and perhaps necessary way to conceptualize the major questions involved in the education of teacher educators. This study mainly suggested that the content of inquiry as a stance should be included in teacher educators' education curriculum.

LOOKING TO THE FUTURE

We synthesized the findings from 65 relevant research articles to understand mathematics teacher educators' learning from their professional practice, defined using the perspective of their teaching-related experiences that include five elements as shown in Figure 9.1. As evidenced by the number of articles for each practice category shown in Table 9.3, preparation of prospective teachers practice attracted the most attention, followed by research practice and professional development for practising teachers practice, with the practices of school teaching and teacher educators' education and professional development attracting the least attention. The prominence of the preparation of prospective teachers practice is understandable considering the fact that most teacher education research is conducted by teacher educators (Adler et al., 2005) and given the dual role of university-based teacher educators as researchers as well as teacher educators who prepare prospective teachers.

Themes in each practice category give a brief indication of the research foci and status quo related to teacher educators' professional practice. Generally speaking, the study results related to the preparation of prospective teachers practice mainly focused on the goals and strategies involved in teaching prospective teachers as well as on knowledge, competencies, and beliefs necessary for fulfilling teacher educators' duties required for this practice. With regard to professional development for practising teachers' practice, the study results pointed to teacher educators' strategies in working with practising teachers and their various perspectives in designing and organizing professional development activities. The study results for research practice concentrated on teacher educators' self-study related to their knowledge, competencies, and beliefs required for being teacher educators as well as on their own teaching strategies, which confirm the patterns found for the preparation of prospective teachers and professional development for practising teachers' practices. Only some studies addressed school teaching practice, focusing on teacher educators' own school-teaching experiences and their views about school teaching. The study results for teacher educators' education and professional development practice suggest that formal preparation programs for preparing teacher educators are rare, with teacher educators' professional development relying heavily on leaning communities of various kinds.

We now turn to comparing the data shown in Tables 9.3 to 9.8 between mathematics teacher educators and other teacher educators to identify any differences. Research about mathematics teacher educators seems to pay more attention to concrete teaching contexts, such as goals and strategies for teaching prospective teachers, various perspectives in designing and organizing professional development activities, and strategies for working with practising teachers, whereas research on other teacher educators appears to attend to more general issues such as teacher educators' general quality and competencies, their identity construction, and their learning in a community.

This chapter has synthesized findings from research about mathematics teacher educators' teaching-related professional practice and provided significant insights on mathematics teacher educators' development through their practices. The number of papers published and the number of issues discussed show that the progress of research related to (mathematics) teacher educators' learning is quite encouraging. However, continuous effort is needed to better understand mathematics teacher educators' development through their own practice. The conceptual framework shown in Figure 9.1, functions well in guiding and synthesizing our review and could also be taken as a research framework to continuing research in this area. In this section, we situated our recommendations for future research related to mathematics teacher educators' professional development within this framework. We will discuss conceptions of teacher educators and methodological considerations in conducting research on mathematics teacher educators first, followed by issues related to the practice box in the left and the development box in the right of the conceptual framework shown in Figure 9.1.

Unified Conception of Teacher Educators

During our review, we came to realize that the group of teacher educators is actually composed of a variety of individuals, including school-based practitioners (e.g., Hopkins & Spillane, 2014; Uibu et al., 2017), those who work simultaneously as teachers in schools and in universities on a part-time basis (e.g., Williams, 2014), university faculty members with rich prior school-teaching experiences (e.g., Trent, 2013), university faculty members with substantial research commitments who also teach disciplinary and pedagogical courses in teacher preparation programs (e.g., Chauvot, 2009; Chen et al., 2018), teaching researchers working in teaching-research institutions (e.g., Gu & Gu, 2016; Huang et al., 2017), and doctoral students involved in university programs for preparing prospective teachers (e.g., Amador, 2016; Masingila et al., 2017). Cochran-Smith (2003) also noticed the complex composition of teacher educators and suggested that "we need a broad answer to the question of who is called a teacher educator in the first place if we are going to consider seriously the education of teacher educators" (p. 22). This issue still remains and there is an urgent need for a unified conception and common language to capture the varied nature of mathematics teacher educators.

One possible approach to resolving this issue is to categorize teacher educators into different types according to their duties and demands, using different terms to refer to different types of teacher educators and providing different modes of preparation and support for them. For example, in the Chinese context, mathematics teacher educators are generally grouped into three types: school-based mathematics mentor teachers, mathematics teaching researchers, and university-based mathematics teacher educators (Wu & Cai, submitted for review). School-based mathematics mentor teachers are master teachers as well as mentors for novice teachers and cooperating teachers for prospective teachers in their practicum. Mathematics teaching researchers, who usually have extensive prior school-teaching experience and excellent classroom teaching expertise, provide guidance to practising mathematics teachers and promote their professional expertise. University-based mathematics teacher educators work to prepare prospective teachers and improve practising teachers' teaching; their other duties include supervising graduate students, conducting research, and participating in the design, implementation, and evaluation of mathematics teacher education programs.

Methodological Considerations for Research Related to Mathematics Teacher Educators

We found that among the 65 research articles, qualitative studies were prominent. As shown in Table 9.2, 57 articles (87.7%) used qualitative methods (de Freitas et al., 2016) for data collection and analysis, involving case study, self-study, action research, interviews, observation, video and audio transcripts, discourse analysis, the phenomenographical approach, and the narrative approach, whereas only three articles used quantitative methods and the remaining five articles used mixed methods.

Scientific research in education uses qualitative and quantitative methods in the modeling and analysis of numerous educational phenomena. Qualitative research aims to produce in-depth and illustrative information to understand various dimensions of the problem under analysis (Queiros, Faria, & Almeid, 2017), and its major characteristics are induction, discovery, exploration, and theory or hypothesis generation (Johnson & Onwuegbuzie, 2004). Quantitative research involves the collection, analysis, and interpretation of numerical data, with the goals of describing, explaining, and predicting phenomena (Ross & Onwuegbuzie, 2012), and its major characteristics are deduction, confirmation, and theory or hypothesis testing (Johnson & Onwuegbuzie, 2004).

Because research on teacher educators in general and mathematics teacher educators in particular is still in its early stages, it is understandable that most of the research studies discussed in this chapter adopted a qualitative approach that was primarily exploratory in nature. To capture the complexities within teacher educators' learning from their professional practice, studies are needed that incorporate quantitative research designs and rely on statistical analysis.

Underscoring the Uniqueness of Mathematics Teacher Educators' Teaching Practice

Teaching responsibilities related to preparation of prospective teachers and professional development for practising teachers are the most prominent responsibilities, which are at the core position among all the five identified elements of teacher educators' professional practices. Many research studies on teacher educators in general and mathematics teacher educators in particular are framed using frameworks that were developed for teachers. For example, Zaslavsky and Leikin (2004) developed a teaching triad of teacher educators who worked with prospective teachers on the foundation of the teaching triad for teachers (Jaworski, 1992). Goodwin et al. (2014) discussed the knowledge essential to teacher educating by drawing to Cochran-Smith and Lytle's (1999) theorizing about relationships of knowledge and practice. Chick and Beswick (2018) extended the construct of pedagogical content knowledge of teachers (Shulman, 1986) to the pedagogical content knowledge of mathematics teacher educators as they work to develop prospective secondary school mathematics teachers' pedagogical content knowledge. Parallel to the construct of Mathematical Knowledge for Teaching (Ball, Thames, & Phelps, 2008), Masingila et al. (2017) explored how mathematics teacher educators use their mathematical knowledge for teaching teachers, a construct defined by Zopf (2010), while helping prospective teachers generate their own mathematics knowledge for teaching by learning mathematics via problem solving. Mathematics teacher educators occupy dual roles in fulfilling both their teaching duties as mathematics teachers and their responsibilities as mathematics teacher educators (Chick & Beswick, 2018; Wu et al., 2017). It is natural to conceptualize mathematics teacher educators from the perspective of mathematics teachers given that mathematics teacher educators sometimes function as mathematics teachers. However, differences in knowledge and practice between mathematics teacher educators and mathematics teachers need to be explicitly elucidated so as to highlight their unique function as teacher educators.

Efforts have begun to emerge in this aspect both theoretically and empirically. For example, Zopf (2010) delineated differences in content, learners, and purpose of instruction between the work of mathematics teacher educators and the work of mathematics teachers. Smith (2003) suggested that the major difference in the professional knowledge of teachers and teacher educators is in the skills related to teaching different learners – in other words, children and adults. Li and Castro Superfine (2018) indicated that the practice of connecting prospective teachers' mathematics learning to school teaching practice is a unique aspect in mathematics teacher educator's teaching of content courses.

Future research could be devoted to identifying unique practices that mathematics teacher educators use in their work. Results from such research would inform the knowledge, competencies, and beliefs involved in becoming a mathematics teacher educator and would provide significant insight on how to support mathematics teacher educators' learning and development. One theme worthy of exploration is teachers' thinking of mathematics and mathematics teaching. Jacobs and Spangler (2017) stated that much of the recent research on mathematics teaching has included attention to students' mathematical thinking. Similarly, mathematics teacher educators need to understand teachers' thinking about mathematics and mathematics teaching so that they can provide relevant guidance and create appropriate opportunities for promoting teachers' learning to teach.

Mathematics Teacher Educators' Learning and Development from the Supportive Practice

In this chapter, we described in detail about the nature of teacher educators' teaching with prospective and practising teachers; strategies that they use to cope with their teaching duties; and how their research practice, school teaching practice, and teacher educators' education and professional development practice could support their teaching. Some cross-cutting features emerged that seem to demonstrate a positive effect on teacher educators' learning and development. These features include self-studies on the knowledge, competencies, and instructional strategies used in their teaching; developing an identity as teacher educators; and building a learning community. More evidence is needed to confirm the effectiveness of these features.

Moreover, how the three supportive practices, especially the practices of school teaching and teacher educators' education and professional development, relate to teacher educators' teaching with prospective and practising teachers are worthy of further exploration considering the fact that only a few relevant studies have been conducted. As reported, only six articles addressed the school teaching practice, which is the smallest number of articles for all of the five elements of teacher educators' professional practice. In fact, in the discussion of the type of knowledge and expertise that mathematics teacher educators should have, Perks and Prestage (2008) proposed that teacher educators' knowledge is based on the knowledge of teachers, which implies that school-teaching experience is an essential component. In a study with Chinese mathematics teacher educators, Wu et al. (2017) found that mathematics teacher educators who had more than five years of prior school-teaching experience and those who had never been school mathematics teachers self-reported different patterns of challenges in teaching secondary school mathematics method courses to prospective teachers. The mathematics teacher educators who had never been school teachers reported more concern about curriculum issues and their own readiness to teach, whereas the mathematics teacher educators with more than five years of prior school mathematics teaching experience reported more challenges related to prospective teachers' engagement in teacher education classrooms. Should direct school-teaching experience be a prerequisite for being a mathematics teacher educator? Can this experience be replaced by experiences from other activities such as observing classroom teaching? What role does school-teaching experience play in fulfilling mathematics teacher educators' teaching responsibilities? All of these questions deserve further investigation.

In addition, our review identified only six studies related to teacher educators' education and professional development practice, with only one of the six about the preparation of mathematics teacher educators. Overall, the programs for teacher educators' education and professional development activities discussed in the studies indicate the content of the programs seemed not systematically planned, although various learning communities among teacher educators within the same subject (e.g., mathematics teacher educators) or across subjects (e.g., mathematics teacher educators and science teacher educators), researchers, subject experts (e.g., mathematicians), and school teachers were set up to facilitate their communications and collaborations. Thus, it is important to explore how we shall develop a formal and systematic preparation program for mathematics teacher educators. Issues such as the goals, content, and implementation of the preparation program require careful consideration and investigation.

As stated earlier, there are different types of mathematics teacher educators with different prior experiences and, therefore, with different learning needs. For university-based mathematics teacher educators, and especially for those without rich prior school-teaching experiences, the preparation program needs to be designed to foster school-based mathematics practical knowledge and knowledge about students (Hadar & Brody, 2016; Psycharis & Kalogeria, 2018). For school-based mathematics teacher educators, theories about learning and teaching in general and about mathematics teaching and learning in particular need to be added into the content of the program because learning theories and applying theories to teaching practice is important though challenging for this set of teacher educators (Cochran-Smith, 2003). In addition, learning within various communities such as communities of inquiry and communities of practice seems to be an effective mode to promote teacher educators' learning (Hadar & Brody, 2010) and to reduce isolation for teacher educators (Hökkä et al., 2017). More empirical studies are needed to explore how to design and implement appropriate programs to better prepare and develop mathematics teacher educators.

Constructing Knowledge for Practice to Support Mathematics Teacher Educators' Learning and Development

We consider practice and development as a spiral process in mathematics teacher educators' learning to become teacher educators. As indicated in Figure 9.1, we view mathematics teacher educators' knowledge, competencies and beliefs as a foundation to support their practice, and practice as a means to develop their knowledge, competencies and beliefs. The changes in their knowledge, competencies and beliefs again lead to improvements in their practice. In this way, mathematics teacher educators' practice and knowledge are promoted and enhanced by each other. This is an ideal vision of mathematics teacher educators' learning and development. In fact, the foundation, though critically important, has not been well-established at present. There is no consensus as to the knowledge and competencies required to be a teacher educator. We know much less than we ought to. However, it is encouraging to observe that some valuable attempts have been made to crack this issue from different perspectives, including theoretical extensions from knowledge and competencies required for mathematics teachers to those for mathematics teacher educators (e.g., Chick & Beswick, 2018; Masingila et al., 2017) and empirical studies on mathematics teacher educators' teaching practices as a means to reflect on the underlying knowledge and competencies required to support their teaching with teachers (e.g., Castro Superfine & Li, 2014; Zazkis & Zazkis, 2011).

Moreover, the knowledge and theory for teaching teachers is not systematically defined. Goodwin et al. (2014) suggested a lack of knowledge-for-teacher educating practice. There exist insufficient attempts to develop teacher-education pedagogies. Based on Hammerness et al.'s (2005) framework for learning to teach and the work by Grossman et al. (2009a, 2009b) on the three key pedagogies of teacher education (representation, decomposition, and approximation of practice), Ghousseini and Herbst (2016) started to explore how mathematics teacher educators could facilitate prospective teachers' learning to lead classroom mathematics discussion. Further work on this endeavor is needed.

NOTES

1 https://link.springer.com
2 http://ccc.calis.edu.cn

REFERENCES

Adler, J., Ball, D., Krainer, K., Lin, F.-L, & Novotna, J. (2005). Mirror images of an emerging field: Researching mathematics teacher education. *Educational Studies in Mathematics, 60*(3), 359–381.

Amador, J. (2016). Professional noticing practices of novice mathematics teacher educators. *International Journal of Science and Mathematics Education, 14*(1), 217–241.

Appova, A., & Taylor, C. E. (2019). Expert mathematics teacher educators' purposes and practices for providing prospective teachers with opportunities to develop pedagogical content knowledge in content courses. *Journal of Mathematics Teacher Education, 22*(2), 179–204.

Ariza, R. P., Pozo, R. M. D., & Toscano, J. M. (2002). Conceptions of school-based teacher educators concerning ongoing teacher education. *Teaching & Teacher Education, 18*(3), 305–321.

Association of Teacher Educators. (2008). *Standards for teacher educators.* Retrieved from http://www.ate1.org/pubs/uploads/tchredstds0308.pdf

Ball, D. L., Thames, M. H., & Phelps, G. (2008). Content knowledge for teaching: What makes it so special? *Journal of Teacher Education, 59*, 389–407.

Berry, A., & Van Driel, J. H. (2013). Teaching about teaching science: Aims, strategies, and backgrounds of science teacher educators. *Journal of Teacher Education, 64*(2), 117–128.

Bleiler, S. K. (2015). Increasing awareness of practice through interaction across communities: The lived experiences of a mathematician and mathematics teacher educator. *Journal of Mathematics Teacher Education, 18*(3), 231–252.

Borg, S., & Alshumaimeri, Y. (2012). University teacher educators' research engagement: Perspectives from Saudi Arabia. *Teaching and Teacher Education, 28*(3), 347–356.

Borko, H., & Livingston, C. (1989). Cognition and improvisation: Differences in mathematics instruction by expert and novice teachers. *American Educational Research Journal, 26*(4), 473–498.

Brock, C. H., Moore, D. K., & Parks, L. (2007). Exploring pre-service teachers' literacy practices with children from diverse backgrounds: Implications for teacher educators. *Teaching and Teacher Education, 23*(6), 898–915.

Brody, D., & Hadar, L. (2011). "I speak prose and I now know it." Personal development trajectories among teacher educators in a professional development community. *Teaching and Teacher Education, 27*(8), 1223–1234.

Bullough Jr., R. V. (2005). Being and becoming a mentor: School-based teacher educators and teacher educator identity. *Teaching & Teacher Education, 21*(2), 143–155.

Caspersen, J. (2013). The valuation of knowledge and normative reflection in teacher qualification: A comparison of teacher educators, novice and experienced teachers. *Teaching and Teacher Education, 30*, 109–119.

Castro Superfine, A., & Li, W. (2014). Exploring the mathematical knowledge needed for teaching teachers. *Journal of Teacher Education, 65*(4), 303–314.

Chauvot, J. B. (2009). Grounding practice in scholarship, grounding scholarship in practice: Knowledge of a mathematics teacher educator–researcher. *Teaching and Teacher Education, 25*(2), 357–370.

Chen, J. C., Lin, F. L., & Yang, K. L. (2018). A novice mathematics teacher educator–researcher's evolution of tools designed for in-service mathematics teachers' professional development. *Journal of Mathematics Teacher Education, 22*(5), 517–539.

Chick, H., & Beswick, K. (2018). Teaching teachers to teach Boris: A framework for mathematics teacher educator pedagogical content knowledge. *Journal of Mathematics Teacher Education, 21*(5), 475–499.

Clarke, D. J., & Hollingsworth, H. (2002). Elaborating a model of teacher professional growth. *Teaching and Teacher Education, 18*(8), 947–967.

Clarke, D. J., Keitel, C., & Shimizu, Y. (Eds.). (2006). *Mathematics classrooms in twelve countries: The insider's perspective.* Rotterdam, The Netherlands: Sense Publishers.

Cochran-Smith, M. (2003). Learning and unlearning: The education of teacher educators. *Teaching & Teacher Education, 19*(1), 5–28.

Cochran-Smith, M., & Lytle, S. (1999). Relationships of knowledge and practice: Teacher learning in communities. In G. Griffin (Ed.), *Review of research in education* (Vol. 24, pp. 249–305). Washington, DC: American Educational Research Association.

Darling-Hammond, L., & Cobb, V. L. (1995). *Teacher preparation and professional development in APEC members: A comparative study.* Retrieved from http://eric.ed.gov/?id=ED383683

Day, C., & Leitch, R. (2001). Teachers' and teacher educators' lives: The role of emotion. *Teaching & Teacher Education, 17*(4), 403–415.

de Freitas, E., Lerman, S., & Parks, A. N. (2016). Qualitative methods. In J. Cai (Ed.), *Compendium for research in mathematics education* (pp. 159–182). Reston, VA: National Council of Teachers of Mathematics.

Dinkelman, T. (2003). Self-study in teacher education: A means and ends tool for promoting reflective teaching. *Journal of Teacher Education, 54*(1), 6–18.

Doecke, B. (2004). Professional identity and educational reform: Confronting my habitual practices as a teacher educator. *Teaching & Teacher Education, 20*(2), 203–215.

Doerr, H. M., & Thompson, T. (2004). Understanding teacher educators and their pre-service teachers through multi-media case studies of practice. *Journal of Mathematics Teacher Education, 7*(3), 175–201.

Ellis, V., McNicholl, J., Blake, A., & McNally, J. (2014). Academic work and proletarianisation: A study of higher education-based teacher educators. *Teaching and Teacher Education, 40*, 33–43.

European Commission/EACEA/Eurydice. (2015). *The teaching profession in Europe: Practices, perceptions, and policies* (*Eurydice report*). Luxembourg: Publications Office of the European Union.

Even, R. (2008). Facing the challenge of educating educators to work with practicing mathematics teachers. In B. Jaworski & T. Wood (Eds.), *The international handbook of mathematics teacher education: The mathematics teacher educator as a developing professional* (pp. 57–73). Rotterdam, The Netherlands: Sense Publishers.

Even, R., & Ball, D. (Eds.). (2009). *The professional education and development of teachers of mathematics – the 15th ICMI study.* New York, NY: Springer.

Even, R., & Krainer, K. (2014). Education of mathematics teacher educators. In S. Lerman (Ed.), *Encyclopedia of mathematics education* (pp. 202–204). Dordrecht: Springer.

Felbrich, A., Müller, C., & Blömeke, S. (2008). Epistemological beliefs concerning the nature of mathematics among teacher educators and teacher education students in mathematics. *ZDM Mathematics Education, 40*(5), 763–776.

Gallagher, T., Griffin, S., Parker, D. C., Kitchen, J., & Figg, C. (2011). Establishing and sustaining teacher educator professional development in a self-study community of practice: Pre-tenure teacher educators developing professionally. *Teaching and Teacher Education, 27*(5), 880–890.

Galman, S., Pica-Smith, C., & Rosenberger, C. (2010). Aggressive and tender navigations: Teacher educators confront whiteness in their practice. *Journal of Teacher Education, 61*(3), 225–236.

Gelfuso, A. (2017). Facilitating the development of preservice teachers' pedagogical content knowledge of literacy and agentic identities: Examining a teacher educator's intentional language choices during video-mediated reflection. *Teaching and Teacher Education, 66*, 33–46.

Ghousseini, H., & Herbest, P. (2016). Pedagogies of practice and opportunities to learn about classroom mathematics discussions. *Journal of Mathematics Teacher Education, 19*, 79–103.

Goodell, J. E. (2006). Using critical incident reflections: A self-study as a mathematics teacher educator. *Journal of Mathematics Teacher Education, 9*(3), 221–248.

Goodwin, A. L., Smith, L., Souto-Manning, M., Cheruvu, R., Tan, M. Y., Reed, R., & Taveras, L. (2014). What should teacher educators know and be able to do? Perspectives from practicing teacher educators. *Journal of Teacher Education, 65*(4), 284–302.

Grossman, P., Compton, C., Igra, D., Ronfeldt, M., Shahan, E., & Williamson, P. (2009a). Teaching practice: A cross-professional perspective. *Teachers College Record, 111*(9), 2055–2100.

Grossman, P., Hammerness, K., & McDonald, M. (2009b). Redefining teaching, reimagining teacher education. *Teachers and Teaching: Theory and Practice, 15*(2), 273–289.

Gu, F., & Gu, L. (2016). Characterizing mathematics teaching research specialists' mentoring in the context of Chinese lesson study. *ZDM Mathematics Education, 48*(4), 441–454.

Gueudet, G., Sacristán, A. I., Soury-Lavergne, S., & Trouche, L. (2012). Online paths in mathematics teacher training: New resources and new skills for teacher educators. *ZDM Mathematics Education, 44*(6), 717–731.

Hadar, L. L., & Brody, D. L. (2010). From isolation to symphonic harmony: Building a professional development community among teacher educators. *Teaching and Teacher Education, 26*(8), 1641–1651.

Hadar, L. L., & Brody, D. L. (2016). Talk about student learning: Promoting professional growth among teacher educators. *Teaching and Teacher Education, 59*, 101–114.

Hammerness, K., Darling-Hammond, L., Bransford, J., Berliner, D., Cochran-Smith, M., McDonald, M., & Zeichner, K. (2005). How teachers learn and develop. In L. Darling-Hammond & J. Bransford (Eds.), *Preparing teachers for a changing world. What teachers should learn and be able to do* (pp. 358–388). San Francisco, CA: Jossey-Bass Educational Series.

Hatton, N., & Smith, D. (1995). Reflection in teacher education: Towards definition and implementation. *Teaching and Teacher Education, 11*(1), 33–49.

Hökkä, P., Vähäsantanen, K., & Mahlakaarto, S. (2017). Teacher educators' collective professional agency and identity – Transforming marginality to strength. *Teaching and Teacher Education, 63*, 36–46.

Hopkins, M., & Spillane, J. P. (2014). Schoolhouse teacher educators: Structuring beginning teachers' opportunities to learn about instruction. *Journal of Teacher Education, 65*(4), 327–339.

Hsieh, F. J., Law, C. K., Shy, H. Y., Wang, T. Y., Hsieh, C. J., & Tang, S. J. (2011). Mathematics teacher education quality in TEDS-M: Globalizing the views of future teachers and teacher educators. *Journal of Teacher Education, 62*(2), 172–187.

Huang, R., Su, H., & Xu, S. (2014). Developing teachers' and teaching researchers' professional competence in mathematics through Chinese lesson study. *ZDM Mathematics Education, 46*(2), 239–251.

Huang, R., Zhang, J., Mok, I., Zhou, Y., Wu, Z., & Zhao, W. (2017). Perceived professional knowledge and competence of teaching research specialists and their development in China. *International Journal for Lesson and Learning Studies, 6*(4), 321–335.

Jacobs, V., & Spangler, D. A. (2017). Research on core practices in K-12 mathematics teaching. In J. Cai (Ed.), *Compendium for research in mathematics education* (pp. 766–792). Reston, VA: National Council of Teachers of Mathematics.

Jaspers, W. M., Meijer, P. C., Prins, F., & Wubbels, T. (2014). Mentor teachers: Their perceived possibilities and challenges as mentor and teacher. *Teaching and Teacher Education*, 44, 106–116.

Jaworski, B. (1992). Mathematics teaching: What is it? *For the Learning of Mathematics, 12*(1), 8–14.

Jaworski, B., & Wood, T. (Eds.). (2008). *The international handbook of mathematics teacher education: The mathematics teacher educator as a developing professional.* Rotterdam, The Netherlands: Sense publishers.

John, P. D. (2002). The teacher educator's experience: Case studies of practical professional knowledge. *Teaching and Teacher Education, 18*, 323–341.

Johnson, R. B., & Onwuegbuzie, A. J. (2004). Mixed methods research: A research paradigm whose time has come. *Educational Researcher, 33*(7), 14–26.

Kalliola, S., & Mahlakaarto, S. (2011). The methods of promoting professional agency at work. In H. Jian, L. Deen, M. Songge, & P. Simin (Eds.), *Proceedings of 7th International Conference on researching work and learning* (pp. 1141–1152). Shanghai: East China Normal University.

Kaufman, D. K. (2009). A teacher educator writes and shares: Student perceptions of a publicly literate life. *Journal of Teacher Education, 60*(3), 338–350.

Kinach, B. M. (2002). Understanding and learning-to-explain by representing mathematics: Epistemological dilemmas facing teacher educators in the secondary mathematics "methods" course. *Journal of Mathematics Teacher Education, 5*(2), 153–186.

Koster, B., Brekelmans, M., Korthagen, F., & Wubbels, T. (2005). Quality requirements for teacher educators. *Teaching and Teacher Education, 21*(2), 157–176.

Krainer, K., Chapman, O., & Zaslavsky, O. (2014). Mathematics teacher educator as learner. In S. Lerman (Ed.), *Encyclopedia of mathematics education* (pp. 431–434). Dordrecht: Springer.

Leikin, R., Zazkis, R., & Meller, M. (2018). Research mathematicians as teacher educators: Focusing on mathematics for secondary mathematics teachers. *Journal of Mathematics Teacher Education, 21*(5), 451–473.

Li, W., & Castro Superfine, A. (2018). Mathematics teacher educators' perspectives on their design of content courses for elementary preservice teachers. *Journal of Mathematics Teacher Education, 21*(2), 179–201.

Lin, F. L., Yang, K. L., Hsu, H. Y., & Chen, J. C. (2018). Mathematics teacher educator-researchers' perspectives on the use of theory in facilitating teacher growth. *Educational Studies in Mathematics, 98*(2), 197–214.

Loughran, J., & Berry, A. (2005). Modelling by teacher educators. *Teaching and Teacher Education, 21*(2), 193–203.

Lunenberg, M. (2010). Characteristics, scholarship and research of teacher educators. In E. Baker, B. McGaw, & P. Peterson (Eds.), *International encyclopedia of education* (3rd ed., Vol. 7, pp. 676–680). Oxford: Elsevier.

Lunenberg, M., Korthagen, F., & Swennen, A. (2007). The teacher educator as a role model. *Teaching and Teacher Education, 23*(5), 586–601.

Lunenberg, M., & Korthagen, F. A. J. (2003). Teacher educators and student-directed learning. *Teaching and Teacher Education, 19*(1), 29–44.

Masingila, J. O., Olanoff, D., & Kimani, P. M. (2017). Mathematical knowledge for teaching teachers: Knowledge used and developed by mathematics teacher educators in learning to teach via problem solving. *Journal of Mathematics Teacher Education*, 1–22. doi:10.1007/s10857-017-9389-8

McGraw, R., Lynch, K., Koc, Y., Budak, A., & Brown, C. A. (2007). The multimedia case as a tool for professional development: An analysis of online and face-to-face interaction among mathematics pre-service teachers, in-service teachers, mathematicians, and mathematics teacher educators. *Journal of Mathematics Teacher Education, 10*(2), 95–121.

McLaughlin, M. W., & Talbert, J. E. (2001). *Professional communities and the work of high school teaching.* Chicago, IL: University of Chicago Press.

Murray, J., & Male, T. (2005). Becoming a teacher educator: Evidence from the field. *Teaching & Teacher Education, 21*(2), 125–142.

Paine, L. W., Fang, Y. P., & Jiang, H. (2015). Reviving teacher learning: Chinese mathematics teacher professional development in the context of educational reform. In B. Sriraman, J. Cai, et al. (Eds.), *The first sourcebook on Asian research in mathematics education: China, Korea, Singapore, Japan, Malaysia and India* (pp. 617–638). Charlotte, NC: The Information Age Publishing House.

Pantić, N., & Wubbels, T. (2010). Teacher competencies as a basis for teacher education–views of Serbian teachers and teacher educators. *Teaching and Teacher Education, 26*(3), 694–703.

Peled, I., & Hershkovitz, S. (2004). Evolving research of mathematics teacher educators: The case of non-standard issues in solving standard problems. *Journal of Mathematics Teacher Education, 7*(4), 299–327.

Perks, P., & Prestage, S. (2008). Tools for learning about teaching and learning. In B. Jaworski & T. Wood (Eds.), *The international handbook of mathematics teacher education: The mathematics teacher educator as a developing professional* (pp. 265–280). Rotterdam, The Netherlands: Sense Publishers.

Ponte, P., Ax, J., Beijaard, D., & Wubbels, T. (2004). Teachers' development of professional knowledge through action research and the facilitation of this by teacher educators. *Teaching & Teacher Education, 20*(6), 571–588.

Psycharis, G., & Kalogeria, E. (2018). Studying the process of becoming a teacher educator in technology-enhanced mathematics. *Journal of Mathematics Teacher Education, 21*(6), 631–660.

Queiros, A., Faria, D., & Almeid, F. (2017). Strengths and limitations of qualitative and quantitative research methods. *European Journal of Education Studies, 9*(3), 369–387.

Ross, A., & Onwuegbuzie, A. J. (2012). Prevalence of mixed methods research in mathematics education. *The Mathematics Educator, 22*(1), 84–113.

Sakonidis, C., & Potari, D. (2014). Mathematics teacher educators'/researchers' collaboration with teachers as a context for professional learning. *ZDM Mathematics Education, 46*(2), 293–304.

Shulman, L. S. (1986). Those who understand: Knowledge growth in teaching. *Educational Researcher, 15*(2), 4–14.

Smith, K. (2003). So, what about the professional development of teacher educators? *European Journal of Teacher Education, 26*(2), 201–215.

Smith, K. (2005). Teacher educators' expertise: What do novice teachers and teacher educators say? *Teaching & Teacher Education, 21*(2), 177–192.

Tatto, M. T., Schwille, J., Senk, S. L., Ingvarson, L., Rowley, G., Peck, R., & Reckase, M. (2012). *Policy, practice, and readiness to teach primary and secondary mathematics in 17 countries: Findings from the IEA Teacher Education and Development Study in Mathematics (TEDS-M)*. Amsterdam: IEA.

Tillema, H. H., & Kremer-Hayon, L. (2002). "Practising what we preach" – Teacher educators' dilemmas in promoting self-regulated learning: A cross case comparison. *Teaching & Teacher Education, 18*(5), 593–607.

Toerner, G., & Arzarello, F. (2012). Grading mathematics education research journals. *EMS Newsletter, 12*, 52–53.

Trent, J. (2013). Becoming a teacher educator: The multiple boundary-crossing experiences of beginning teacher educators. *Journal of Teacher Education, 64*(3), 262–275.

Troyer, M. B. (1986). A synthesis of research on the characteristics of teacher educators. *Journal of Teacher Education, 37*(5), 6–11.

Twombly, S. B., Wolf-Wendel, L., Williams, J., & Green, P. (2006). Searching for the next generation of teacher educators: Assessing the success of academic searches. *Journal of Teacher Education, 57*(5), 498–511.

Tzur, R. (2001). Becoming a mathematics teacher-educator: Conceptualizing the terrain through self-reflective analysis. *Journal of Mathematics Teacher Education, 4*(4), 259–283.

Uibu, K., Salo, A., Ugaste, A., & Rasku-Puttonen, H. (2017). Beliefs about teaching held by student teachers and school-based teacher educators. *Teaching and Teacher Education, 63*, 396–404.

Vanassche, E., & Kelchtermans, G. (2014). Teacher educators' professionalism in practice: Positioning theory and personal interpretative framework. *Teaching and Teacher Education, 44*, 117–127.

Wang, J., Spalding, E., Odell, S. J., Klecka, C. L., & Lin, E. (2010). Bold ideas for improving teacher education and teaching: What we see, hear, and think. *Journal of Teacher Education, 61*, 3–15.

Willemse, M., Lunenberg, M., & Korthagen, F. (2005). Values in education: A challenge for teacher educators. *Teaching & Teacher Education, 21*(2), 205–217.

Williams, J. (2014). Teacher educator professional learning in the third space: Implications for identity and practice. *Journal of Teacher Education, 65*(4), 315–326.

Wu, Y., & Cai, J. (submitted for review). Supporting secondary mathematics teacher educators in China: challenges and opportunities. In M. Goos & K. Beswick (Eds.), *The learning and development of mathematics teacher educators: International perspective and challenges*. New York, NY: Springer.

Wu, Y., Hwang, S., & Cai, J. (2017). Being a mathematics teacher educator in China: Challenges and Strategic Responses. *International Journal of Science and Mathematics Education, 15*(7), 1365–1384.

Xiao, L., & Chen, L. (2005). CALIS and development of academic digital libraries in China. *Library and Information Service, 49*(11), 9–14. [in Chinese]

Yang, K. L., Hsu, H. Y., Lin, F. L., Chen, J. C., & Cheng, Y. H. (2015). Exploring the educative power of an experienced mathematics teacher educator-researcher. *Educational Studies in Mathematics, 89*(1), 19–39.

Zaslavsky, O. (2008). Meeting the challenges of mathematics teacher education through design and use of tasks that facilitate teacher learning. In B. Jaworski & T. Wood (Eds.), *The international handbook of mathematics teacher education: The mathematics teacher educator as a developing professional* (pp. 93–114). Rotterdam, The Netherlands: Sense Publishers.

Zaslavsky, O., Chapman, O., & Leikin, R. (2003). Professional development of mathematics educators: Trends and tasks. In A. Bishop, M. A. Clements, C. Keitel, J. Kilpatrick, & F. S. Leung (Eds.), *Second international handbook of mathematics education* (pp. 877–917). Dordrecht: Kluwer Academic Publishers.

Zaslavsky, O., & Leikin, R. (2004). Professional development of mathematics teacher educators: Growth through practice. *Journal of Mathematics Teacher Education, 7*(1), 5–32.

Zazkis, R., & Zazkis, D. (2011). The significance of mathematical knowledge in teaching elementary methods courses: Perspectives of mathematics teacher educators. *Educational Studies in Mathematics, 76*(3), 247–263.

Zeichner, K. (2005). Becoming a teacher educator: A personal perspective. *Teaching & Teacher Education, 21*(2), 117–124.

Zopf, D. (2010). *Mathematical knowledge for teaching teachers: The mathematical work of and knowledge entailed by teacher education* (Unpublished doctoral dissertation). Retrieved from http://deepblue.lib.umich.edu/bitstream/handle/2027.42/77702/ dzopf_1.pdf

APPENDIX: DETAILS OF THE 65 RESEARCH ARTICLES

Authors	*Year published*	*Journal*	*Practice type*	*Method*
Amador, J.	2016	International Journal of Science and Mathematics Education	(3)	qualitative
Appova, A., & Taylor, C. E.	2019	Journal of Mathematics Teacher Education	(1)	qualitative
Ariza, R. P., Pozo, R. M. D., & Toscano, J. M.	2002	Teaching & Teacher Education	(1)	qualitative
Berry, A., & Van Driel, J. H.	2013	Journal of Teacher Education	(1)	qualitative

(*cont.*)

Authors	*Year published*	*Journal*	*Practice type*	*Method*
Bleiler, S. K.	2015	Journal of Mathematics Teacher Education	(1)	qualitative
Borg, S., & Alshumaimeri, Y.	2012	Teaching & Teacher Education	(6)	quantitative
Brock, C. H., Moore, D. K., & Parks, L.	2007	Teaching & Teacher Education	(1)	qualitative
Brody, D., & Hadar, L.	2011	Teaching & Teacher Education	(5)	qualitative
Bullough, R. V., Jr.	2005	Teaching & Teacher Education	(6)	qualitative
Caspersen, J.	2013	Teaching & Teacher Education	(3)	quantitative
Castro Superfine, A., & Li, W.	2014	Journal of Teacher Education	(1)	qualitative
Chauvot, J. B.	2009	Teaching & Teacher Education	(1)(4)	qualitative
Chen, J. C., Lin, F. L., & Yang, K. L.	2018	Journal of Mathematics Teacher Education	(2)	qualitative
Chick, H., & Beswick, K.	2018	Journal of Mathematics Teacher Education	(1)(4)	qualitative
Cochran-Smith, M.	2003	Teaching & Teacher Education	(5)	qualitative
Day, C., & Leitch, R.	2001	Teaching & Teacher Education	(2)(4)	qualitative
Doecke, B.	2004	Teaching & Teacher Education	(1)(4)	qualitative
Doerr, H. M., & Thompson, T.	2004	Journal of Mathematics Teacher Education	(1)	qualitative
Ellis, V., Mcnicholl, J., Blake, A., & Mcnally, J.	2014	Teaching & Teacher Education	(1)	mixed method
Felbrich, A., Müller, C., & Blömeke, S.	2008	ZDM – Mathematics Education	(4)	quantitative
Gallagher, T., Griffin, S., Parker, D. C., Kitchen, J., & Figg, C.	2011	Teaching & Teacher Education	(1)(4)	qualitative
Galman, S., Picasmith, C., & Rosenberger, C.	2010	Journal of Teacher Education	(1)(4)	qualitative

(*cont.*)

Authors	*Year published*	*Journal*	*Practice type*	*Method*
Gelfuso, A.	2017	Teaching & Teacher Education	(1)	qualitative
Goodell, J. E.	2006	Journal of Mathematics Teacher Education	(1)(4)	qualitative
Goodwin, A. L., Smith, L., Souto-Manning, M., Cheruvu, R., Tan, M. Y., Reed, R., & Taveras, L.	2014	Journal of Teacher Education	(1)	mixed method
Gu, F., & Gu, L.	2016	ZDM – Mathematics Education	(2)	qualitative
Gueudet, G., Sacristán, A. I., Soury-Lavergne, S., & Trouche, L.	2012	ZDM – Mathematics Education	(2)	qualitative
Hadar, L. L., & Brody, D. L.	2016	Teaching & Teacher Education	(5)	qualitative
Hadar, L., & Brody, D.	2010	Teaching & Teacher Education	(5)	qualitative
Hökkä, P., Vähäsantanen, K., & Mahlakaarto, S.	2017	Teaching & Teacher Education	(5)	mixed method
Hopkins, M., & Spillane, J. P.	2014	Journal of Teacher Education	(2)	mixed method
Hsieh, F. J., Law, C. K., Shy, H. Y., Wang, T. Y., Hsieh, C. J., & Tang, S. J.	2011	Journal of Teacher Education	(1)	quantitative
Huang, R., Su, H., & Xu, S.	2014	ZDM – Mathematics Education	(2)	qualitative
John, P. D.	2002	Teaching & Teacher Education	(1)	qualitative
Kaufman, D. K.	2009	Journal of Teacher Education	(1)(4)	qualitative
Kinach, B. M.	2002	Journal of Mathematics Teacher Education	(1)(4)	qualitative
Koster, B., Brekelmans, M., Korthagen, F., & Wubbels, T.	2005	Teaching & Teacher Education	(4)	qualitative
Leikin, R., Zazkis, R., & Meller, M.	2018	Journal of Mathematics Teacher Education	(1)	qualitative

(*cont.*)

Authors	*Year published*	*Journal*	*Practice type*	*Method*
Li, W., & Castro Superfine, A. C.	2018	Journal of Mathematics Teacher Education	(1)	qualitative
Lin, F. L., Yang, K. L., Hsu, H. Y., & Chen, J. C.	2018	Educational Studies in Mathematics	(2)	qualitative
Loughran, J., & Berry, A.	2005	Teaching & Teacher Education	(1)(4)	qualitative
Lunenberg, M., & Korthagen, F. A. J.	2003	Teaching & Teacher Education	(1)	qualitative
Lunenberg, M., Korthagen, F., & Swennen, A.	2007	Teaching & Teacher Education	(1)	qualitative
Masingila, J. O., Olanoff, D., & Kimani, P. M.	2017	Journal of Mathematics Teacher Education	(1)(4)	qualitative
McGraw, R., Lynch, K., Koc, Y., Budak, A., & Brown, C. A.	2007	Journal of Mathematics Teacher Education	(3)	qualitative
Murray, J., & Male, T.	2005	Teaching & Teacher Education	(1)	qualitative
Pantić, N., & Wubbels, T.	2010	Teaching & Teacher Education	(6)	qualitative
Peled, I., & Hershkovitz, S.	2004	Journal of Mathematics Teacher Education	(1)(2)(4)	qualitative
Ponte, P., Ax, J., Beijaard, D., & Wubbels, T.	2004	Teaching & Teacher Education	(2)	qualitative
Psycharis, G., & Kalogeria, E.	2018	Journal of Mathematics Teacher Education	(5)	qualitative
Sakonidis, C., & Potari, D.	2014	ZDM – Mathematics Education	(2)(4)	qualitative
Smith, K.	2005	Teaching & Teacher Education	(4)	qualitative
Tillema, H. H., & Kremer-Hayon, L.	2002	Teaching & Teacher Education	(1)	qualitative
Trent, J.	2013	Journal of Teacher Education	(6)	qualitative
Twombly, S. B., Wolfwendel, L., Williams, J., & Green, P.	2006	Journal of Teacher Education	(6)	qualitative

(*cont.*)

Authors	*Year published*	*Journal*	*Practice type*	*Method*
Tzur, R.	2001	Journal of Mathematics Teacher Education	(1)(2)(3)(4)(5)	qualitative
Uibu, K., Salo, A., Ugaste, A., & Rasku-Puttonen, H.	2017	Teaching & Teacher Education	(3)	qualitative
Vanassche, E., & Kelchtermans, G.	2014	Teaching & Teacher Education	(1)	qualitative
Willemse, M., Lunenberg, M., & Korthagen, F.	2005	Teaching & Teacher Education	(1)	qualitative
Williams, J.	2014	Journal of Teacher Education	(1)	qualitative
Wu, Y., Hwang, S., & Cai, J.	2017	International Journal of Science and Mathematics Education	(1)	mixed method
Yang, K. L., Hsu, H. Y., Lin, F. L., Chen, J. C., & Cheng, Y. H.	2015	Educational Studies in Mathematics	(2)(4)	qualitative
Zaslavsky, O., & Leikin, R.	2004	Journal of Mathematics Teacher Education	(2)	qualitative
Zazkis, R., & Zazkis, D.	2011	Educational Studies in Mathematics	(1)	qualitative
Zeichner, K.	2005	Teaching & Teacher Education	(1)(2)(3)	qualitative

Note: (1) refers to preparation of prospective teachers practice, (2) refers to PD for practising teachers practice, (3) refers to school teaching practice, (4) refers to research practice, (5) refers to teacher educators' education and professional development practice, (6) refers to others.

Yingkang Wu
School of Mathematical Sciences
East China Normal University

Yiling Yao
College of Education
Hangzhou Normal University

Jinfa Cai
Department of Mathematical Sciences
University of Delaware

ALAN H. SCHOENFELD, EVRA BALDINGER,
JACOB DISSTON, SUZANNE DONOVAN, ANGELA DOSALMAS,
MICHAEL DRISKILL, HEATHER FINK, DAVID FOSTER,
RUTH HAUMERSEN, CATHERINE LEWIS, NICOLE LOUIE,
ALANNA MERTENS, EILEEN MURRAY, LYNN NARASIMHAN,
COURTNEY ORTEGA, MARY REED, SANDRA RUIZ,
ALYSSA SAYAVEDRA, TRACY SOLA, KAREN TRAN,
ANNA WELTMAN, DAVID WILSON AND ANNA ZARKH

10. LEARNING WITH AND FROM TRU

Teacher Educators and the Teaching for Robust Understanding Framework

The Teaching for Robust Understanding Framework delineates five essential aspects (dimensions) of classroom practice. Research indicates that students who emerge from classrooms that do increasingly well along the five dimensions of Teaching for Robust Understanding are increasingly knowledgeable and resourceful thinkers and problem solvers. The framework, along with tools developed to support its use, are used in a range of teacher learning communities. Berkeley's teacher preparation programs now use Teaching for Roust Understanding as a foundation for entering the profession. Teaching for Robust Understanding is the basis of ongoing work with practising teachers. This chapter describes the framework, the tools, and their uses; it describes the ways in which those who helped develop the tools and work with teacher learning communities themselves have developed deeper understandings of ways to support teachers at various stages in their development.

INTRODUCTION

The Teaching for Robust Understanding (TRU) framework delineates five essential dimensions of classroom practice. The framework and tools developed to support its use are used in a range of teacher learning communities, both prospective and practising. TRU is not prescriptive, leaving great latitude for teacher educators in its implementation – and thus for teacher and teacher educators' learning. This chapter begins with an outline of the framework and its affordances. The bulk of the chapter contains descriptions by teacher educators who have developed and worked with TRU of the ways in which they have, by virtue of their involvement with the framework, developed deeper understandings of ways to support teachers, and their own conceptions of teaching and teacher education.

© KONINKLIJKE BRILL NV, LEIDEN, 2020 | DOI: 10.1163/9789004424210_011

The key idea undergirding TRU is that in order for students to emerge from mathematics classrooms as knowledgeable and flexible thinkers and problem solvers, the classrooms must offer significant opportunities along the dimensions described in Figure 10.1. Those five dimensions are necessary and sufficient. If a classroom does well along all of the dimensions in Figure 10.1, students will emerge as powerful disciplinary thinkers; if there are significant difficulties in any of the five dimensions, many students will not (Baldinger, Louie, & The Algebra Teaching Study and Mathematics Assessment Project, 2016; Schoenfeld, 2013, 2014, 2015, 2018; Schoenfeld and the Teaching for Robust Understanding Project, 2016).

The Five Dimensions of Powerful Classrooms				
The Content	**Cognitive Demand**	**Equitable Access to Content**	**Agency, Ownership, and Identity**	**Formative Assessment**
The extent to which classroom activity structures provide opportunities for students to become knowledgeable, flexible, and resourceful disciplinary thinkers. Discussions are focused and coherent, providing opportunities to learn disciplinary ideas, techniques, and perspectives, make connections, and develop productive disciplinary habits of mind.	*The extent to which students have opportunities to grapple with and make sense of important disciplinary ideas and their use. Students learn best when they are challenged in ways that provide room and support for growth, with task difficulty ranging from moderate to demanding. The level of challenge should be conducive to what has been called "productive struggle."*	*The extent to which classroom activity structures invite and support the active engagement of all of the students in the classroom with the core disciplinary content being addressed by the class. Classrooms in which a small number of students get most of the "air time" are not equitable, no matter how rich the content: all students need to be involved in meaningful ways.*	*The extent to which students are provided opportunities to "walk the walk and talk the talk" – to contribute to conversations about disciplinary ideas, to build on others' ideas and have others build on theirs – in ways that contribute to their development of agency (the willingness to engage), their ownership over the content, and the development of positive identities as thinkers and learners.*	*The extent to which classroom activities elicit student thinking and subsequent interactions respond to those ideas, building on productive beginnings and addressing emerging misunderstandings. Powerful instruction "meets students where they are" and gives them opportunities to deepen their understandings.*

Figure 10.1. The five dimensions of powerful classrooms

In essence, TRU provides *principles* for powerful learning environments. TRU does not prescribe particular ways of teaching; rather, it provides a growing set of tools for problematizing and reflecting on teaching, with an eye toward enhancing teaching practices and classroom environments along the five dimensions described in Figure 10.1.

As such, TRU provides educators of prospective and practising teachers a great deal of latitude – the central question being, "how can I use the ideas in TRU to help teachers, coaches, and teacher learning communities develop richer understandings of teaching that enable them to produce more powerful learning environments for their students?" Addressing this question provides multiple learning opportunities. In using TRU, teacher educators are called upon to reflect on the nature of powerful learning environments, and to develop mechanisms for helping teachers learn more deeply. Because there is latitude in implementing TRU (there is no prescribed "right way" to teach, and there are many ways to enhance current teaching practices along the TRU dimensions), teacher educators working with TRU have the latitude to adapt TRU to their local contexts and to build tools for them. Indeed, "early adopters" – better called "early adapters and developers" – have helped to shape the collection of available TRU tools[1] and our collective understandings of ways to support teacher development using TRU. At the same time, TRU's openness places demands on teacher educators. The challenge of figuring out how to make effective use of the principles in any particular professional environment can be substantial.

In this chapter, six sets of early adapters/developers briefly describe their work with TRU and reflect on their learning as a result of that work. Each group was asked to address the following:

a. Who you are, and what your responsibilities are,
b. How you came to encounter TRU,
c. What your learning trajectory has been, including
   i. what you find easy or challenging, in theoretical terms. (How has your understanding of TRU developed? This may well include how you use TRU as a tool with teacher learning communities.)
   ii. what you find easy or challenging, in practical terms. What issues do you face as you try to help members of a teacher learning community get their heads around TRU?
d. Other issues you want to raise/discuss.

Following those descriptions, we discuss the challenges and opportunities of working within this kind of framework. We note that many of our discussions end with questions or unresolved issues. That is because we as teacher educators are grappling with those issues as we move forward. Our learning is an ongoing process.

EARLY DEVELOPMENT AND SUBSEQUENT TENSIONS (NICOLE LOUIE AND EVRA BALDINGER)

We were part of the research team that initially developed the TRU framework. As the team's attention turned to teacher professional development, we were charged with developing tools that would be more conducive to supporting teacher learning and less likely to be misused for high-stakes teacher evaluation than the rubrics we had been working with for research purposes. Drawing on our experiences as mathematics teachers and instructional coaches as well as current research, we developed the TRU Math Conversation Guide[2] (Baldinger, Louie, & The Algebra Teaching Study and Mathematics Assessment Project, 2014).

Our primary goal in designing the Guide was to support teachers and the professionals who work with them (e.g., coaches, administrators, and colleagues) to leverage TRU to nurture collaborative relationships, building on teachers' own concerns, goals, and strengths to advance collective learning. We drew from the premise that teachers' most meaningful learning toward ambitious teaching occurs when they work with others, coordinating diverse perspectives and expertise to investigate problems of practice (Cabana, Shreve, & Woodbury, 2014; Horn & Little, 2010). As a result, we framed the Guide as a set of key questions for a teacher learning community (or a teacher and coach) to think about during planning, execution, and review of a lesson. For example, the core questions for Dimension 4 of the TRU Framework, *Agency, Ownership, and Identity*, are: "What opportunities do students have to see themselves and each other as powerful doers of mathematics? How can we create more of these opportunities?" These are elaborated as points of discussion for teachers with a set of questions that begins as follows:

- Who generates the ideas that get discussed?
- What kinds of ideas do students have opportunities to generate and share
- (strategies, connections, partial understandings, prior knowledge, representations)?
- Who evaluates and/or responds to others' ideas?
- How deeply do students get to explain their ideas? Etc.

Here, we discuss two considerations that came into our design of the Conversation Guide: (1) supporting focused and coherent learning over time, and (2) supporting teacher agency and choice. Although many other ideas influenced our design, we focus on these two because of their relationship to questions that have emerged for us since the Guide's initial release. Our aim is less to explain or justify our choices than to bring readers into some of the spaces in which we ourselves are still wondering and learning.

Supporting Focus and Coherence over Time

Research has consistently identified focus, coherence, and duration as essential for effective professional development (Darling-Hammond, Hyler, & Gradiner, 2017; Desimone, 2009). These qualities can be difficult to achieve, however. Setting aside structures that promote one-off workshops and trainings, the complexity of teaching itself can lead teachers and their partners to jump from one idea to another, without making clear progress in any particular direction. In the midst of "the blooming, buzzing confusion" of classroom life (Sherin & Star, 2011, p. 69), we have seen that a TRU focus can counter that tendency. A coach in the Oakland Unified School District described an instance of this in her work with a teacher:

> She [the teacher] had a lot of thoughts swirling in her head. … [The Conversation Guide] focused her and she was able to pick some questions and some ideas

> that she wanted to talk about. … When she started to fly off and get in her head, we were able to reground her and call her back to these questions.

In addition to focusing individual conversations on issues at the heart of teaching, the dimensions have the potential to serve as a backbone that organizes teacher learning and lends it coherence over an extended period of time, across multiple contexts. For example, we have seen a district adopt a focus on Agency, Authority, and Identity (Dimension 4), and invest substantial resources over the course of several years to support teacher collaboration around creating opportunities for students to see themselves as powerful doers of mathematics. Spending this amount of time with a consistent focus has been essential for building a shared vision and capacity to enact that vision.

Naturally, we experienced tensions associated with focus and coherence. One is the potential for tunnel vision, losing sight of important aspects of teaching and their connections in favor of one particular piece of the puzzle. TRU's five dimensions span a broad range of concerns, and in the Conversation Guide, we attempted to craft questions that draw out connections and overlap between areas. Whether or when this works well is an open question. There are also tensions between fostering focus and coherence and supporting teacher agency and choice, as we discuss below.

Supporting Teacher Agency and Choice

Professional development (not to mention public discourse) often positions teachers as deficient and in need of fixing. We are committed to a contrasting perspective that highlights the knowledge, goals, and strengths that teachers bring to their work. The TRU Conversation Guide reflects this in a number of ways, including open-ended questions that invite educators with diverse experiences, background knowledge, and roles to all contribute to collaborative learning. We also aimed to give teachers room to direct the focus of their learning, which TRU's range facilitates. Teachers can choose a dimension (or part of a dimension, or an intersection between multiple dimensions) that speaks *to them*. Some teachers might choose to focus on developing "more meaningful connections" between facts, procedures, and important ideas and practices (see Dimension 1: The Mathematics); others might work on creating more opportunities for students "to see themselves and each other as powerful doers of mathematics" (see Dimension 4: Agency, Ownership, and Identity). Honoring those choices is an important way of respecting teachers' goals, commitment, and intelligence, and of developing the trust necessary to support relationships that in turn support ongoing learning. As one coach who has used the Conversation Guide has said,

> [Teachers] feel like they're constantly being told, "You're doing it wrong. Do this. Why isn't this strategy being implemented?" [Conversations with the Guide can be] more about … things that they want for their kids and things that are important to them.

Supporting teachers to focus on "things that they want for their kids" may be in tension with supporting system-wide coherence, however. In some cases, we have seen teachers with varied concerns find space to exercise agency within a TRU focus that has been externally defined (e.g., a focal dimension chosen by their district). But in other cases, teachers have experienced mandates to connect to TRU as yet another directive, inconsistent with their sense of professional autonomy.

One challenge we continue to face, both in the Guide and in professional development sessions, is how to orient conversations in productive directions without being directive. In the guide we can point to important issues with our questions (see the examples above). But, because TRU is not prescriptive, the challenge has been to take teacher concerns and frame them in ways that can be addressed productively. For example, the problem "my students don't persevere" can be re-framed as an issue of formative assessment and cognitive demand: "What do you know about their thinking (formative assessment)? Can you offer them challenges within their capacity for productive struggle (cognitive demand)? With some experience succeeding at this, support in reflecting on it, they might come to persevere more." Developing and refining the capacity to do such framing is an ongoing issue. We are getting better, individually, at recognizing and managing such tensions – but we do not yet have ways to build such support into our materials.

Additionally, TRU does not include everything teachers might legitimately wish to focus on. Teachers might, for example, want to focus on developing strong relationships with students, partnering with families, and supporting students to understand and critique social injustices using mathematics, none of which are easy to locate in TRU. It has become clear that fleshing out TRU tools to make such connections will be an ongoing challenge.

These tensions surrounding agency also raise questions with implications beyond the learning interactions that we had in mind when we wrote the Conversation Guide. How might a focus on teacher agency and choice on a larger scale – for example, in selecting, adapting, or authoring frameworks for teaching and learning, or in structuring the work day – create opportunities for teachers to build on their commitments and experiences to inspire and nurture their personal growth as professionals, and the growth of their profession?

When and How to Name Oppression

A third area that has raised tensions and questions for us regards when and how to name oppression. When we first wrote the Conversation Guide, supporting teachers to investigate and transform inequitable classroom interactions was very much on our minds. Yet there is only one sentence in the Guide that makes oppressive power dynamics explicit. That sentence is part of a suggestion to ground conversations in specific, detailed evidence of student thinking and strengths. It reads:

> Attending to particular students can help us think about patterns of marginalization in society at large (e.g., fewer resources for ELLs [English language learners], or stereotypes that link race, gender, and mathematics ability), and how our classrooms might work to replicate or counter those patterns for our own students. (p. 5).

This sentence is buried in the Guide's front matter, which many users look at infrequently if ever. In other parts of the Guide, we posed questions asking, "which students" participate and how, and how more opportunities could be created for "each student." Our observations suggest that this language fails to support teachers and others to engage in conversations about race, class, gender, and other lines along which privilege and oppression are organized (Martin, 2003). Instead of encouraging these conversations, our uses of "which students" and "each student" seem to perpetuate the taboo of naming social hierarchies. We have observed instances of teachers talking around these hierarchies, seemingly uncomfortable with confronting them, as well as instances of teachers casually reproducing damaging assumptions about who is capable of doing what. How could explicit and frequent naming of our concern for Black, Latinx, and indigenous students, in the main text of the Conversation Guide, shape teachers' opportunities to examine oppressive power dynamics in their classrooms? What would be the tradeoffs in using words like "minoritized" or "nondominant," which are more adaptable to local conditions of privilege and oppression but also more easily taken up in ways that do not challenge those conditions? How could explicit naming of oppression be done in ways that are consistent with foregrounding teachers' questions, strengths, and goals and nurturing collaborative, collegial relationships between teachers, their colleagues, and those who are often in positions to evaluate them (such as administrators and coaches)?

Concluding Thoughts

We have always known that the real value of the TRU Conversation Guide would rest not in the document itself but in the hands of the educators who would pick it up, adapt it, and use it. It has been exciting and thought-provoking to see what people in Oakland, San Francisco, Chicago, and elsewhere have done with it. These experiences have prompted new learning for us as teacher educators and scholars of teacher learning. We look forward to continuing to learn about the tool, how it interacts with different educational systems, and how it might be productively adapted to support more powerful work. Three main take-aways from our work as teacher educators have been that: (1) collaboration with those "in the trenches" – both district coaches and teachers – is essential to build and refine tools that have ecological validity; (2) theory can help drive practice in productive ways, if the two live in synergy; and (3) extended work of this type is very much context-driven and context-sensitive, requiring sensitivity to the needs of teacher learning communities while remaining focused on key aspects of professional development. We are still learning, individually and collectively, how to deal with the tensions identified here.

TRU IN THE MASTER'S AND CREDENTIAL IN SCIENCE AND MATHEMATICS EDUCATION PROGRAM (JACOB DISSTON)

I direct the Master's and Credential in Science and Mathematics Education (MACSME) secondary school teacher education program at the University of California, Berkeley, a program that supports prospective teachers in earning both a post-baccalaureate teaching credential and a Master of Arts degree in Education.[3] Our aim with TRU is to develop a shared structure and language for talking about teaching and learning, both in planning for instruction and reflecting on and analyzing instruction. Here I describe ways we have worked to integrate TRU and TRU-related tools into the parts of the program that primarily support students' fieldwork: our Supervised Teaching Seminar, our Supervisor Student Teaching Observation Form, and our Teaching Methods Course.

The most meaningful and transformative professional development experiences I experienced in my 17 years as a middle school mathematics teacher were those that involved collaboration with colleagues, and those that supported my development over an extended time period – specifically, working with student teachers and Lesson Study. Both positioned me as a collaborator and co-learner, working to investigate teaching and learning in real classrooms. Partnering with prospective teachers called for being explicit about teaching decisions, in planning and in the moment; student teachers and I could examine the intentions and understand the implications of choices we make on student engagement and learning. We explored how the collection of individual decisions come together to support structures and classroom norms, and examined whether and how these structures support the types of student engagement and learning that we wanted. Lesson Study creates opportunities to collaborate in depth on lesson planning, and to gather data to understand the impact of the lesson.

My experience in working as a supervising teacher and in Lesson Study prior to having TRU is that a lot of time and effort is spent searching for words and structures to describe what is taking place. With the TRU framework and vocabulary, the work becomes sorting what is going on into the five dimensions and building a consensus about the effects of moments and events within the lesson on student learning with regards to the five dimensions.

What follows describes three structures we have developed to introduce and integrate the TRU framework into our work with prospective teachers.

TRU Video Jigsaw

Our first challenge has been to orient prospective teachers to TRU quickly. Over time, we evolved the following "immersion" strategy. The first day of the MACSME program prospective teachers watch a 5-minute video clip of a classroom lesson. We begin by having prospective teachers examine the task that is the focus of the video, asking them on to work on it and then to anticipate all the ways that students might approach it. They reflect on approaches that lead to complete and correct answers and those that do not – all of which illuminate student thinking and understanding. We then introduce TRU and the Observation Guide, splitting the students into five 'home groups,' each focusing on one TRU dimension. Prospective teachers reflect individually on how ideas in their focal dimension connect to their experiences as school students and their early experiences tutoring or teaching. They discuss their experiences within groups to see how their understandings do or do not align, and what they anticipate seeing in the video of the teaching episode.

When they watch the video, the prospective teachers write down everything they see that is connected to their chosen dimension. They are asked to avoid judgmental observations and to focus on how to sort observations into one or more of the TRU dimensions, identifying the evidence from the video supporting their decisions (e.g., "The point at which the teacher turned her attention to the female student who was leaning back and not participating in the group's conversation seemed important for Agency/Ownership/Identity because …"). Members of each dimension home group then take turns sharing what they noticed, supporting observations with evidence from the video. They add other group members' ideas to their own lists so that they can represent their group's complete set of observations once we switch to jigsaw groups.

Jigsaw groups composed of at least one member from each of the home groups share observations and evidence one at a time, in the order given in Figure 10.1. The group then opens up the discussion, comparing what was noticed through each of the lenses. It becomes clear that different events and moments in a lesson stand out and have different significance depending on which TRU lens you are looking through, and that sharing these different perspectives helps everyone in the group deepen their understanding of classroom dynamics.

The class then reflects collectively on the experience, focusing on what TRU seemed to surface that might not otherwise have been noticed and what, if anything, happened in the clip that was left out of the discussion and was missed – whether there are aspects of teaching and learning that may not be accounted for in the TRU framework. This surfaces tensions we are still dealing with. Prospective teachers commonly point out that issues related to classroom management and time management do not get discussed, even though they seem to feature prominently in the classrooms in which our prospective teachers observe and teach. Negotiating the tension between key points of focus and their everyday concerns is a challenge. (Of course, the sooner they think big picture, the sooner management issues begin to get resolved.) We have also noticed that issues related to how race and gender play out in classrooms do not feature in the video jigsaw activity discussions. A question for us to consider is whether these issues are absent in the jigsaw discussions because the TRU dimensions somehow focus our attention elsewhere, or because the video clips we have picked either do not feature these issues or video does not portray these issues accurately enough. The video and live classroom observations create vantage points and opportunities to explore aspects of teaching that ultimately deepen our ability to analyze events in the classroom so that we can better support student learning. There are still questions: Does TRU guide or force us to see certain things? Or do we only see certain things, and then try to sort them into the TRU dimensions?

The MACSME-TRU Supervisor Observation Form

Another challenge was to support observations of practice teaching in a way that keeps TRU central. We developed a TRU-based[4] form to organize observation notes and feedback to student teachers. The form includes identification information about the placement, a description of the teaching practice goal the prospective teacher is currently working on, a description of the lesson activities and links to the lesson plan. These are filled out by the prospective teacher and shared with the supervising teacher prior to the observation visit. The observation section of the form, used by the supervisor to record observations during the lesson, is divided into six cells: one for each dimension and a cell for general comments.

During the lesson, the supervising teacher makes notes and sorts them into the cells according to which TRU dimension or dimensions are most appropriate. Depending on the student teacher's goals for the day and what feedback she has asked for, the supervisor may focus on one, a select few, or all TRU dimensions. The supervising teacher makes decisions in real time about what to record, and where, in the observation sheet. This act of sorting significant moments and aligning them to specific TRU dimensions puts the supervisor in a more active mode of observation.

Afterward, the supervising teacher and prospective teacher meet to reflect on the lesson. Typically, the prospective teacher begins by reflecting on how she feels about the lesson generally, and then what aspects of the lesson stand out as interesting or important, and how these aspects align to and are informed by the TRU dimensions. The supervising teacher can respond by building on the prospective teacher's reflections, adding details and evidence they collected which supports the alignment the student teacher has made. Or, the supervising teacher can ask questions or make statements that surface important aspects of the lesson which the prospective teacher has not mentioned.

The TRU framework grounds the reflection in common vocabulary and provides a structure for reflecting on a lesson, both in terms of what worked well and what could be improved. But because the supervising teacher is the one sorting the significant moments into the TRU dimensions, we added a step in the process to include a response to feedback from the prospective teacher following the debrief discussion: prospective teachers write responses in the shared observation document, sharing their thoughts about the five dimensions with regard to the lesson itself, the debrief discussion, and the supervising teacher's written notes. We have also experimented with a process for debriefing an observation in which the supervising teacher presents a set of interesting and significant moments with the prospective teacher, one at a time, and asks the prospective teacher which dimension(s) of TRU she feels it aligns to. In this way, the prospective teacher participates in the process of examining the notes from the lesson enactment and exploring the possible alignment to the TRU dimensions, rather than accepting the alignment that the supervising teacher felt was strongest.

TRU as a Focus for Weekly Reflective Journals

All MACSME students are enrolled in the Supervised Teaching Seminar in which we discuss issues that arise in their field placements. Each week, prospective teachers respond to a journal prompt on our online class portal by Friday night. The group reads and comments on each other's posts over the weekend, and the collection of posts serves as the foundation for our weekly discussion.

The weekly journal topics help to guide what our prospective teachers focus on during fieldwork, in observing lessons their Supervising Teachers teach, and during lessons the prospective teachers lead themselves. By focusing on each of the TRU dimensions for a week or two, prospective teachers become experienced in looking through a specific lens to identify the moments in a lesson that align to that dimension. We also ask them to focus on connections between the dimensions, and aspects of teaching and learning that fall within the intersection between dimensions, as a way to explore how certain lesson structures or teacher moves might be leveraged to achieve particular teaching practice goals, or might help identify potential pitfalls and compromises that can inhibit student engagement and learning. For example, one of our journal prompts in the fall semester asked prospective teachers to observe their classrooms through the lens of Agency/Ownership/Identity, and to make connections to the Access dimension:

> Last week we focused on looking for moments in the classroom that connect to Agency/Ownership/Identity …
>
> This week continue to observe/reflect on issues related to Agency/Ownership/Identity – especially in how those issues relate to Equitable Access: if we are working to structure things so that all students have access, why aren't they all engaging?

In this way prospective teachers can question what specific elements of lesson structure and teacher moves, besides those related to creating equitable access for students, might be necessary for teachers to consider in order to support all students in developing productive identities as learners. Similarly, we can examine through the intersection of Access and Cognitive Demand how structures that support improved access might serve either to diminish or maintain the cognitive demand of an activity and the potential for students to engage in productive struggle. These intersections between dimensions have proven to be a productive way to develop a deeper shared understanding of the TRU dimensions, and to identify aspects of a classroom episode that otherwise might be missed.

What We Have Learned about TRU, and the Questions That Have Arisen

TRU provides a useful structure for investigating teaching and learning within a professional learning community like MACSME. The TRU framework as experienced through the structures described in this chapter provides a means to establish a common vocabulary and a set of lenses for identifying what is important to notice and examine in an episode of teaching and learning, and for planning and reflecting on instruction in ways that supports the development of effective teaching practices.

The structures described above resulted from some years of experimenting with how best to introduce TRU to prospective teachers whose main experience of teaching has been through the "apprenticeship of observation." They do help prospective teachers re-orient to classroom phenomena, but they are still a work in progress. Questions we will examine in future work include: What, if anything, do we miss when we look at teaching and learning through TRU? What is the difference between looking at a classroom through a specific TRU lens, where we only pay attention to moments that seem aligned to the one dimension, and looking through all of them at once, picking out moments that seem important, and sorting them into the TRU dimensions? And in what ways do the individual TRU lenses, and the intersections or combinations of multiple TRU lenses, help us notice and understand more nuanced aspects of teaching and learning?

TRU MATH IMPLEMENTATION BY THE SILICON VALLEY MATHEMATICS INITIATIVE (DAVID FOSTER AND TRACY SOLA)

Supporting productive shifts in mathematics instruction is challenging. Dominant belief systems, counterproductive federal, state, district and school policies, a long tradition of "demonstrate and practice" pedagogy, and historically low expectations for under-represented students are just some of the obstacles that must be overcome.

The Silicon Valley Mathematics Initiative[5] has been working to support improved mathematics teaching and learning since 1996. Participating districts receive year-round professional learning, a formative and summative performance assessment system, funding to support district mathematics coaching, and a network including meetings and workshops with mathematics teachers, leaders and administrators.[6] Our purpose has been to describe a new vision of teaching and learning and to share innovative instructional methodologies to improve instruction. Until recently, communicating all the necessary elements for a program of sustainable change has been an immense challenge. We had long lists describing mission statements, goals, a range of loosely connected programs, and a set of strategies and improvement plans. To create a common vision, Silicon Valley Mathematics Initiative facilitators provided readings, cited important research, developed lengthy bullet points of actions, and created bibliographies of books and papers for reference.

As an illustration of the challenge consider National Council of Teachers of Mathematics (2014) newest guide to practice, *Principles to Actions*. The book has tons of lists – lists for mathematics teaching practices (8), beliefs about teaching and learning mathematics (6), establishing mathematics goals to focus learning (7), implementing tasks (9), using mathematical representations (12), levels of classroom discourse (24 cells), facilitating mathematical discourse (8), posing purposeful questions (8), building procedural fluency from conceptual understanding (9), supporting productive struggle (8), etc. For anyone without extensive experience, these lists are overwhelming and hard to embrace. Creating coherent professional learning activities out of these lists is at best challenging and at worst unwise, since teaching appears fragmented when described in lists. What we needed was a comprehensive and concise vision of mathematics teaching and learning.

TRU came about at an opportune time. American education was focused on shifting to the *Common Core State Standards for Mathematics* (CCSSM) and districts and schools were intent on learning about CCSSM. Standards allowed teachers and leaders to focus on mathematics content but offered little help regarding ways to describe or create mathematically powerful classrooms. That is what TRU does, in efficient and coherent form. Being able to organize our work around the five TRU dimensions, with confidence that all key issues can be addressed, allows for much more efficient professional development. Silicon Valley Mathematics Initiative has rolled out TRU Math in in Northern California, Southern California, the Greater Chicago Metro Area, and New York City. It is core to our work.

Introducing TRU

A TRU launch begins with formal presentations at our Math Network Meetings, where mathematics leaders, mathematics coaches, and teachers on special assignment meet regularly for professional development. We show a video of a highly engaged classroom and ask participants to list the attributes they observed in the video lesson. One or more of five scribes at easels in the front of the room record the comments the participants make. Once the chart papers are full, the scribes reveal the categories they were using to record comments – the five dimensions of TRU. Since every attribute landed on at least one of the chart papers, the five dimensions encompass the entire instructional domain. The fact that some attributes landed on multiple papers, indicated the differences between dimensions but some overlap. The overlaps show connection between dimensions such as Access and Agency or Cognitive Demand and Mathematics.

TRU underpins all Silicon Valley Mathematics Initiative professional development. When working with district superintendents and curricular administrators at our tri-annual meetings, addressing site administrators during our *Principal as Instructional Leader Meetings*, working with mathematics leaders at our *Math Network Meetings*, and presenting professional development to teachers during summer institutes or school year follow-up professional development workshops, the five dimensions of TRU frame professional learning and provide the lens to reflect upon student learning outcomes. For administrators, TRU highlights the kinds of things that need their support. For teachers, TRU focuses on the aspects of instruction over which they have control.

Going into Depth

As we continued working with educators to deeply understand TRU, we realized the usefulness of focusing sessions on a single dimension. An early favorite was Agency, Authority/Ownership, Identity. We assisted educators in exploring those terms by focusing on the root words of the terms. Unpacking what it means to be an agent and identifying attributes of an agent, helps teachers make sense of agency (being self-reliant, a self-starter, responsible for others, being in the role of a leader …). The root word of authority is author, as in author of ideas. Shifting the term authority from merely the concept of a central power to the creator of thought invites educators to see students in a different but very important role – facilitating students as creators of thought is challenging and nuanced, inviting and requiring a different role for the teacher. The term identity was often more accessible to the participants, especially with recent emphasis on growth versus fixed mindset. Using these characteristics, the role of students in mathematically powerful classrooms began to take shape. We then engaged the educators with videos of classrooms where students were actively discussing important mathematical ideas, challenging each other's thinking, sharing ideas and strategies, and clarifying their understandings. The use of classroom video was an important tool for promoting collegial discussions and assisting teachers and leaders in deeply defining Agency, Authority, and Identity. The next levels of discussions, prompted by videos, were about how to create a classroom culture to foster these important characteristics in students. This opened the door to sharing routines and activities that promote and sustain classroom discourse. Mathematics or number talks, cooperative logic activities, group discussion protocols, group work quizzes, think-pair-share routines, sentence frames and sentence starters, are just some of the instructional techniques we introduced and promoted to build Agency, Authority and Identity in our students. These techniques shift the heavy lifting of learning from a teacher who is expounding to a facilitator who fosters learning. The students become the owners of their own learning and resources for one another.

The Equitable Access dimension promotes a social justice agenda, taking on achievement and opportunity gaps directly. At the heart of this dimension are belief systems and student expectations. Our professional learning in this dimension began with conversations and activities that surfaced and confronted belief systems. Using readings and discussions raised awareness. Observing classroom videos that illustrate students, whose capabilities are underestimated, struggling, persevering, and succeeding helped to confront traditional beliefs. Engaging teachers in closely examining student work and conducting consensus scoring sessions created a space to share and negotiate common values about student expectations. Setting goals and creating actionable plans was often a next step for teachers and leaders.

We read about and discussed the pernicious impact of tracking. Another factor that often denies students' access centers on language. Traditionally, English Learners are often prevented or "protected" from engaging in rich tasks, or tasks that require negotiating a written or real-life context. Instead of moving away from language-rich mathematical problems and tasks, teachers need to create opportunities for English Learners to engage productively with mathematically and linguistically rich tasks. We introduced routines such as three reads, problem stems, close reads, and video contexts to create accessible strategies enabling English Learners to tackle rich mathematical tasks. Our professional development included an emphasis on students' explanations and justifications. Status posters, student work analysis, reengagement lessons, peer editing and review, are just some of the instructional techniques we emphasized in our mathematics workshops.

What TRU has allowed us to do is to frame individual issues like tracking as part of the larger picture. Tracking is now framed as an issue of access, and potential remedies point to the domain of Agency, Ownership, and Identity (Dimension 4). That is, we now know we need to do more than just give students access (a partial "solution" to the issue of equity), and to do so in ways that allow students to see themselves as mathematical thinkers and problem solvers. TRU has helped to frame our professional development in more coherent ways.

Productive Uses of Assessment

Perhaps counterintuitively, richer and deeper mathematical tasks (especially those amenable to multiple approaches and/or using multiple representations) provide *greater* access to important mathematics and support rich classroom discourse (and thus possibilities for greater agency). To counter more than a decade of California's emphasis on skills-oriented high stakes tests, starting in 1997, Silicon Valley Mathematics Initiative formed the Mathematics Assessment Collaborative. We made a strategic decision to invest in a mathematics performance assessment test that assessed higher cognitive levels, that needed to be hand-scored, and that produced rich examples of student work. We commissioned the Mathematics Assessment Collaborative/Mathematics Assessment Resource Service (MARS) performance assessment tests in 1999. These assessments, used for both summative and formative purposes in classrooms, provide rich mathematical content, generate student experiences with high levels of cognitive demand, assess students' ability to be productive in high cognitive demand situations, and develop teachers' skills in implementing formative assessment practices. Three dimensions of TRU Math – Mathematical Content, Cognitive Demand, and Formative Assessment – provide the framework for our deep professional learning using the Mathematics Assessment Collaborative/Mathematics Assessment Resource Service Performance Assessment tasks. Our collective debriefs on student work are now framed in the language of TRU.

The mathematics of tasks used in the performance assessments is designed along a learning progression, where the initial questions are accessible to nearly all students and they can demonstrate what they know. Additional questions probe whether students can demonstrate that they are meeting grade/course level standards. The tasks assess student thinking at higher cognitive levels to measure conceptual understanding, applications, generalizations and/or justifications. Our professional development, using performance assessments, strengthens teacher knowledge and focuses on both the mathematical content being taught and the levels of cognition in which the students are engaged. Teachers learn that a mathematically powerful program includes a balanced diet of the levels in Norman Webb's Depth of Knowledge approach (Webb, 2007) or the levels of Cognitive Demand characterized by Smith and Stein (2011).

In addition to the Mathematics Assessment Resource Service performance tasks, Silicon Valley Mathematics Initiative often engages educators in the Formative Assessment Lessons. These lessons, produced by the same team that produced the TRU framework, support all five dimensions of TRU in the classroom. Educators experience the lessons as learners.

Finally, TRU Dimension 5 (Formative Assessment) guides Silicon Valley Mathematics Initiative's essential work to elicit student thinking and use that thinking to promote further learning. Teachers select a Mathematics Assessment Resource Service performance task focused on the mathematics content of the unit they are teaching. During the unit, the teachers administer the task to their classes. Collectively, teachers score and analyze their students' work, identifying common errors, misconceptions, reasoning flaws, varied approaches and representations, successful explanations, and other artifacts of student thinking. The teachers then use actual student work samples to design a lesson, called a reengagement lesson. The lesson is taught by presenting these mined student gems to pose learning disequilibrium or cognitive conflict. Students are asked to critique, analyze, or explain one another's thinking, arriving at correct solutions, reasoning about varied approaches, or improving mathematical explanations or justifications.

Reflections

TRU Mathematics has changed our thinking in several ways. Prior to TRU, we would awkwardly attempt to describe the role of the student in the classroom. To address the varied aspects of their role we would discuss the classroom environment and the culture that needed to be established. We discussed the student role in group work and aspects of status, accountability, inter-personal skills, and self-reliance. Then we would attempt to address classroom discourse, including different talk moves, good questioning strategies such as funneling versus focus questions, strong explanations and justifications, and students' perseverance. Then we would focus on academic and mathematical language, with special treatment for students whose first language is not English. This professional development "to-do" list got longer and longer as more research, such as growth mindset and students' disposition toward learning mathematics came to light. When TRU Math introduced the dimension of Agency, Authority and Identity, lights came on for us. All the descriptions and discrete categorization, formerly described, are captured and condensed into targeted work to develop student with agency, authority and identity. TRU provided both concise language and the needed focus on the core essence of the student role in learning mathematics. This was enlightening and liberating.

At the same time, we still face significant challenges. One is to help build self-sustaining Teacher Learning Communities. It is still an open question as to how to guarantee the longevity and purposefulness of Teacher Learning Communities, and how to make TRU so natural a part of a teaching learning community's work that it automatically frames issues through the lens of TRU. A second is how to secure administrative buy-in at both the school and district levels. It is easy for an administrator to undermine the work of a teaching learning community by, for example, mandating skills testing, not providing adequate time or resources for the teaching learning community to work effectively, or trying to implement so many "helpful" initiatives that coherence is lost. We have begun working on tools that support administrators in supporting teaching learning communities.

TRU IN CHICAGO: SUPPORTING SYSTEMIC CHANGE (RUTH HAUMERSEN, ALANNA MERTENS, AND LYNN NARASIMHAN, WITH NICOLE LOUIE)

We are writing as a group of mathematics educators who have come together to look at TRU more deeply as it has come into the work of the Chicago P12 (Pre-kindergarten through Grade 12) Mathematics Collaborative. The Collaborative began in 2012 as a partnership between the Department of Science Technology, Engineering and Mathematics (STEM) of the Chicago Public Schools and local institutions of higher education, with the goal of strengthening instructional practice and increasing student success in mathematics. We created and implemented a mathematics professional learning model with a district-wide reach (some 1300 teachers across 500 schools) and additional support for a subset (110) of the district's schools. The model included cycles of teacher workshops, administrator institutes, cross-site professional learning communities for teachers and school teams, and both individual and collaborative coaching.

TRU entered the Collaborative's work in 2014. When we were first introduced to the framework at a meeting hosted by the Silicon Valley Mathematics Initiative, we saw great potential for the five dimensions to powerfully summarize the shifts in instruction that we had been working toward in our professional learning model, and for the framework to provide a common language for teachers and administrators to have meaningful conversations around high-quality mathematics instruction across a very large and diverse district. We, therefore, began to ground all professional learning in mathematics in the TRU framework. At each district-wide teacher and administrator session, we looked deeply at one dimension of TRU and explored its significance through participants' own engagement in rich mathematical experiences, as well as through observation of and reflection on how the dimension might look in Chicago Public Schools classrooms. Time was also given to consider next steps and to do some collaborative planning in light of that session's highlighted TRU dimension. We placed a significant emphasis on the use of the TRU Conversation Guide, especially the core questions and "Think About" sections, in all mathematics professional learning, but particularly in the professional learning community sessions and during collaborative coaching.

Since we began our work with TRU, we have seen remarkable changes in how teachers talk about their vision for mathematics instruction. One teacher reflected, "I used to think that [this work] was beyond the capabilities of not only my teaching practice but also my students' ability level. Now I think that … I am more than capable and my students are more than able." We have also seen shifts in how teachers relate to one another. As another teacher described, "I feel that teachers are becoming more open about sharing practices as a result of these trainings …. Observing other teachers' practices has been the most valuable to me."

TRU, and the Conversation Guide in particular, has had a major role in facilitating these shifts. To be clear, it is not a silver bullet that has solved all of our problems. But it has supported us to enact – and learn about enacting – a vision of teacher learning that emphasizes (1) a robust vision of powerful mathematics instruction, (2) building social capital alongside individual human capital, and (3) teacher agency, authority, and professionalism. We discuss each of these points below.

A Robust Vision of Powerful Mathematics Instruction

When the Collaborative was launched, we focused on instructional strategies that had great potential to increase students' opportunities to make sense of big mathematical ideas and explain and justify their thinking – that is, to support the vision of rich, powerful mathematics instruction that *we* had. These strategies included Math Talks,[7] Three Reads,[8] and Formative Assessment Lessons.[9] However, we found that teachers often focused on the "what" of the strategies instead of the "why." Instead of reasoning about their instructional decisions in terms of rich, powerful goals, they were concerned with following protocols and ticking checkboxes.

TRU helped us make the vision in our heads more explicit for teachers, and it helped us develop a shared language that everyone in the Collaborative could use to articulate and reinforce their goals for their instruction. When teachers ask whether the protocol says students should have two minutes or five minutes of independent think time, TRU helps us bring them back to a vision of powerful mathematics instruction with questions like, "How do you think that would affect students' access to the mathematics in this task?" Teachers themselves – including some who have never attended a Collaborative workshop but have colleagues at their schools who have – also bring up these kinds of questions during planning meetings and peer observations.

We knew that it was important for teachers to connect instructional strategies to bigger ideas, goals, and principles – the same way it is important for students to connect procedures and algorithms to underlying concepts. What TRU did was show us how powerful it could be to support those connections with a framework and language that teachers could return to again and again, developing personal meanings and connections that further fueled Collaborative work.

Building Social Capital

Early in the project, our focus was on improving individual teachers' practice. We worked to develop teachers' human capital, deepening their content knowledge and pedagogical content knowledge and providing classroom resources and instructional strategies. In early professional learning community meetings, however, we began to see the value of bringing teachers together in a safe environment where they could share problems of practice and successful classroom experiences. Teachers who initially said things like "I can't do this in my classroom with my students" were trying new strategies after hearing success stories from colleagues at neighboring schools.

As the project evolved, we shifted our focus to building *social capital* – trust and collaboration *between* teachers – through increasing opportunities for teachers to make their practice public within and across schools (Leana, 2011). This shift towards increased collaboration and public practice was reinforced as we began to see that the TRU framework not only describes dimensions of powerful learning for students, but also dimensions of powerful professional learning for adults (Schoenfeld, 2015). Thus, the use of the TRU framework at the scale of professional learning for teachers provides consistency: teachers who are focused on creating powerful learning environments for their students are experiencing such environments in their own learning. It also supports the idea that to be effective, learning environments – whether for teachers or students – need to be shared and collaborative in nature.

As this shift took hold, we saw teachers engaging with one another around problems of practice related to the vision set forth in TRU. One principal described the transformation she was seeing: "Our whole staff is coming kind of to a threshold where they're becoming a collaborative staff. They are trusting each other, to take [one another's] criticism and also to do something positive with it."

Teacher Agency, Authority, and Professionalism

In tandem with working to support strong teacher communities and a strong instructional vision, we have learned to support teachers' sense of agency, authority, and professionalism, in parallel to asking them to support students' agency, authority, and identity. Just as there are many "right" ways to solve rich mathematics problems, there are many "right" ways to teach for robust understanding of mathematics, and teachers have clearly communicated that they appreciate having space to take on leadership in developing their own practice – without being told what to do, and without being evaluated.

When we started using TRU, we saw possibilities for it to empower teachers. As we were introducing instructional strategies, TRU provided room for teachers to continue to use their own strategies or modify ours and connect them to the Collaborative via the larger goals articulated in the dimensions. The questions in the Conversation Guide also provided opportunities for teachers to engage in deeper thinking, as they made their *own* sense of the dimensions, connections between the dimensions and their current practice, and ways *they* wanted to improve or grow. The core questions in the Conversation Guide also created a safety net for teachers engaging in peer observations. Because they are open-ended with no right or wrong answers, it became easier for every teacher to participate in discussing them. Additionally, they made focusing on a particular area less threatening, especially when teachers themselves had picked the focal question for the day. Instead of picking apart an individual teacher's practice because it fell short on a rubric or checklist, we could think together about a question the lead teacher had shared to develop not only her practice but our collective practice as teachers of mathematics.

In practice, teachers took on responsibility not only for trying new strategies and analyzing their effects on student learning but also for organizing and sustaining collegial collaboration at their schools. TRU has helped them to develop a shared focus that every team member could find a personal stake in, and to which everyone had something to contribute. At one school, a teacher described this process as creating

> a system … to not check what people are doing, but to get ourselves into each other's lives, our teaching lives. So we started meeting together as a mathematics team to figure out how can we set up a schedule so we can get into each other's classrooms.

The mathematics team, which consisted of teachers from pre-Kindergarten to middle school, used the TRU dimensions to sharpen the focus of these observations and created a document to support the peer collaboration process. A team member said that the purpose of the document was to help give:

> a focus to the discussion. And to have that same conversation happening across [the] entire school … There's power in being able to ask the same questions, look for the same things, talk about how is this giving Agency and Authority to the students while we're doing a lesson.

A teacher outside of the mathematics team who began to take part in these peer observations reflected that:

> after you do it a couple times, I think it becomes much easier to … realize that … your colleagues are simply there to help … it opens up, it's funny, because

> it kind of opens up the same door that we want for the kids. The comfort level is there so we can really share where we're at with [instructional strategies].

It is through this kind of shared ownership and trust that teachers become empowered to grow their own practice using the common language and vision provided by the TRU Dimensions as they work to create mathematically powerful classrooms for their students.

Concluding Remarks

Although we have treated them as separate, a robust vision for instruction, building social capital, and teacher agency, authority, and professionalism are deeply intertwined. To build social capital, we have leveraged our instructional vision as an organizing tool, as have teachers. And teacher agency, authority, and professionalism both stem from and contribute to strong social capital and a vision that is clear and coherent without being prescriptive.

At a time when teachers are constantly bombarded with resources and strategies and are faced with constant pressure to raise student achievement, we suggest that less is more. Having fewer tools can be immensely generative, when those tools are open-ended enough to support teachers to make their own sense of them, take ownership, work with others to solve problems of practice, and promote a collaborative culture of transparency, reflection, and growth.

TRU AND MULTIPLE TEACHER LEARNING COMMUNITIES (MICHAEL DRISKILL, EILEEN MURRAY AND DAVID WILSON)

We are a research-practice partnership between two teacher-leadership organizations, Math for America (MfA) and the New York State Master Teacher Program (NYSMTP), and two universities, Montclair State University and the State University of New York, Buffalo State.

Math for America is a nonprofit organization based in New York City with a mission to improve mathematics instruction in the United States. To accomplish this mission, Math for America works to retain talented and experienced teachers through selective, four-year fellowships that provide ongoing professional and leadership opportunities. New York State Master Teacher Program is an independent, publicly funded program explicitly based on Math for America's model aimed at improving STEM teacher retention across the state of New York.

Fellowships at Math for America and New York State Master Teacher Program provide teachers opportunities to work with one another, outside of school hours, to pursue a variety of self-selected learning opportunities. These opportunities fall into different categories and include teacher-led learning teams dedicated to understanding how to use high-quality instructional materials effectively.

An example of the type of high-quality instructional materials teachers explore in both programs is the *Classroom Challenges* Formative Assessment Lesson collection. *Classroom Challenges*, often simply called FALs, are a set of 100 free mathematics lessons developed by a team at the University of California at Berkeley and the Shell Centre for Mathematics Education at the University of Nottingham. The lessons support teachers' formative assessment in important mathematical ideas and practices articulated in the *Common Core State Standards for Mathematics*.

Our partnership between teacher-leadership programs and universities is creating a repository of video cases based on Formative Assessment Lessons taught by teachers at Math for America and New York State Master Teacher Program in a diverse set of classrooms across New York State. The video cases are intended as objects of study for communities of practice (Wenger, 1998) dedicated to understanding how to teach Formative Assessment Lessons.

The communities of practice we support foster interactions within and across populations of prospective, early career, and practising teachers; this later group includes teachers who have been awarded Master Teacher fellowships by Math for America and New York State Master Teacher Program, as well as others such as colleagues in their schools.

Each case includes a video segment of secondary instruction of a Formative Assessment Lesson taught by a Math for America or New York State Master Teacher Program fellow along with supporting materials that provide context. Each case also includes a set of discussion prompts, based on the TRU Framework, that support teachers, coaches, professional development leaders and teacher educators in facilitating discussions about mathematics teaching and learning.

The video cases are not intended as exemplars. Rather we understand them within the communities of practice framework as objects of study that make it possible to develop collective knowledge about how to use Formative Assessment Lessons effectively in different contexts. Of particular interest are emerging, collective understandings about how students understand specific mathematical ideas (e.g. sample spaces, domain and range) and teaching moves that can support student thinking in these areas.

As the video cases are used in different contexts (professional learning communities, methods courses, etc.) our team collects commentary (e.g., discussions, mathematical solutions) on the case and adds these artifacts to the materials. This commentary supports the shared repertoire that allows different members of the community to deepen their understanding of the teaching and learning generally and as it pertains to specific lessons.

From the beginning of our research, all partners believed it was important to situate the case study materials in the context of a research-based framework characterizing the dimensions of high quality instruction. We chose TRU because of its accessibility, comprehensiveness, readability, and abundance of open-source support materials.

In what follows we will discuss two challenges we faced related to TRU in developing and using the video cases. The first is deciding how to select video clips that can support rich discussions for teachers in different contexts and at different stages of their careers. The second is which TRU tools we should include with the case materials to foster ideas that can be used collectively not only to improve the teaching of specific lessons, but to deepen our understandings of practice generally.

Theoretical and Practical Challenges

There is a gap in research about how the same records of teaching, such those in our video case studies, can be used effectively for learning in different contexts (Ball, Ben-Peretz, & Cohen, 2014). This relates to our project particularly with respect to the video clips we select that are essential ingredients in the video cases. We have approached this issue by using the TRU Framework to guide our selection of video, and studying the ensuing discussions as teachers mutually engage in analyzing and discussing the video, and iteratively improving our selection over time. We see the TRU Framework, the rationale for selecting video, and the video itself, all as integral in developing the shared repertoire that sustains our communities of practice in a joint enterprise of learning how to teach *Classroom Challenges* Formative Assessment Lessons.

Our initial thinking was that we could start with a particular dimension of TRU, for example *Agency, Ownership, and Identity*, and look for video that we believed might support rich discussions in that area. We imagined that for a particular *Classroom Challenges* Formative Assessment Lesson, we might end up with several video cases each centered on a particular dimension of TRU. We found, however, that video clips selected in this way would often support some users but not others. A certain clip selected that supported a rich discussion for a group of prospective teachers, for example, might fall flat with practising teachers. Over time we learned that the heart of this difficulty related to *The Mathematics*. Specifically, if the clip did not allow for exploration of a rich mathematical activity (either because aspects of the activity were not clear on tape, or because the lesson activity was not particularly rich), the clip would not work across communities. This changed our approach, and we now use *The Mathematics* as a starting point for selecting the clips and engaging in the video case materials. This approach is theoretically consistent with the nonlinear representation of TRU that places The Mathematics at the center (Figure 10.2).

Another challenge we faced was in deciding which TRU tools to use when discussing the video. As in selecting the video, we proceeded by iteratively testing different approaches. We used tools individually and in combination, pulling from the TRU Conversation Guide (Baldinger & Louie, 2014), the Observation Guide (Schoenfeld and the Teaching for Robust Understanding Project, 2016), *On Target* (Schoenfeld and the Teaching for Robust Understanding Project, 2018), and the Framework itself. We found that each of these tools appeared to produce meaningful learning opportunities for practising teachers and in prospective teacher classrooms. We were particularly impressed with the learning opportunities afforded by *On Target*; practising teachers were able to characterize and unpack complex teaching situations by locating teaching episodes on a target with various descriptors corresponding to a particular dimension. They found using the tool particularly helpful for their

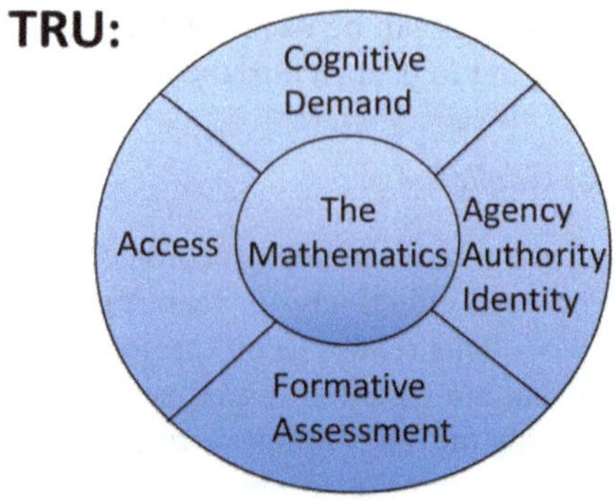

Figure 10.2. A non-linear representation of TRU, representing the interconnections of the five TRU dimensions, with mathematics at the core (from Schoenfeld, 2016, reproduced with permission)

individual growth, often remarking that they knew exactly where they fell on the targets in their own instruction. Watching video of others and thinking about how to move closer to the bullseye gave them ideas for their own practice. This tool also successfully scaffolds discussion of complex moments for prospective teachers; the descriptors, along with the Framework, gave them an accessible language to notice what matters.

And yet something was missing in terms of our project's overall goals. We use communities of practice as a way to both describe and characterize the learning process we are studying. The case studies and TRU-based discussions taking place as various teacher groups engage with the materials allow us to capture the development of teachers' thinking about mathematics instruction, and one of our aims is to reify teachers' ideas about the video in ways that establish a "community memory" (Orr, 1990) about teaching and learning Formative Assessment Lessons. With *On Target*, as with other tools, the great flexibility in the tools led to considerable variation in the analysis by different groups, and often failed to unearth common themes that could be refined and built on over time.

We made progress in this area by simplifying our approach somewhat and adapting one of the most basic TRU tools: a description of the Framework written from the students' point of view in the Observation Guide. Here the dimensions are framed as questions – for example *The Mathematics* is introduced with the question: "What's the big mathematical idea in this lesson? How does it connect to what I already know?"

As we had already decided to select clips based on *The Mathematics*, we wrote the following questions that are now answered by users of the video cases after doing the task in the lesson and before they watch the video:

1. What are different ways to solve the problem?
2. How do the different ways to solve the problem illuminate the big mathematical idea(s) in this lesson?
3. How do different ways to solve the problem connect to one another?
4. What approaches are students likely to take when trying to solve the problem?

For the videos, we created questions for each dimension, again based on the school students' point of view. For *Agency, Ownership, and Identity* we used the characterization: "*Do I get to explain, to present my ideas? Are they built on? Am I recognized as being capable and able to contribute in meaningful ways?*" to write the following questions:

1. What do the students' different explanations tell us about how they might be thinking and what they might understand?
2. Imagine we could go back in time to this part of the lesson and put ourselves in the teacher's shoes. What questions might we ask or what moves might we make to build on the students' thinking?

While these questions are fewer than those posed in *On Target*, we have found that they still support rich discussions around teaching and learning. Critically to our project's goals, these questions have led to shared noticings across groups that focus on students' mathematical understandings. As different conversations are recorded, summarized, and built upon, we see this as reification of our evolving and collective understanding about to teach for robust understanding.

Discussion

Our intent is to produce video cases that give prospective and early career teachers the opportunity to think about complex situations in the classroom and put themselves in the position of decision maker. We hope that focusing on teaching practices rather than the teacher will allow these viewers to consider teachable moments, what one might do next in a lesson, how to handle particular events during the lesson, and the discourse in the classroom. These considerations will build understanding of teaching practices. For practising teachers, the videos provide opportunities for discussion across the five dimensions that build on their own experiences and help foster a vision of classrooms that feature mathematically rich, accessible, and cognitively demanding learning environments.

The TRU Framework is supported by an impressive array of high-quality tools. In our experience all the tools support the learning experiences described above, and some have worked better than others to develop a shared repertoire of practices and orientations across prospective and practising teachers' environments. Our observations that support this position, are still based on a relatively small number of users. Our next step is to build and maintain a micro-site for the cases. This micro-site will be open to any individual or group of educators, and we are planning to reflect on feedback from the users for continued refinement, revision, and development.

TRU AND LESSON STUDY (ANGELA DOSALMAS, HEATHER FINK, SANDRA RUIZ, ALYSSA SAYAVEDRA, ALAN SCHOENFELD, ANNA WELTMAN, AND ANNA ZARKH; SUZANNE DONOVAN AND KAREN TRAN; COURTNEY ORTEGA AND MARY REED; CATHERINE LEWIS)

The TRU-Lesson Study partnership (UC Berkeley, the Strategic Education Research Partnership (SERP) Institute, Mills College, and the Oakland Unified School District) is a National Science Foundation-funded effort to enhance teacher professional development in ways that combine the strengths of the TRU framework and Lesson Study, a collaborative teacher learning program with origins in Japan. TRU-Lesson Study, like Lesson Study, engages teachers in inquiry and reflection around important problems of practice through inquiry cycles of studying, planning, enacting, and reflecting that culminate in live research lessons (Lewis & Hurd, 2011). Every step of the TRU-Lesson Study inquiry cycle, from design to enactment to reflection, is framed by the vision of mathematics teaching and learning provided by the TRU Framework. Hence, Lesson Study provides the overarching activity structure while TRU provides a theoretical and structural frame for professional learning content (Schoenfeld, Dosalmas, Fink, Sayavedra, Weltman, Zarkh, & Zuniga-Ruiz, 2019).

Ultimately, the goal of TRU-related professional development is for individual teachers and teacher learning communities to "own" and to internalize TRU – for the principles underlying TRU to frame both lesson planning and in-the-moment enactment of lessons, as mechanisms to support powerful instruction. This raises significant tensions. On the one hand there is a body of knowledge to be internalized (not simply "learned" – the goal is not to talk about or recognize TRU dimensions, but to think with them); on the other hand, there are issues of building and teacher autonomy to be respected.

As planners and facilitators of professional development, we often found ourselves wondering: How do we as teacher educators set learning goals that inspire but do not constrain or impose? How do we present teachers with a framework meant to structure their work, but also support dialogue that allows teachers to engage with issues that are meaningful for them and their community? How do we avoid making "learning TRU" the goal, rather than "learning to use TRU as a means to inquire into one's practice, foster communication, and improve instruction"? Such issues arise in all professional development, of course, especially in work that tries to be respectful of and support teachers' professionalism. We were not always successful in negotiating these tensions; some sessions were too much about TRU and some seemed to push our agenda more than might have been helpful. But over time, with feedback and review after every session with teachers, we learned how to better integrate the TRU Framework into our teachers' inquiry projects in ways that provided teachers with an ambitious horizon of mathematics teaching and learning to aim for, but also nurtured teachers' own valued problems of practice and ownership over them as a community.

In the remainder of this section we highlight two decisions that we faced as teacher educators designing and implementing TRU-Lesson Study in which the tension between structure and agency was salient.

Decision 1: How Should Teacher Educators Integrate the TRU Framework into the Lesson Study Activity Structure?

TRU can be used in generative fashion as well as for reflection. However, there is a lot to learn before one can be fluent with it. An early issue, then, was how we might integrate TRU with the more classic lesson study processes. Lesson study provides an activity structure within which teachers can pursue collaborative inquiry projects while the TRU Framework, on the other hand, comes with no such structure for activity.

Initially, we used TRU as a lens for reflecting on videos of practice. For a particular video, what could we say about (for example) students' agency? What opportunities did the students have to develop productive mathematical identities, and how might the space of opportunities be opened up? Similarly, when we integrated TRU and Lesson Study, TRU played a natural role in the formal lesson commentaries – How rich was the mathematics; when and where were the students engaged in productive struggle; which students participated, in which ways; what opportunities were there for agency, etc.

As intended, the TRU Framework supported teachers in noticing features of lessons, students' engagement, and teachers' decisions as experienced through the eyes of a student. However, using the TRU Framework primarily as a reflection tool had some drawbacks related to the balance between agency and structure. We began to see some teachers understanding the framework as a static, canonical entity to be used as a reference for checking whether the lesson they observed or planned had all of the features "required" by TRU. They used the framework to label their observations, with the result that their reflections tended to stay in the territory of *what* was noticed or planned, rather than *how* what was noticed or planned came about or *why* it mattered for students. Using the TRU Framework often became the endpoint of conversations – an evaluation rather than the beginning of deeper reflection. This led us to develop a new activity structure aimed at enhancing teachers' own agendas.

The new activity structure, the TRU Inquiry Cycle, was incorporated into the study phase of Lesson Study. As in Lesson Study, teachers began by setting a shared goal based in a current problem of practice. Then, to explore that goal, they choose a pedagogical strategy to try in their classrooms.[10] The teachers collected classroom artifacts demonstrating how students reacted to the pedagogical strategy, brought them to department meetings, and reflected on how the strategies had played out, using the TRU Framework. Typically, those reflections spurred revisions of their statements of their problem of practice and goals; those, in turn, led to the selection of new or revised pedagogical strategies, launching another round of the inquiry cycle. As the teachers came to use the TRU Framework to pose questions for inquiry and to design teaching experiments, they were positioned as initiators of the work rather than as consumers. This gave them authority over the meaning of the framework, with the TRU team playing a supportive role rather than dictating meaning from a position of authority.

Decision 2: How Should Mathematics Teacher Educators Reference and Leverage the TRU Framework in Conversation with Teachers?

A second decision that we faced in balancing the structure provided by the TRU Framework with our commitment to teachers having agency over what and how they learned concerned in-the-moment decisions of when and how to incorporate the TRU Framework into teacher discussions. Within a given conversation, we asked ourselves: when was a good time to push discussions in particular direction, and how should such moves be best articulated? We have found that these in-the-moment decisions cue facilitators' and TRU's positioning in ways that influence teachers' sense of agency, authority and identity.

Here an analogy to classroom instruction may be useful. The classic "demonstrate and practice" form of instruction in mathematics (Lappan & Phillips, 2009) has the virtue of clarity: students know what they are supposed to be doing. However, it denies them agency: they are doing "other people's mathematics." In contrast, problem-based learning starts with issues (admittedly, typically chosen by instructors) and then builds on student thinking. Our work takes this approach one step further. We find that it is best to start with goals and problem statements that come from teachers. Then, appropriately timed interventions using TRU as a tool can help demonstrate its value, in service of the teachers' goals. To give one example, teachers at a particular site were concerned that their students did not persevere when working on the problems the teachers had designed. They had created resources for the students; why weren't the students using them?

The challenge with regard to framing things in terms of perseverance is that it can be a dead end: "what can we do if the students won't persevere?" When this issue arose, the TRU facilitators helped re-frame the question. Perseverance is a function of agency (Dimension 4 of TRU): students are likely to persevere if they think they have a chance of success. How do they develop that sense of agency? By being successful. How does that happen? When instruction offers students challenges that are within reach (a matter of formative assessment and cognitive demand, Dimensions 5 and 2), giving students an opportunity to make legitimate progress and build agency.

This kind of re-framing helped teachers pursue their own goals ("We need to craft tasks and lessons in which students can experience legitimate success"), both supporting teacher agency and demonstrating the ways in which TRU can facilitate their own agendas. We have found that if communities of teachers develop routines around using the TRU Framework to problematize their inquiry goals, and around actively negotiating how the TRU Framework can support their inquiry, that in-the-moment integrations of the TRU Framework into conversation are more likely to be taken up as inviting, rather than constraining, teachers' agency. However, we also found that each decision to bring the TRU Framework into a teacher conversation already-in-progress must be carefully considered. What impact will the TRU Framework have on the trajectory of conversation? How will its introduction position the teachers and facilitators with respect to each other and the broader professional community? What shared meanings have teachers and facilitators begun to develop around the TRU Dimensions, and how will invoking the dimensions constrain or open up dialogue about those meanings? Any introduction of the framework will necessarily redirect conversation, reposition teachers and facilitators, and assert meanings for key terms.

Discussion

The non-prescriptive nature of the TRU Framework – TRU does not tell teachers or teacher educators what to do – is a significant virtue, but it may also be its greatest challenge. We have found ourselves struggling to balance the need to bring TRU forward when we see it can help and the need to respect teacher agency and authority. There are, we suspect, no easy solutions to this dilemma – although we hope that the construction of additional tools will provide more resources for teacher educators as they deal with this challenge.

CONCLUDING DISCUSSION

The TRU framework was designed to focus on what counts – to the degree that a framework with five dimensions *can* focus. Any distillation of a phenomenon as complex as teaching into five dimensions necessarily foregrounds some critical concerns and backgrounds others, issues of race and power being examples raised in an earlier section of the chapter. It is not that such issues are not implicated; one cannot reasonably consider issues of equitable access and agency/ownership/identity without dealing with issues of race and power head on. But, there is a lot of unpacking to be done to help TRU deal adequately with such issues, and to provide useful tools for addressing them. Doing so with any degree of success will require a significant program of research and development. Learning how to bring such concerns naturally into TRU-based professional development will take some learning on the part of teacher educators.

The non-prescriptive character of TRU provides essential opportunities and in doing so raises a set of challenges and tensions. Ultimately, there is a need for powerful and self-sustaining teacher learning communities – communities that continue to refine their members' understandings and practices in ongoing ways. This is essential for two reasons: (1) learning communities are the best "growth medium" in which understanding can take hold and grow, and (2) as a matter of scale, there simply are not enough teacher educators to provide the relevant support for teachers, on their own. Teacher educators can serve as catalysts, but part of our role needs to be to help the communities we help foster become increasingly independent (but supported with good tools, of course.)

In the United States, at least, this means that community *building* is a critical part of the challenge – Lortie's (1975) invocation of the "egg crate" to describe teacher isolation is still a significant reality. Bringing teachers together and telling them what to do is deprofessionalizing – and it does not work. As a matter of respect and because meaningful attention to teacher support is context-dependent, there must be substantial flexibility. But with such flexibility come tensions related to focus and coherence. We have found that there are certain patterns of teachers' perceived needs: in the United States, Dimension 4 (agency, ownership, and identity) is often perceived as a needed expansion of a focus on equitable access (Dimension 3), and a good place to dig in at first; after some time with Dimension 4, it becomes clear that efforts will be more effective if one understands how to help students engage in productive struggle (Dimension 2), and teacher learning communities often turn to that. This, of course, necessitates attention to student thinking (Dimension 5) – all the time, with content worth engaging with (Dimension 1).

What we have just outlined is one possible order of a curriculum for professional development. Its effectiveness would depend, of course, on community; on issues being meaningful and workable for participants; on their making it their own. If that sounds familiar, it should. TRU is a theory of productive learning environments, and if teacher educators are to help teachers shape powerful learning communities, those communities themselves should do well along the dimensions of TRU (see Schoenfeld, 2015, for more detail).

There is one further challenge to community building. Point (2) above was that even if teacher educators in the United States were familiar with and predisposed toward using TRU, there is not an adequate number of teacher educators to provide the relevant support. Thus, further work needs to be done along at least two dimensions: helping communities that have made significant progress to become self-sustaining, so that teacher educators can be freed to have broader impact, and building networks in ways that teachers themselves can become ambassadors of change, and mentors to other teachers.

Making this happen is a significant challenge. But as the discussions in this chapter indicate, taking on that challenge is a source of significant learning for both teacher educators and the teachers they work with.

ACKNOWLEDGEMENT

This chapter was produced with support from the National Science Foundation grant 1503454, "TRUmath and Lesson Study: Supporting Fundamental and Sustainable Improvement in High School Mathematics Teaching," a partnership between the Oakland Unified School District, Mills College, the SERP Institute, and the University of California at Berkeley.

NOTES

1 See http://TRUframework.org
2 See https://truframework.org/tru-conversation-guide/
3 In California, most teachers earn their Teaching Credential subsequent to earning their undergraduate (Bachelor's) degree. The MACSME program combines this professional credential with an academic program that leads to the Master's degree.
4 See https://docs.google.com/document/d/1NtkzSgL0LRWeU-nG6TTgdvLfzSAKm7VT2GGwjpBwk1k/edit
5 See http://www.svmimac.org/home.html
6 Space limitations preclude a discussion of SVMI's history and contributions. For partial documentation of impact see Boaler and Foster (2018), Foster, Noyce, and Spiegel (2007), Foster and Paek (2012), Foster and Poppers (2009), and Ridgway, Crust, Burkhardt, Wilcox, Fisher and Foster (2000).
7 http://www.sfusdmath.org/math-talks-resources.html
8 http://www.sfusdmath.org/3-read-protocol.html
9 http://map.mathshell.org/lessons.php
10 A TRU tool we created offered a list of strategies. Teachers were not constrained to this list, but they typically used it as a resource and selected strategies from it.

REFERENCES

Baldinger, E., Louie, N., & The Algebra Teaching Study and Mathematics Assessment Project. (2014/2016). *TRU conversation guide: A tool for teacher learning and growth.* Berkeley, CA & E. Lansing, MI: Graduate School of Education, University of California, Berkeley & College of Education, Michigan State University. Retrieved from http://ats.berkeley.edu/tools.html and/or http://map.mathshell.org/materials/pd.php

Ball, D., Ben-Peretz, M., & Cohen, D. (2014). Records of practice and the development of collective professional knowledge. *British Journal of Educational Studies, 62*(3), 317–335.

Ball, D., Thames, M. H., & Phelps, G. (2008). Content knowledge for teaching: What makes it special? *Journal of Teacher Education, 59*(5), 389–407.

Boaler, J., & Foster, D. (2018). *Raising expectations and achievement. The impact of wide scale mathematics reform giving all students access to high quality mathematics.* Retrieved June 10, 2018, from http://www.svmimac.org/images/Raising-Expectations_Paper_by_Jo_and_David.pdf

Cabana, C., Shreve, B., & Woodbury, E. (2014). Building and sustaining professional community for teacher learning. In N. S. Nasir, C. Cabana, B. Shreve, E. Woodbury, & N. Louie (Eds.), *Mathematics for equity: A framework for successful practice* (pp. 175–186). New York, NY: Teachers College Press.

Darling-Hammond, L., Hyler, M., & Gardner, M. (2017). *Effective teacher professional development.* Palo Alto, CA: Learning Policy Institute. Retrieved from https://learningpolicyinstitute.org/product/teacher-prof-dev

Desimone, L. (2009). Improving impact studies of teachers' professional development: Toward better conceptualizations and measures. *Educational Researcher, 38*(3), 181–199.

Foster, D., Noyce, P., & Spiegel, S. (2007). When assessment guides instruction: Silicon Valley's mathematics assessment collaborative. In A. Schoenfeld (Ed.), *Assessing mathematical proficiency* (pp. 137–154). Cambridge: Cambridge University Press.

Foster, D., & Paek, P. (2012, April 15). *Improved mathematical teaching practices and student learning using complex performance assessment tasks.* Paper presented at the annual meeting of the National Council on Measurement in Education (NCME), Vancouver, Canada.

Foster, D., & Poppers, A. (2009). *Using formative assessment to drive learning.* Retrieved June 10, 2018, from http://www.svmimac.org/images/Using_Formative_Assessment_to_Drive_Learning_Reduced.pdf

Horn, I. S., & Little, J. W. (2010). Attending to problems of practice: Routines and resources for professional learning in teachers' workplace interactions. *American Educational Research Journal, 47*(1), 181–217.

Lappan, G., & Phillips, E. (2009). A designer speaks. *Educational Designer, 1*(3). August 7, 2013, http://www.educationaldesigner.org/ed/volume1/issue3/article11

Leana, C. R. (2011, Fall). The missing link in school reform. *Stanford Social Innovation Review*, 30–35.

Lortie, D. (1975). *School teacher: A sociological study.* Chicago, IL: University of Chicago Press.

Martin, D. B. (2003). Hidden assumptions and unaddressed questions in mathematics for all rhetoric. *Mathematics Educator, 13*, 7–21.

National Council of Teachers of mathematics. (2014). *Principles to actions: Ensuring mathematical success for all.* Reston, VA: National Council of Teachers of Mathematics.

Orr, J. E. (1990). Sharing knowledge, celebrating identity: Community memory in a service culture. In D. Middleton & D. Edwards (Eds.), *Collective remembering* (pp. 169–189). Newbury Park, CA: Sage Publications.

Ridgway, J., Crust, R., Burkhardt, H., Wilcox, S., Fisher, L., & Foster, D. (2000). *MARS report on the 2000 tests.* San Jose, CA: Mathematics Assessment Collaborative.

Schoenfeld, A. H. (2013). Classroom observations in theory and practice. *ZDM, the International Journal of Mathematics Education, 45*, 607–621. doi:10.1007/s11858-012-0483-1

Schoenfeld, A. H. (2014, November). What makes for powerful classrooms, and how can we support teachers in creating them? *Educational Researcher, 43*(8), 404–412. doi:10.3102/0013189X1455

Schoenfeld, A. H. (2015). Thoughts on scale. *ZDM, the International Journal of Mathematics Education, 47*, 161–169. doi:10.1007/s11858-014-0662-3

Schoenfeld, A. H. (2016, May 15). *Creating classrooms that produce powerful mathematical thinkers.* Keynote presentation, 8th International Conference on Technology and Mathematics education and workshop on mathematics teaching, Taipei, Taiwan.

Schoenfeld, A. H. (2018). Video analyses for research and professional development: The Teaching for Robust Understanding (TRU) framework. In C. Y. Charalambous & A.-K. Praetorius (Eds.), *Studying instructional quality in mathematics through different lenses: In search of common ground* (An issue of ZDM). https://doi.org/10.1007/s11858-017-0908-y

Schoenfeld, A., Dosalmas, A., Fink, H., Sayavedra, A., Weltman, A., Zarkh, A, Tran, K., & Zuniga-Ruiz, S. (2019). Teaching for Robust understanding with lesson study. In R. Huang, A. Takahashi, & J. P. Ponte (Eds.), Theory and practices of lesson study in mathematics: An international perspective. New York, NY: Springer.

Schoenfeld, A. H., & The Teaching for Robust Understanding Project. (2016). *An introduction to the Teaching for Robust Understanding (TRU) framework.* Berkeley, CA: Graduate School of Education. Retrieved from http://map.mathshell.org/trumath.php or http://tru.berkeley.edu

Schoenfeld, A. H., & The Teaching for Robust Understanding Project. (2016). *The Teaching for Robust Understanding (TRU) observation guide: A tool for teachers, coaches, administrators, and professional learning communities.* Berkeley, CA: Graduate School of Education, University of California, Berkeley. Retrieved from http://TRUframework.org

Schoenfeld, A. H., & The Teaching for Robust Understanding Project. (2018). *On target: A TRU guide to crafting richer learning environments for our students.* Berkeley, CA: Graduate School of Education, University of California, Berkeley. Retrieved from http://TRUframework.org

Alan H. Schoenfeld
University of California, Berkeley

Evra Baldinger
University of California, Berkeley

Jacob Disston
University of California, Berkeley

Suzanne Donovan
SREP Institute, Washington DC

Angela Dosalmas
University of California, Berkeley

Michael Driskill
Math for America, New York

Heather Fink
University of California, Berkeley

David Foster
Silicon Valley Mathematics Initiative

Ruth Haumersen
Chicago P12 Mathematics Collaborative
DePaul University

Catherine Lewis
Mills College

Nicole Louie
University of Wisconsin-Madison

Alanna Mertens
Chicago P12 Mathematics Collaborative
DePaul University

Eileen Murray
Montclair State University

Lynn Narasimhan
Chicago P12 Mathematics Collaborative
DePaul University

Courtney Ortega
Oakland Unified School District

Mary Reed
Oakland Unified School District

Sandra Ruiz
University of California, Berkeley

Alyssa Sayavedra
University of California, Berkeley

Tracy Sola
Silicon Valley Mathematics Initiative

Karen Tran
SREP Institute, Oakland, CA

Anna Weltman
University of California, Berkeley

David Wilson
SUNY Buffalo State

Anna Zarkh
University of California, Berkeley

JAMES A. MENDOZA ÁLVAREZ, KATHRYN RHOADS AND THERESA JORGENSEN

11. MATHEMATICS TEACHER EDUCATORS LEARNING FROM EFFORTS TO FACILITATE THE LEARNING OF KEY MATHEMATICS CONCEPTS WHILE MODELLING EVIDENCE-BASED TEACHING PRACTICE

Many mathematics teacher educators regularly assess their efforts to promote critical mathematical insight and understandings in a classroom environment that models evidenced-based teaching practice. Mathematics teacher educators may face challenges in linking key mathematical insights to school mathematics and maintaining a learning environment that engages prospective or practising teachers in a manner that reflects the expectations for teaching practice in pre-tertiary classrooms. This chapter describes the evolution of mathematical tasks used by mathematics teacher educators as they learn from practice when emphasizing the learning of important mathematical concepts while simultaneously firmly grounding advanced mathematics topics in school mathematics. For each task, we will discuss mathematics teacher educators' learning resulting from the process of task revision, research on their own practice, and focusing on teacher learning and on facilitating teacher learning.

INTRODUCTION

Facilitating the learning of key mathematical concepts while modelling evidence-based teaching practices provides many opportunities for mathematics teacher educators to learn from their practice. In mathematics courses designed for teachers and offered in departments of mathematics, finding appropriate ways to connect pre-tertiary mathematics topics to advanced mathematics also contributes to mathematics teacher educators' understanding of teacher learning and facilitating teacher learning.

We write as three mathematics teacher educators who work with both prospective and practising secondary mathematics teachers in a university setting. The courses we teach are mathematics courses, but these courses are designed specifically for mathematics teachers. The process of designing tasks for these courses has taught us about teaching, learning, and mathematics. We designed tasks both for immediate

© KONINKLIJKE BRILL NV, LEIDEN, 2020 | DOI: 10.1163/9789004424210_012

use in our own courses and for use by novice mathematics teacher educators. As such, for many tasks, we have developed facilitation notes for mathematics teacher educators that intend to capture some of the instructional practices that we have found to be effective from our research and our experiences.

In this chapter, we describe the evolution of tasks from our practice and the mathematics teacher educator learning that occurred both from interactions with prospective and practising teachers and efforts to improve the facilitation of the tasks. The tasks discussed fall under three topics: Functions and Equations, Visualising Complex-valued Zeros, and Building Functions. Each topic was addressed in an undergraduate course for prospective secondary school mathematics teachers and a graduate course for practising secondary school mathematics teachers. The evolution of each task followed our assessment of how the task promotes critical mathematical understandings, how its facilitation responds to the use of evidence-based teaching and learning practices, how it is grounded in school mathematics, and how well its use can be aligned with learning environments that model learning environment expectations in secondary school classrooms. For each task, we frame our learning as mathematics teacher educators following Zaslavsky's (2008) seven themes (described later in the Conceptual Framework) which "represent qualities and kinds of competence and knowledge that mathematics teacher education seeks to promote in prospective and practising teachers in a broad sense" (p. 95) and can be linked to how mathematics teacher educators may use carefully designed tasks to address them. Our growing understandings of the tasks through practice and research on our practice, the prospective and practising teachers in our courses, and the mathematics in this context contribute to our view of the effectiveness of a task in meeting the challenges related to the themes described by Zaslavsky (2008).

BACKGROUND

Although mathematics teacher education is a rapidly growing area of research, there remain few resources for mathematics teacher educators as they develop, adapt, and implement tasks for future or practising teachers (Zaslavsky, 2008; Koichu, Zaslavsky, & Dolev, 2016), and we have found the resources to be especially limited for mathematics teacher educators working with secondary school teachers. Yet, some recent research has contributed to a growing understanding of using tasks in mathematics teacher education, and we draw on this research in our work.

We use Epperson and Rhoads' (2015) definition of *task*, which is "a mathematical problem, prompt, or guided exploration that is posed to learners" (p. 38). Like Epperson and Rhoads (2015) and Watson and Sullivan (2008), we consider a task to be "a starting point of mathematical activity" (Watson & Sullivan, 2008, p. 109). Drawing from a review of the literature on tasks in mathematics teacher education as well as their own experiences in mathematics teacher education, Epperson and Rhoads (2015) shared three guiding characteristics for choosing or developing high-yield tasks for practising secondary teachers. Tasks can aim to build deeper and more flexible understanding of mathematical concepts, yet tasks can be grounded in school mathematics so that teachers can see how they will use the concepts in their work with secondary students. In addition, tasks can reinforce mathematical habits of mind, such as conjecturing, justifying, and making connections among multiple representations. Epperson and Rhoads' characteristics are broad, but research has also shown that more specific details of a task influence what teachers learn from it. Koichu et al. (2016) described three iterations of a geometry task used with practising secondary school mathematics teachers. The authors cautioned that too many technical components of a task can detract from the conceptual goals of the task, and sequencing of activities within a task is also important to ensure that teachers' decisions and thinking are developing in productive ways as they complete the task.

As mathematics teacher educators develop and implement tasks for mathematics teachers, they are also developing and strengthening their own knowledge – particularly their pedagogical content knowledge. As introduced by Shulman (1986), pedagogical content knowledge blends content and pedagogy and is not needed in contexts outside of teaching. For example, pedagogical content knowledge includes knowledge of how to explain content, ways of representing the content, and what aspects of the content may pose challenges to learners. Since Shulman introduced the term, a great deal of research has focused on the pedagogical content knowledge needed by mathematics teachers, but research on pedagogical content knowledge needed by mathematics teacher educators is a much more recent but growing field.

For example, Chauvot (2009) systematically analysed her experiences as a mathematics teacher educator to create a knowledge map of the knowledge she uses as a mathematics teacher educator-researcher. The map included content knowledge, curriculum knowledge, and pedagogical content knowledge. Within pedagogical content knowledge, Chauvot highlighted her knowledge of college students/adults as learners as well as her knowledge of teacher learning, including teachers' beliefs about mathematics, mathematics teaching, and mathematics learning. More recently, Chick and Beswick (2018) presented a framework for MTEPCK (mathematics teacher educator pedagogical content knowledge), which built on existing research in pedagogical content knowledge as well as the work of teacher education. The authors argued that each aspect of schoolteachers' pedagogical content knowledge has a related component of MTEPCK. For example, whereas knowledge of examples that illustrate a particular mathematical concept or procedure is a part of school mathematics teachers' pedagogical content knowledge, knowledge of examples that illustrate pedagogical content knowledge concepts for school mathematics teachers is part of MTEPCK. In other words, mathematics teacher educators must have knowledge of how to develop school mathematics teachers' pedagogical content knowledge.

A different but related concept was introduced by Even (2008), in which she discussed the challenges in educating practising mathematics teachers. Even argued that effective mathematics teacher educators draw on both knowledge and practice – the concept of *knowtice* – when working with practising mathematics teachers.

Knowtice includes both knowledge and practice in the areas of mathematics education, mathematics content, teacher education, and practices of teacher education.

A key method by which mathematics teacher educators develop may be through research on their own practice. Chapman (2008) proposed four conditions that enable mathematics teacher educator learning from research on their practice. First, mathematics teacher educators should not judge the instructional approach being used in the research. Second, mathematics teacher educators should be focused on teacher thinking as they engage in the instructional approach. Third, mathematics teacher educator learning is enabled when mathematics teacher educators experience conflict between what was expected and what actually happened in their instructional approach. Fourth, mathematics teacher educators critically examine teachers' learning. These conditions were met in a lesson experiment described by Chamberlin and Candelaria (2018). The authors shared what they learned through the implementation and revisions of a lesson with prospective elementary teachers, focusing both on how their instruction affected the teachers' understanding of the content as well as what they learned from the lesson experiment process. However, Chamberlin and Candelaria's paper is somewhat rare: Chapman (2008) argued that (at the time the chapter was written) very few studies on mathematics teacher educator learning through research addressed mathematics teacher educator learning explicitly, and she recommended that future studies articulate "how the teacher-educator-researchers reflected, what practical knowledge they acquired, and how this knowledge impacted or is likely to impact their future behaviour in working with their students" (p. 132). In this chapter, we respond to Chapman's call by discussing our mathematics teacher educator learning from practice, much of which came as a result of our research on our practice.

CONCEPTUAL FRAMEWORK

The work of mathematics teacher educators can be framed by the goals that mathematics teacher educators have as facilitators of teacher learning. Zaslavsky (2008) summarized these goals in seven themes, which represent both goals for mathematics teacher education and the challenges inherent in them. Each theme can be considered from both a mathematical and a pedagogical perspective. For each goal, mathematics teacher educators must both demonstrate the theme and provide opportunities for prospective teachers and practising teachers to experience it. The themes are:

- *Developing adaptability*, which includes developing in teachers an orientation to being adaptable with regards to tasks, curriculum, approaches, etc.,
- *Fostering awareness to similarities and differences,* which includes helping teachers to develop a state of mind that includes a tendency to notice and identify similarities and differences,

- *Coping with conflicts, dilemmas and problem situations,* which includes preparing teachers to be problem solvers who can deal with conflicting constraints, both in mathematical problem situations and within the uncertainties and complexities of decision making in their classrooms,
- *Learning from the study of practice*, which includes developing in teachers a career long orientation to learning from the study of their own teaching and the teaching of others,
- *Selecting and using (appropriate) tools and resources for teaching,* which includes enhancing teachers' competence for selecting and effectively using tools, while being sensitive to teachers' general reluctance to use unfamiliar innovations in their teaching,
- *Identifying and overcoming barriers to students' learning*, which includes educating teachers on the existence and sources of barriers, and developing ways to engage all students in meaningful mathematics, and
- *Sharing and revealing self, peer, and student dispositions*, which includes helping teachers become aware of their own beliefs and their students' dispositions, and the impact those can have on opportunities to learn mathematics (Zaslavsky, 2008).

Implementing appropriate mathematical tasks is a key way through which mathematics teacher educators transform the learning of prospective teachers and practising teachers, and thus meet these challenges of mathematics teacher education. Typically, a mathematics teacher educator is involved in the process of designing such tasks, because the availability of resources for mathematics teacher educators to draw upon when structuring learning for their teachers is limited. This process often features the interplay between mathematics teacher educator research and practice. The reflective process of designing, implementing, and modifying tasks is a vehicle for mathematics teacher educator learning. Zaslavsky (2008) argued that a natural way to track mathematics teacher educator growth is to use the evolution of a well-designed mathematical task as a platform for understanding how mathematics teacher educators use and construct their own knowledge in the process of facilitating teachers' learning.

Following Zaslavsky's (2008) framework, we will document our own growth as mathematics teacher educators by connecting the design of tasks and the seven themes above. We will show how these themes are interwoven in the process of task design and adjustment, and how this process tracks our professional development as mathematics teacher educators, going beyond the confines of each task.

SETTINGS

In this chapter, we will describe our learning – as mathematics teacher educators – through the practice of mathematics teacher education in two settings: one setting in which we work with undergraduate prospective teachers of secondary school mathematics and one in which we work with graduate practising teachers of secondary school mathematics. Both populations attend the same highly diverse, urban university in the southwestern United States with over 42,000 students. Our work is situated in a Department of Mathematics, and we teach mathematics content courses. As we plan and implement lessons, teachers' pedagogical development is an important consideration, but our primary focus is on teachers' mathematical learning. As such, the MTEPCK that we develop through practice is mainly focused on teachers' learning of mathematics and secondarily focused on their learning about *teaching* mathematics.

At the same time, a major challenge of our work is ensuring that we model evidence-based teaching practices. That is, our teaching practices are grounded in research on engaging and productive learning environments. For example, we aim for all mathematics courses for teachers to have a strong inquiry component. The inquiry-based learning community in mathematics describes inquiry-based learning in mathematics as engaging students in sense-making activities. Mathematics teacher educators take the role of guide or mentor, and key components of our activities include, "deep engagement in rich mathematical activities" and "opportunities to collaborate with peers" (Academy of Inquiry-Based Learning, n.d.). In science education, the inquiry continuum includes *confirmation inquiry* in which students confirm known results, *structured inquiry* which includes teacher-presented and teacher-scaffolded questions, *guided inquiry* which includes teacher-presented questions but student-selected approaches and procedures, and *open inquiry* in which questions are student-formulated and students design and select procedures (Brachi & Bell, 2008). Consistent with the inquiry-based learning community in Mathematics and the science education inquiry continuum, we define inquiry-based instructional materials as classroom tasks that engage students in sense-making, foster making rich mathematical connections, and generate opportunities for collaboration among peers.

As such, we aim to embed collaborative learning as a component of all mathematics courses for teachers. We strive to implement Stein, Engle, Smith, and Hughes' (2008) *five practices* for orchestrating productive mathematical discussions around cognitively demanding tasks: anticipating, monitoring, selecting, sequencing, and connecting (p. 322; see also Smith & Stein, 2011, p. 8). Other evidence-based practices we employ include setting ambitious learning goals for lessons, building on teachers' existing knowledge and skills, focusing on conceptual understanding (as opposed to only skill proficiency), paying attention to teachers' ways of thinking as they engage in mathematics, and reflecting on our practice (e.g., Hiebert, 2003; Watson & Mason, 2007).

Setting 1: Undergraduate Prospective Secondary School Mathematics Teachers

The university uses the UTeach (UTeach Institute, n.d.) teacher preparation program for students majoring in science, mathematics, and computer science, a program replicated at over 40 universities in the United States. The tasks we describe for undergraduate prospective teachers come from a mathematics course, usually called Functions and Modeling, that is designed for all mathematics-intended UTeach majors. This is a required course that typically follows a second-semester calculus course. The intent of the course is to deepen prospective teachers' experiences with functions and immerse them in an inquiry-based learning environment. The UTeach curriculum materials include the course manuscript, Functions in Mathematics (Armendariz & Daniels, 2011), which is disseminated by the UTeach Institute. The materials contain 23 lessons, each consisting of several explorations that are meant to be implemented using an inquiry-based approach.

The authors are conducting research on the *Functions and Modeling* course, as principal investigators for the *Enhancing Explorations in Functions for Preservice Secondary Mathematics Teachers* Project. The project is partially funded by the United States National Science Foundation. Our goal is to develop research-based tasks and explorations for use in mathematics courses for prospective teachers of secondary school mathematics, as well as to develop mathematics teacher educator materials that assist mathematicians and other mathematics teacher educators in using the tasks and explorations in an inquiry-based, active learning environment. All authors of this chapter have experience in teaching this course.

Our project uses a design experiment framework. We aim to create research-based materials for prospective teachers of secondary school mathematics while simultaneously studying prospective teachers' processes of learning and how the materials and classroom environment can support their learning (e.g., Cobb, Confrey, di Sessa, Lehrer, & Schauble, 2003). Following recommendations by the Design-Based Research Collective (2003), we follow a cyclic process as we engage in "design, enactment, analysis, and redesign" (p. 5). Our research is both "prospective and reflective" (Cobb et al., 2003, p. 10) in that we draw on existing theory and research to design and implement instructional materials, and we simultaneously collect data and reflect on the success of the materials in an iterative process. As the mathematics teacher educators engaged simultaneously in instruction and research, our learning drove the evolution of tasks.

Although all the authors of this chapter have practitioner experience on which they draw, we have collected formal research data from three semesters of the *Functions and Modeling* course, each iteration taught by one of the authors. The class met 30 times over the semester (twice per week for 80 minutes each meeting). A graduate assistant was present in all class meetings and video recorded these meetings (except for three exam days). During the second iteration, at least one non-teaching author of this chapter observed each class meeting in which new materials were implemented and took notes on the class environment, lesson structure, implementation of the lesson, and mathematics content. The course was inquiry-oriented, and students worked in groups of three to four to learn concepts through inquiry-based tasks. The mathematics teacher educator facilitated small group work and some whole-class discussion, but there were very few lectures.

Approximately 20–30 prospective teachers were enrolled in *Functions and Modeling* each semester. A copy of all written work was collected for each participant who consented to participate in the research. This included written explorations, a daily journal, homework, hands-on experiments, exams, and a midterm project. All prospective teachers also completed a pre- and post-assessment on function knowledge that we developed, and 20 participants (approximately 10 each semester) completed individual, task-based interviews with a graduate assistant to further explore their understanding of functions.

Before each class meeting, the mathematics teacher educators completed a written log documenting his or her plans, and after each class, the mathematics teacher educator completed an interview with the graduate assistant to debrief and discuss her implementation choices.

Setting 2: Graduate Practising Secondary School Mathematics Teachers

The tasks we describe for graduate practising teachers come primarily from a graduate mathematics program designed for practising secondary school teachers in a department of mathematics. The goal of the program is for practising teachers to deepen their mathematical knowledge of high school concepts from an advanced standpoint. Participants earn a Master of Arts (M.A.) degree in mathematics, extending their undergraduate mathematics knowledge in the area of specialized content for secondary school teaching.

Approximately 10–25 practising teachers were enrolled in each course. A one-semester course in the M.A. program meets once per week for 15 weeks, with each class meeting lasting three hours. The courses in the M.A. program are all inquiry oriented and incorporate extensive group work. The mathematics teacher educator acts as a facilitator for small-group and whole-class discussion.

All authors of this chapter have experience teaching courses for the M.A. program. Although we have not conducted formal research on the design and re-design of tasks for the M.A. courses, we draw on our collective experience in teaching these courses for over 15 years and our reflection through observing practising teachers' discussions and questions, writing informal teacher logs, assessing written work from practising teachers, and discussing teaching ideas and issues with other mathematics teacher educators.

TASKS FROM PRACTICE

In this section, we share examples of tasks from our practice and discuss how the tasks have evolved as we learn from practice. We present tasks in relation to three main topics: Functions and Equations, Visualising Complex-valued Zeros, and Building Functions. Within each topic, we discuss two related tasks: one used with undergraduate prospective teachers and one used with graduate practising teachers.

For each task, we also discuss our learning as mathematics teacher educators from the facilitation, development, and revisions of the task.

Topic 1: Functions and Equations

Functions are a key element of the structure of secondary school and undergraduate mathematics. A modern definition for function is a relation that uniquely associates the elements of one set to elements of another set. Inherent to this definition are the underlying concepts of the associated domain and codomain, and the requirement that for each value in the domain, there is exactly one associated value in the range. The definition of function allows for examples that are not numeric, that cannot necessarily be graphed, or that cannot be defined by an algebraic formula, though such function examples are not typical in grades K-12 mathematics curricula.

Despite extensive exposure to functions in their K-12 and undergraduate studies, prospective teachers' conceptions of functions do not always align with modern definitions. Throughout high school and undergraduate mathematics, students are accustomed to working with functions that can be defined by algebraic formulas, and students often use formulas to identify the functions they discuss (Cooney, Beckmann, & Lloyd, 2010). This can be very useful in courses such as calculus, and such courses can reinforce students' concept image of functions as being defined by formula. Students' conceptions of functions can be limited by such thinking. For example, some prospective teachers believe that a function can *always* be represented by an algebraic formula, and others believe that the terms function and equation are interchangeable (Álvarez, Jorgensen, & Rhoads, 2018; Even, 1993; Hitt, 1998).

Secondary school curricula emphasize that zeros of a function f are the solutions to the equation $f(x) = 0$ (National Governors Association Center for Best Practices and Council of Chief State School Officers [CCSSM], 2010). Although this connection is important, students sometimes misinterpret this relationship. For example, in a study in the United States, Carlson (1998) found that students earning A's in College Algebra, "do not make a distinction between the zeros of functions and solutions to equations" (p. 141). In a 1999 study, Carlson also reported that second-semester calculus students had similar confusions between solutions to equations and zeros of functions. Many prospective teachers have incorrect conceptions about the relationships between functions and equations. For example, in Even's (1993) study, some prospective teachers provided definitions of function in which they claimed a function was an equation or expression. Breidenbach, Dubinsky, Hawks, and Nichols (1992) found that some prospective teachers described a function as "a mathematical equation with variables" (p. 252). In similar fashion, Chazan and Yerushalmy (2003) documented that learners have difficulty in distinguishing between functions and equations.

Functions and Equations in an Undergraduate Course

Evolution of the task. In the *Functions and Modeling* course designed for undergraduate prospective teachers, we had seen evidence of prospective teachers missing the important but under-regarded distinctions between functions and equations. For example, when describing their own thinking, prospective teachers would make statements such as "I solved the function" or "this equation is a function."

To better understand the conceptions that our prospective teachers had of the relationships between function and equation, we used as a pre- and post-test a written instrument consisting of ten items and corresponding sub-items targeting the prospective teachers' understanding of function and equation. The items on the assessment required the prospective teachers to explain their reasoning and provide multiple representations, when appropriate, for example, "Can the terms *function* and *equation* ever be used interchangeably? Why or why not?" The pre-test was administered during the first week of the course and the post-test was completed following the unit on functions. It took the prospective teachers approximately one hour to complete each assessment.

We used qualitative methods to analyse the written responses from the assessments. Using the principles of grounded theory method (Strauss & Corbin, 1990), the data were coded through the lens of emerging themes. The data were then grouped into similar conceptual themes to characterize the prospective teachers' descriptions contrasting function and equation.

We learned from this assessment that the predominant concept image for functions among the prospective teachers entailed the idea that a function establishes a relationship between inputs and outputs, regardless of whether their description of an equation also used the idea of a relationship between quantities. No prospective teachers attempted to contrast equations and functions by referring to solution sets or domain and range, respectively, in either the pre- or the post-test (Álvarez et al., 2018).

Based upon the initial results of the pre-test, we developed an inquiry-based lesson with the goal of developing prospective teachers' understanding that mathematical language in algebra, specifically uses of the terms equation and function, is important and has underlying implications related to student understanding. The lesson focused on the meaning of the term equation and the different meanings associated with the equal sign. Here, constructed meanings refer to meanings constructed by individual learners that may differ from the concept definition (see Noss & Hoyles, 1996, as cited in Kieran, 2007, p. 711). The focus on equations in this lesson linked to generational and transformational activities that involve variables, expressions, and symbols (Kieran, 2007).

This lesson had components that required the prospective teachers to generate their own examples of equations and come to a consensus on a list of such examples. The prospective teachers were then presented with a formal definition of equation (An equation is a mathematical statement that asserts the equivalence between two quantities). The prospective teachers had little problem with understanding this definition and were able to apply it to their equation examples.

The second exploration in the lesson required that the students identify and discuss the different constructed meanings that "=" conveys in various expressions/statements. For instance, prospective teachers were asked to identify the constructed meaning of "=" in $3 + 5 =$ ___, $f(x) = x^2 + 5$, or $x^2 + 3x = x - 1$. (The meanings we were expecting were akin to "*compute*," "*is defined to be*," and "*is equivalent to*" respectively.) This group of prospective teachers was unexpectedly not open to considering situations in which an equal sign could be utilized to define an object, such as $f(x) = x^2 + 5$. They declared the constructed meaning of "=" in that statement to be the same as the constructed meaning in $A = \pi r^2$. Mathematics teacher educator questioning during whole class discussion led to a tenuous agreement among the prospective teachers that there exist different constructed meanings of the equal sign, but there remained dissent about the distinctions between those meanings.

The prospective teachers then considered situations arising from students' work and made connections between that work and the students' understanding of the use of the equal sign. For instance,

> David has no problem with computations such as $3 + 5 =$ ___, but has trouble solving $3 + 5 = 2 +$ ___. David may have a limited understanding of the use of the equal sign. Which meaning might David be missing? How would you know?

The final exploration in the lesson was also situated in teaching practice and connected to school mathematics. Prospective teachers were asked to consider some homework and assessment questions and select the appropriate word or words to use in the question. For instance, (i) *Evaluate/Simplify/Solve* $5x + 2$ when $x = 2$, or (ii) Given functions g and h, *evaluate/simplify/solve* $g(x) = h(x)$. The prospective teachers then created guidelines for a high school mathematics teacher to use for determining when it is appropriate to use the instructions "solve," "evaluate," or "simplify" on her homework assignments and assessments in exercises or tasks involving functions and equations.

In a similar fashion to the whole group discussion of the constructed meanings of the equal sign, there was disagreement among the prospective teachers as to which situations required the use of the instruction "solve" versus "evaluate." Some prospective teachers were adamant that the distinction between the two instructions was meaningless and arbitrary, perhaps showing a misconception in their understanding of the difference between a solution set of an equation and the domain of a function. Mathematics teacher educator questioning was not successful in guiding the prospective teachers to see value in distinguishing between the verbs based upon the situation.

This first iteration of the implementation of this lesson on equations occurred at the end of the *Functions and Modeling* course, and we saw little subsequent evidence of change in the prospective teachers' conceptions of function and equation on the post-test that semester. The prospective teachers, even while they were engaged in the lesson on equation, expressed the view that the ideas of equation, although interesting, did not seem connected to the focus of the course. In individual interviews with prospective teachers at the end of the course, they noted that they still were confused about the difference between functions and equations and did not feel as though they had had sufficient opportunity to understand the distinctions.

Based upon the data and lessons learned from the first implementation of the lesson on equation, we made changes in the timing of the lessons in the *Functions and Modeling* course. In particular, we reordered the lessons in the first unit, which is focused on developing and applying the concept of function, to include the lesson on equations directly after the lesson on the definition of function. We also developed mathematics teacher educator notes to provide facilitation questions for the mathematics teacher educator and to highlight potential caveats and conceptions that the prospective teachers may hold, for instance, that an equation must be a true statement, or confusion between the contexts in which one considers a solution set versus a domain.

In the second implementation of the lesson on equations, the mathematics teacher educator (different from the first mathematics teacher educator) was able to guide the prospective teachers through the activities on constructed meanings of the equals sign and the subtle differences between the terms “solve” and “evaluate” in a way that led smoothly to consensus among the prospective teachers. This contrasted with the frustration exhibited by the prospective teachers in the first iteration of the lesson. The change in prospective teachers’ dispositions toward the task may have been due to the revised placement of the lesson in the scope and sequence of the course. Since it directly followed the lesson on the definition of function, it was a natural time for the prospective teachers to ponder the distinctions between function and equation.

To directly address the potential conflict between the “defining” constructed meaning of the equal sign and the “equating” constructed meaning, the mathematics teacher educator implemented another exploration in the lesson that had not been used in the first iteration of the lesson, due to time constraints. The exploration required the prospective teachers to consider the functions of two variables, $f(x, y) = y$ and $g(x, y) = 2x + 1$. The graphs of the functions were provided to them for reference (both graphs are planes). The two questions shown in Figure 11.1 were then posed.

a. What is the meaning of $f(x,y) = g(x,y)$? Explain.
b. Discuss why $y = 2x + 1$ is an equation (part of your discussion should involve the definition of equation). Is it also a function? Explain.

Figure 11.1. Probing “defining” and “equating” contructed meanings of the equal sign

The prospective teachers, most of whom had previously completed a course in multivariable calculus, had great difficulty in understanding the setting, and contrasting the resulting linear equation and its solution set with a linear function that arises from a context in which the domain and range are specified. To connect to a situation in which the prospective teachers had more prior experience, the mathematics teacher educator incorporated an example that equated two functions of one variable, and had the prospective teachers consider the resulting equation. This analogy was helpful to some of the prospective teachers. The resulting whole class discussion culminated with the prospective teachers deciding that whether a given object is a function or equation depends very much on the context and associated assumptions with which it arises.

These classroom experiences led to another revision of the lesson. A two-dimensional example (see Figure 11.2), akin to the one used by the mathematics teacher educator has been written into the lesson, preceding the 3-dimensional task, so that the prospective teachers have the opportunity to consider the situation in that more familiar setting before contemplating the three-dimensional example.

Consider the functions $h(x) = x^2 - 3$ and $r(x) = 2x$.

a) What is the meaning of $h(x) = r(x)$? Explain.
b) Discuss why $x^2 - 3 = 2x$ is an equation. Is it also a function? Explain.

Figure 11.2. Scaffolding for the task shown in Figure 11.1

The mathematics teacher educators' learning. The evolution of this task for prospective teachers illustrates key ways in which we as mathematics teacher educators developed. One of the themes of the prospective teacher task was *fostering awareness to similarities and differences* (Zaslavsky, 2008). The challenge for us as mathematics teacher educators was to design problem situations for the prospective teachers that naturally led to contrasting and comparing functions and equations so that they could identify the distinctions. In working to foster this kind of noticing, we had to think about prospective teachers' underlying conceptions of function and equation, and develop situations that caused them to question their assumptions. Our own mathematics teacher educator pedagogical content knowledge grew as we developed the associated mathematics teacher educator questions, teaching moves, and facilitation plans that would support this tendency for prospective teachers to attend to similarities and differences between these connected concepts.

A second theme that emerged from the evolution of the task was *identifying and overcoming barriers to students' learning* (Zaslavsky, 2008). In these activities, the epistemological facets of distinguishing function and equation were a substantial barrier to the prospective teachers' engagement in the task. Prospective teachers exhibited a level of comfort with their somewhat informal understanding of the equal sign that needed to be overcome. As mathematics teacher educators, we needed to think about the nature and cause of this barrier to the prospective teacher learning, and develop productive interventions. In this task, that led to including activities that directly highlighted pitfalls in the thinking of grade K-12 students that can arise out of a non-precise understanding of the equal sign, as well as structuring the prospective teacher learning experience in a way that brought conversations about the importance of addressing these barriers to learning to the fore.

Another example of how this theme of *identifying and overcoming barriers to students' learning* led to mathematics teacher educator growth was in the realization that the prospective teachers needed scaffolding in the activity in which they were asked to consider if the equation $y = 2x + 1$ that arose from equating two functions of two variables is also itself a function.

A third theme of the functions and equations task for the prospective teachers was *sharing and revealing self, peer, and student dispositions* (Zaslavsky, 2008). The whole-class discussions of different versions of this task revealed prospective teachers' frustration in facing conflict with their own conceptions of function and equation. As mathematics teacher educators, we were stretched to help the prospective teachers become aware of their dispositions in a productive way, while exhibiting our own positive dispositions and enthusiasm for the process of the sorting out of ideas that is a necessary step in deepening the prospective teachers' understanding of these subtle mathematical concepts.

Finally, the evolution of this task was situated within the context of our research, so the Zaslavsky's (2008) theme of *learning from the study of practice* is interwoven in our resulting mathematics teacher educator learning. The prospective teachers were informed participants in this research, so they have a start on developing the orientation to learn from the study of their own teaching. The research project provided opportunity and motivation for us as mathematics teacher educators to reflect upon our practice in a way that allowed us to document, analyse, and collaborate on teaching moves and curricular changes in a systematic way.

Functions and Equations in a Graduate Course

Evolution of the task. The graduate course *Concepts and Techniques in Precalculus* is a course for practising teachers that develops the foundations for functions and explores functions as a unifying theme from an advanced standpoint. The course connects and extends the mathematics based in the high school mathematics curriculum with an emphasis on transformations, inverses, and solving equations related to exponential, polynomial, power, trigonometric, and rational functions, and polar and parametric relationships.

The consideration of the interplay between functions and equations fits naturally into this course, but the mathematics teacher educator did not always have a task designed purposefully to have the practising teachers examine the concepts in tandem. One mathematics teacher educator of the course had had discussions with practising teachers who expressed consternation at students from their own classes who consistently represented a square root with an "attached" plus/minus, disregarding the context. This led to a discussion about the distinction between "the" square root function and the use of the square root function to solve an equation.

Because of these discussions, the mathematics teacher educator inserted a task in which practising teachers considered various mathematical statements and decided whether they were true or false. If false, practising teachers had to provide a counter example, and if true, they had to explain why. Together, the statements in Figure 11.3 led to a rich discussion of functions and equations.

Consider whether the following are TRUE or FALSE.
If false, give a counterexample. If true, explain why.

(a) $(\sqrt{x})^2 = x$

(b) $\sqrt{x^2} = x$

Figure 11.3. Excerpt of task addressing questions about plus/minus and square root

Each time the mathematics teacher educator implemented this task, there were practising teachers who claimed both statements were true, and cited the rules of exponents as their justification. It was rare that the practising teachers would think to note the solution set of the equations, but when they did, typically they would say the statements were true for all non-negative real numbers, because "you can't take the square root of a negative number." That is, they were specifically thinking about the standard domain of square root function when considered as a function mapping from the real numbers to the real numbers.

When prompted to think about what happens if one allows x to be a negative real number in the equations in Figure 11.3, many of the practising teachers were surprised to realize that statement (b) is false (in fact, the right-hand side of the equation should be $|x|$), and moreover, that the real solution set of (b), when corrected, is all real numbers, not just the non-negative real numbers. In prior iterations of this lesson, the mathematics teacher educator did not push the practising teachers much beyond this, thinking that this realization would be sufficient for practising teachers to put together the connection between their students' understanding of the square root function and its use in solving equations. However, informed by her experience with undergraduate prospective teachers and their struggle to see the relationship, the mathematics teacher educator changed her questioning to ask questions about the assumptions that the practising teachers inherently had about the context of the equations.

In contrast to the way that prospective teachers thought about the notion of functions and equations in their undergraduate course, the practising teachers were able to consider mathematical questions that were richer and more directly tied to their practice. Questions that the practising teachers considered that arose out of this task included: Are there restrictions to the rules of exponents? That is, if we think of the rules of exponents as outputs of function composition, what domain issues or assumptions need to be considered? What is the best way to help secondary students understand the connection between the square root function with its associated domain and range and the use of "taking the square root of both sides" to solve an equation? How does one explain, using function concepts, why a plus/minus "appears" when a student uses a square root to isolate the variable in an equation such as $(x - 3)^2 = 9$?

Mathematics teacher educator learning. The task relating functions and equations in the graduate practising teacher course arose out of a different context than did the task used with prospective teachers, but both tasks exhibit connected underlying mathematical ideas of the key distinctions between functions and equations. The themes that guided the development and enactment of the tasks and associated mathematics teacher educator learning are similarly related.

When working with the practising teachers, the Zaslavsky's (2008) theme of *coping with conflicts, dilemmas, and problem situations* was a key feature of the task. Based on secondary student claims or over-generalizations that the practising teachers had noted in their practice, the task was designed so that practising teachers were presented with a problem situation in which their own mathematical understanding of the rules of exponents conflicted with their knowledge about the relationship between functions and their inverses. The mathematics teacher educator's MTEPCK was deepened in the challenge of facilitating the mathematical discussion to focus on the assumptions the practising teachers were making about the solutions sets, contexts of the equations, and domains and ranges of the association functions.

Another theme of the practising teacher functions and equations task was *learning from the study of practice* (Zaslavsky, 2008). For the practising teachers, many of the questions that they fruitfully considered in thinking about the mathematics of the statements in Figure 11.3 were intertwined with thinking about how their students learn concepts in secondary school mathematics, providing evidence that they were engaged in the process of learning from their own practice. The evolution of this practising teacher task led to growth in the mathematics teacher educator's MTEPCK by bridging ideas about how to facilitate the task with the knowledge acquired from better understanding how prospective teachers struggled with thinking about the distinctions between functions and equations. That is, the mathematics teacher educator was able to take the insights about student learning gained from the implementation of the functions and equations tasks with prospective teachers and connect that understanding with how the practising teachers needed to have mathematical problems posed so that they could mirror that same layered thinking about their own students' assumptions and conceptions of domain of a function contrasted with solution set of an equation.

Topic 2: Visualising Complex-Valued Zeros

The topic of quadratic functions is ubiquitous in secondary school in the United States and in many other countries. In secondary school, the graph of a quadratic function can help students to visualise the zeros of a quadratic function, and this visualisation offers a powerful connection between graphical and algebraic forms. However, in many cases, it is assumed that the quadratic functions in question have real numbers as the domain, without explicitly stating so. In one task that we use, we challenge teachers to consider quadratic functions whose domain is the set of all complex numbers.

We have explored the concept of complex zeros of quadratic functions with the two populations described above: (a) undergraduate prospective teachers of secondary school mathematics and (b) graduate practising teachers in a mathematics program designed specifically for practising teachers. The task used in the undergraduate setting differed from the task used in the graduate setting, in response to different curricula as starting points and what we learned about the different background and needs of the teachers being served. However, in both cases, tasks evolved according to our growing understanding of the mathematics and the learning environment.

Complex Zeros in an Undergraduate Course

Evolution of the task. In the *Functions and Modeling* course designed for undergraduate prospective teachers, we began exploring complex-valued zeros through a task taken from Armendariz and Daniels (2011). In the task, prospective teachers were provided with a quadratic function that has no real-valued zeros. They were then asked to show (a) that a given complex-valued domain value is a zero for the function and (b) that some complex-valued domain values yield real-valued function values, whereas others yield complex values. Prospective teachers were then prompted to conjecture and prove the complex-valued domain values for which the function yields real-valued range values and represent this on a three-dimensional coordinate system in which the axes are x (real parts of domain values), y (real-valued codomain values), and i (imaginary parts of domain values).

As mathematics teacher educators, we found Armendariz and Daniels' (2011) complex zeros task to provide a nice extension of concepts that prospective teachers will teach in secondary school, as well as opportunities for conjecturing, proving, and illuminating visual representations. However, interviews with prospective teachers who had completed the *Functions and Modeling* course revealed that although many prospective teachers enjoyed the task, they remembered it as some type of "neat graph" of quadratics but were unable to explain the concept in much detail. Understanding prospective teachers' perspectives prompted us to revise the task for future use. We made revisions both to the written task and to the implementation of the task, which we captured in our mathematics teacher educator notes.

For example, we sought to connect the task to other ideas in the course to better illuminate the main concepts. Armendariz and Daniels' (2011) initial task was offered immediately after an introduction to functions that included tasks in which prospective teachers discussed various definitions of function and identified functions and non-functions in several examples. However, the connection between definitions of functions and the complex zeros task was unclear, and we wondered if this contributed to prospective teachers' limited understandings of the concept. In future iterations of the introduction to functions tasks, we made revisions and wrote mathematics teacher educator notes to prompt prospective teachers to discuss why the domain must be specified when deciding whether a given relation is a function (e.g., see Functions and Equations section). An enhanced discussion of the domain in the functions task led to, in the complex zeros task, more natural emphases on the complex domain of the quadratic functions (a domain not commonly considered in secondary school) and the restricted domain when considering only real-valued outputs. We were explicit about the domain consideration in the written background for the task given to prospective teachers, and we added text to emphasize the domain under consideration in each part of the task. In addition, we wrote mathematics teacher educator notes recommending class discussion about the connections according to the domain.

We also revised the task by better connecting to prospective teachers' knowledge. Through implementing the complex zeros task, we became more familiar with prospective teachers' existing mathematical knowledge. The prospective teachers in our course had had limited exposure to complex numbers prior to the course, and a great deal of time was spent introducing complex numbers and operations with complex numbers during the task. This left limited time for prospective teachers to do the mathematical work of conjecturing, proving, and exploring representations of the mathematics. To address this issue, the next time that we implemented the task, we asked prospective teachers to complete some homework prior to the class meeting in which they evaluated a quadratic function for three different complex domain values. Not only did the preparation work offer the opportunity for prospective teachers to learn or refresh basic skills before class, but the complex zeros task also employed all three of the calculations included in the preparation work. Our intent was to allow prospective teachers more time to engage deeply in the mathematics with their peers during class time. This is similar to Koichu et al.'s (2016) finding that too many technical aspects of a task can detract from the conceptual theme of the task.

Although adding preparation work certainly improved the use of class time, we noticed that prospective teachers had a difficult time connecting this preparation work to the task. For example, in the preparation work, prospective teachers were asked to evaluate $f(-2 + \sqrt{3}i)$ for $f(z) = z^2 + 4x + 7$ (the result is 0). The task prompted prospective teachers to show that $-2 + \sqrt{3}i$ is a zero for f, which prospective teachers had essentially already shown in the preparation work. However, very few prospective teachers recognized this connection, even after the mathematics teacher educator prompted the class by saying that the calculations in their preparation work could be used for the task. This is one example of a place where mathematics teacher educator questioning became important in helping prospective teachers to make meaning of the task. The mathematics teacher educator could ask questions such as, "What does it mean for a value to be a zero for a function?" Rather than include these questions in the written task for prospective teachers, we added these questions to mathematics teacher educator notes that could be used by other novice mathematics teacher educators. We also reminded prospective teachers of the definition of *zero of a function* in the written background of their copy of the task.

An important consideration for our team was balancing opportunities for open-ended inquiry with the explicit instructions and background given to prospective teachers. That is, we wanted tasks to be scaffolded enough that prospective teachers could pull out important concepts and were not struggling with meaningless details, but we did not want tasks to be so structured that all productive struggle and inquiry was gone from the task. As such, we did not change the overall structure of the task, which prompted prospective teachers to form their own conjectures and then prove them. We also added questions prompting prospective teachers to consider the range of the function, which highlighted a key difference between the function with domain of real numbers and the function with domain of all complex numbers.

Finally, recent advances in technology allow many options to help prospective teachers visualise complex zeros. In class, our prospective teachers shared their 2-dimensional sketches of the 3-dimensional visualisation, and some prospective teachers created 3-dimensional models with transparencies. After the prospective teachers had come to visualise what the 3-model would look like, the mathematics teacher educator shared a GeoGebra file representing the 3-dimensional graph in which prospective teachers could explore the graph of the function with real number outputs.

Mathematics teacher educator learning. The revised prospective teacher complex zeros task not only led to deeper prospective teacher learning, but also deepened our own learning as mathematics teacher educators. First, one of our themes in the task was *fostering awareness to similarities and differences* (Zaslavsky, 2008). We aimed for prospective teachers to understand how the complex zeros task connected to the previous classwork on the domain of functions. To achieve this goal, we devised teaching strategies, explanations, mathematics teacher educator questions, and approaches that would highlight the structure and connections within the content, thereby extending our mathematics teacher educator pedagogical content knowledge.

A second theme of the prospective teacher complex zeros task was *coping with conflicts, dilemmas, and problem situations* (Zaslavsky, 2008). We created skills-based preparation homework for prospective teachers so they were better prepared to engage in mathematical dilemmas and problem situations with their peers during class time. In addition, we aimed to maintain a high level of inquiry during the lesson, and we added questions to the task to prompt additional mathematical connections. Facilitating an inquiry-based lesson strengthened our mathematics teacher educator pedagogical content knowledge because it required that we were able to respond to prospective teachers' (sometimes-unexpected) questions, ideas, or confusion in-the-moment.

A third theme of the prospective teacher complex zeros task was *selecting and using (appropriate) tools and resources for teaching* (Zaslavsky, 2008). We aimed to enhance prospective teacher learning by allowing them to create physical three-dimensional models as well as showing how dynamic software could be used to visualise the situation in three dimensions. At the same time, we strengthened our mathematics teacher educator pedagogical content knowledge related to resources and tools for teaching mathematics as we considered tools that would represent the concept well and raise prospective teacher awareness of ways to use tools in their future classrooms.

An overarching theme of the prospective teacher complex zeros task was *identifying and overcoming barriers to students' learning* (Zaslavsky, 2008). Although this was not an explicit goal that we had for prospective teachers in the task, it was a goal that we had as mathematics teacher educators. That is, we aimed to better understand prospective teachers' barriers to their learning of the topic and determine ways to overcome those barriers. For example, each implementation of the task strengthened our own mathematics teacher educator pedagogical content knowledge concerning prospective teachers' background knowledge, as we learned of areas in prospective teachers' knowledge of skills and concepts related to complex numbers that needed particular attention. As discussed above, we then had to determine ways to revise the task in order to address these difficulties.

We reiterate that the research context in which we worked afforded us the opportunity to develop our own learning. In fact, the revisions of this task were strongly motivated by what we learned about prospective teachers' understanding (and lack of understanding) of the task through conducting research interviews with prospective teachers at the end of the course as well as analysis of video-recordings of class meetings. In essence, we were enacting Zaslavsky's (2008) theme of *learning from the study of practice* as we collaboratively studied our own practice and deepened our mathematics teacher educator pedagogical content knowledge.

In addition, our understanding of the purpose of the task grew as we needed to defend our choice to continue to include the task in the course. As part of the process of developing tasks for the *Functions and Modeling* course, we have consulted an expert panel and an advisory board team, composed of experts in the preparation of secondary school mathematics teachers, including mathematics teacher supervisors, mathematics teacher educators, and mathematics education researchers. Both the expert panel and advisory board have questioned the usefulness of this lesson in a course focused on understanding the fundamentals of functions and modelling with functions. Some of our experts commented that they found the task confusing, and others saw it as superfluous to secondary school mathematics. However, our research team chose to keep the task as part of the course, for several reasons. Notably, in more recent interviews about their experience in the course, prospective teachers said that the complex zeros task was one of the most memorable lessons in the course and a lesson that opened their eyes to the power of visualisation. Also, the *Mathematical Education of Teachers II* (Conference Board of the Mathematical Sciences, 2012) report states that, "Complex numbers can fall into the chasm between high school and college, with high school teachers assuming they will be taught in college and college instructors assuming they have been taught in high school" (p. 64). The report goes on to recommend that students be given opportunities to see the use of complex numbers in context and explore complex numbers with meaning. As secondary school teachers, prospective teachers will be required to teach complex-valued zeros, or, at the very least, teach complex numbers as they relate to the quadratic formula. As mathematics teacher educators, we see that the complex zeros lesson has the potential to add depth to prospective teachers' understanding of this content as well as provide novel ways to connect graphical representations of functions to assumptions about the domain and restriction of the range. In addition, as stated above, we believe that our complex zeros task has evolved to reduce unproductive confusion and increase its meaningfulness for prospective teachers.

Complex Zeros in a Graduate Course

Evolution of the task. The premise of the complex zeros task was also used in a graduate course for practising teachers, but the specific approach to the task differed. The graduate course, called *Concepts and Techniques in Algebra,* was designed to deepen the understanding of secondary school mathematics practising teachers in the domain of algebra. The textbook for the course was Usiskin, Peressini, Marchisotto, and Stanley (2003): *Mathematics for High School Teachers: An Advanced Perspective.*

This text addressed the notion of complex-valued zeros of quadratic functions in a chapter focused on real and complex numbers. The approach was to consider the solutions to the equations $x^2 + bx + c = 0$, where b and c are real numbers, b is constant, and c varies. The mathematics teacher educator offered a worked example in which $b = 2$ and three specific values for c are chosen for $x^2 + 2x + c = 0$. The book showed each of the three graphs of the corresponding functions in the xy-plane as well as the solutions for each of the three equations $x^2 + 2x + c = 0$ on a separate complex plane. (The solutions were graphed as points in the complex plane.) The text went on to justify the graphical representations, using an analytic approach with the quadratic formula. The exploration of $x^2 + bx + c = 0$ where c is constant and b varies was a homework problem.

In her first time teaching the *Concepts and Techniques in Algebra* course, one mathematics teacher educator assigned textbook reading for outside of class that included the worked example addressing complex solutions to $x^2 + bx + c = 0$ as c varies. During the following class, the mathematics teacher educator revisited the example, leading a class discussion about the purpose of the example and what practising teachers gained from it. The exploration of $x^2 + bx + c = 0$, where c is constant and b varies, was assigned for homework, as suggested in the text. However, the mathematics teacher educator recognized that this approach did not allow for much inquiry from the practising teachers. Rather than discovering the properties of the solutions to $x^2 + bx + c = 0$ as c varies, practising teachers were told the properties through the example. The homework exercise then became largely a replication that lacked depth.

In the next iteration, the mathematics teacher educator modified the task to assign it as an in-class problem for group work, *before* assigning the textbook reading. The modified task is provided in Figure 11.4.

Consider $x^2 + 5x + c = 0$.

i. Choose a value for c such that the solutions to the equation are two, distinct real numbers. (Hint: How can the quadratic formula help you to choose these numbers?)
ii. Choose a value for c such that there is only one solution to the equation.
iii. Choose a value for c such that there are only complex solutions to the equation.
iv. Graph all the solutions found in parts (i - iii) in one complex plane, using a different color for each pair of solutions. Make observations about the visual representations of these solutions.
v. Graph the equations $y = x^2 + 5x + c$ for your values of c in one x-y ($\mathbb{R}^2$) plane. Compare and contrast these graphs.
vi. Use the quadratic formula to express the solution set of the general equation $x^2 + 5x + c = 0$.
vii. Make observations about the connections between parts (iv) – (vi).
viii. What did you learn from this working on this problem? How might your findings help you in teaching?

Figure 11.4. Modified complex zeros task for graduate practising teachers (adapted from Usiskin et al., 2003)

A major consideration in the revision of this task was raising the level of inquiry for the task. Rather than have practising teachers read a worked example, they were required to develop their own example by choosing different values for c, making observations, and justifying their observations. As with the undergraduate complex zeros task, a major consideration was to make the problem inquiry-based rather than shown as a worked example. In addition, the mathematics teacher educator's intent was to provide enough scaffolding that practising teachers could complete the problem in a meaningful way and be prompted to think deeply about the mathematical connections among representations. After all practising teachers had the opportunity to do so, the mathematics teacher educator then assigned the textbook reading as a follow-up to the task.

We have learned that practising teachers are capable of rigorous justification, when they have the appropriate mathematical knowledge and are pushed to do so. For example, in (vii) in Figure 11.4, practising teachers may make superficial observations at first, but with mathematics teacher educator prompting, practising teachers can be quite specific in their justifications of the physical location of solutions to $x^2 + 5x + c = 0$ as they compare to the graphs of $y = x^2 + bx + c$. In both this problem and the homework problem, practising teachers can go so far as to describe and justify not only the location of the solutions to the given quadratic equations, but also the rate of change of the distance between the solutions with respect to the parameter.

In addition, the mathematics teacher educator added the final question (viii in Figure 11.4) prompting practising teachers to consider the purpose of the problem in the context of the course and in the context of their teaching. In working with practising teachers, we have learned that practising teachers are constantly seeking connections between their coursework and their practice. Without prompting questions, practising teachers sometimes do not consider the benefits of learning concepts that they will not directly teach to their students. Nevertheless, many practising teachers in this course commented that they would use a version of this problem as an extension for secondary school students.

After exploring the in-class problem, practising teachers were assigned Usiskin et al.'s (2003) simply-stated homework problem, "Track the solution set in the complex plane of the quadratic equation $x^2 + bx + 2 = 0$ as the value of the real coefficient b varies" (p. 53). With a thorough mathematical understanding of the in-class problem, practising teachers were well equipped to explore this homework problem, which offered an additional opportunity to explore related concepts.

Much of the mathematical power of the practising teacher task lies in the dynamic nature of the problem. In fact, practising teachers in this graduate program are often familiar with dynamic graphing software, and in working this problem, several practising teachers have naturally extended the problem to create dynamic graphs to illustrate the concepts being explored. It is especially powerful when practising teachers share their dynamic graphs through presentation to their peers.

Mathematics teacher educator learning. Although the complex zeros task used with the practising teachers had different origins than the one used with prospective teachers, each task highlights (related) mathematical features of complex zeros. In addition, although there are similarities in the themes emphasized in the two tasks, the ways in which these themes were enacted varied according to the population of teachers served. In many cases, when working with practising teachers, the mathematics teacher educator's MTEPCK was challenged and strengthened in deeper ways than it was when working with prospective teachers.

For example, a major theme of the practising teacher complex zeros task was *coping with conflicts, dilemmas, and problem situations* (Zaslavsky, 2008). Much like the prospective teacher task related to complex zeros, a major consideration in revising the practising teacher task was implementing a high level of inquiry. The task was revised so that practising teachers were required to develop their own examples, make observations, and provide rigorous justifications. The process of writing a new task strengthened the mathematics teacher educator's MTEPCK in relation to teaching strategies and approaches to promote inquiry. In addition, much like the prospective teacher lesson, the mathematics teacher educator was required to respond to practising teachers' mathematical productions in-the-moment as they worked through the inquiry lesson. However, practising teachers' mathematical questions, observations, solutions, and ideas went far beyond those of the prospective teachers, challenging and strengthening the mathematics teacher educator's MTEPCK to a greater extent than in the prospective teacher lesson.

Another theme of the practising teacher complex zeros task was *selecting and using (appropriate) tools and resources for teaching* (Zaslavsky, 2008). Like the prospective teacher task, the practising teacher complex zeros task was enhanced through the use of dynamic graphing software. However, unlike the prospective teacher task, many of the practising teachers were familiar with dynamic graphing software and its uses, and ideas from the prospective teachers strengthened the pedagogical content knowledge of both their classmates and the mathematics teacher educator in the area of selecting and using resources for mathematics instruction.

In addition, like the prospective teacher complex zeros task, the mathematics teacher educator grew in understanding the purpose of the practising teacher complex zeros task as she considered how to open the task for inquiry and situate it within the course. However, the mathematics teacher educator's learning went beyond understanding the purpose of the task, as she also sought to understand the purpose from the point of view of practising teachers and facilitate discussion on the connections between the task and practising teachers' teaching.

Topic 3: Building Functions

The notion of building functions from existing functions using transformations or building a function that models a relationship between quantities are common high school mathematics topics (see for example, CCSSM, 2010, p. 70). The proliferation of mathematics-specific technologies such as Geometer's Sketchpad®, GeoGebra, or Desmos facilitates experimentation on existing functions by enabling dynamic views of the effects on the graph of a function when various parameters change. However, anticipating the dynamic changes in the graph or relating the dynamic views to expected changes based upon, say the algebraic form of a transformation, is not necessarily straightforward for practising teachers and prospective teachers.

Because the tasks used in the undergraduate setting evolved from the tasks and mathematics teacher educator's experiences in the graduate setting, the evolution of the task for practising teachers is presented first for this topic. The use of dynamic software to explore transformations introduces a level of noticing similarities and differences that is not emphasized in prospective and practising teachers' standard mathematics courses and requires that mathematics teacher educators understand the strengths and weaknesses arising from these experiences to address them adequately.

Building Functions in a Graduate Course

Evolution of the task. The graduate course, called *Mathematics-specific Technologies,* focuses on learning mathematics-specific technologies – those used specifically for the teaching and learning of mathematics – via tasks grounded in secondary school mathematics. The class meets in a computer lab where each practising teacher has access to a computer and the mathematics teacher educator has the equipment to project the mathematics teacher educator's computer screen for the entire class to view. Practising teachers work in pairs on inquiry-based tasks that are designed to learn the technology based upon the need to resolve a mathematical task or to design a mathematical task for their students.

The topic from the course highlighted here involves practising teachers learning to use the power of dynamic visualisation software such as Geometer's Sketchpad®, GeoGebra, and Desmos to teach the effects of basic function transformations. Practising teachers learn to use the technology as they design tasks and lessons that focus on using linear, quadratic, power, trigonometric, exponential, and logarithmic functions to build other functions. The first iteration of the introductory classroom experiences that lay the foundation for practising teachers to complete the project shown in Figure 11.5, involved the mathematics teacher educator projecting a GeoGebra sketch of the graph of $f(x) = x$ and the graph of $y = kf(x)$ with k dynamically changing, and asking the practising teachers to describe their observations. Next, the mathematics teacher educator projected another GeoGebra sketch of the graph of graph of $f(x) = x$ and the graph of $y = f(x) + k$ with k dynamically changing and asked the practising teachers to describe their observations. The mathematics teacher educator followed with a similar dynamic illustration and questioning sequence for $g(x) = x^2$.

> Design an investigation or aid for teaching transformations using GeoGebra. Specifically, your project should include investigation that requires learners to identify the effect on the graph of replacing $f(x)$ by $kf(x)$, $f(kx)$, $f(x+k)$, and $f(x)+k$ for specific values of k and a given function f. You should include examples of the following functions: $f(x) = x$, $g(x) = x^2$, $h(x) = \sqrt{x}$, $s(x) = e^x$, $v(x) = \log x$, $r(x) = \frac{1}{x}$, and $a(x) = |x|$. You should also include ways to animate parameters and pop up text boxes that explain the effects of the parameters (e.g. refection across x-axis, vertical stretch) or pose questions to learners as the animation occurs. In addition, your project should include investigation that requires learners to explore the effect on the graph of replacing $f(x)$ by $|f(x)|$ and $f(|x|)$.

Figure 11.5. Mathematics-specific Technologies course project assignment

During discussion, many of the practising teachers asserted that they "knew" that the effect on the graph of $f(x)$ when replaced by $kf(x)$ should be a vertical dilation of $f(x)$ when $k > 0$, but after viewing the dynamic sketch related to $f(x) = x$, it could also be thought of as a rotation for this function. Also, after viewing the dynamic sketch for determining the effect on the graph of $f(x) = x$ when the graph is replaced by $f(x) + k$, many of the practising teachers grappled with the fact that the latter should be a vertical translation of the former, but they were also seeing a horizontal translation. For $g(x) = x^2$, they "knew" that the graph of $kg(x)$ should be a vertical dilation of $g(x)$

when $k > 0$, but some questioned why the dynamic sketch seemed to be a horizontal dilation. The mathematics teacher educator asked the practising teachers to resolve the apparent discrepancies by asking them to attend to precision (e.g., How do we define a rotation?) and asking them to think about other representations (e.g., How may the form of a defining expression for a function inform features of the graph of the function?). As such, for the mathematics teacher educator, this generated an opportunity to have practising teachers reflect on the mathematical connections and on the pedagogical reasons why certain examples may be preferable to others.

Visualisation may powerfully expand the resources for learning mathematics and may challenge traditional methods of teaching mathematics (Cruz, Febles, & Diaz, 2000; Villarreal, 2000). However, research also suggests that disadvantages may arise from uncontrollable visual imagery and from sole reliance on visual information (Aspinwall, Shaw, & Presmeg, 1997; Boulter & Kirby, 1994). Based upon class discussion and observations, the mathematics teacher educator wondered whether the dynamic sketches were influencing or generating misconceptions for the practising teachers and whether the classwork toward resolutions resulted in the practising teachers having a firm understanding of the underlying mathematics and possible pedagogical issues when using dynamic sketches to investigate transformations of functions. The next time the mathematics teacher educator taught the course, he created a set of six pre-assessment items (Figure 11.6) that the practising teachers completed individually before viewing the dynamic sketches. Then, after the practising teachers viewed and discussed the sketches, they completed the same set of six questions.

(1) For the function $f(x) = x$, we create a new function g by multiplying f by a parameter a where $a > 0$. Identify the effects on the graph of $f(x)$ when replacing the graph of $f(x)$ by the graph of $g(x)$;

(2) For the function $f(x) = x$, we create a new function g by adding a parameter b where $b > 0$. Identify the effects on the graph of $f(x)$ when replacing the graph of $f(x)$ by the graph of $g(x)$;

(3) For the function $f(x) = x^2$, we create a new function g by multiplying f by a parameter a where $a > 1$. Identify the effects on the graph of $f(x)$ when replacing the graph of $f(x)$ by the graph of $g(x)$;

(4) A student in your class claims that the graph of $g(x) = ax$ is a rotation of the graph of the function $f(x) = x$. What do you tell him? Why?;

(5) A student in your class claims that the graph of $g(x) = f(x) + b$ is a horizontal translation of the graph of the function $f(x) = x$. What do you tell her? Why?; and

(6) A student in your class claims that the graph of $g(x) = ax^2$ is a horizontal compression of the graph of the function $f(x) = x^2$. What do you tell him? Why?

Figure 11.6. Pre- and post-assessment items (from Epperson, 2009)

The mathematics teacher educator was most interested in the answers to Questions 4–6 from the assessment because Questions 1–3 were included to establish a baseline knowledge that most of the practising teachers already possessed. Although almost no practising teachers used the word "rotation" in their answer to Item 1, after viewing the sketches the mathematics teacher educator observed that some included the "rotation" in their response to Question 1 on the post-assessment. The responses to Questions 4–6 on the pre-assessment typically focused on explaining the "rule" and do not address why the student's observation may have some validity. On the post-assessment, many responses to Questions 4–6 attended to the student's reasoning and correctly drew upon class discussion to explain the nature of their response to the student. However, the mathematics teacher educator felt that more structure in the class discussion might help those practising teachers whose responses remained persistently weak or weakened after viewing the sketches. Thus, the instructional sequence now involves practising teachers specifically discussing "The graph of a constant multiple of $f(x) = x$ looks like a rotation of f when we view it dynamically. Is it? Explain your reasoning. What would you tell a student?" and "The graph of a constant multiple of the $f(x) = x^2$ looks like both a vertical and a horizontal dilation. Is it both? Explain your reasoning. How would you explain this observation to a student?" Subsequently, they experimented with other functions such as $f(x) = e^x$ regarding the effect on the graph of replacing $f(x)$ by $kf(x)$, $f(x + k)$, and $f(x) + k$ for specific values of k and asked, in a similar way, to resolve why the dynamic sketch that "should be a horizontal translation" appears also to be a vertical dilation.

In this case, the activities in class focused on experiencing dynamic representations of function transformations and then connecting possibly discrepant observations to the corresponding algebraic representations. Using visualisation was aimed at developing better mathematical understanding and encouraging experimentation and discovery (Zimmerman & Cunningham, 1991). The focus on flexible reasoning, attending to precision, and plausible student questions also aimed at having practising teachers delve deeper into their own understandings and meanings of the mathematics.

Mathematics teacher educator learning. The mathematics teacher educator became aware of the attractive, possibly erroneous, conclusions that arise from viewing dynamic function transformations. At a state-level school mathematics meeting, the quandary had arisen from a textbook that claimed that given $f(x) = x$ the graph of $y = af(x)$ is a rotation of the graph of $y = f(x)$. The fact that this had made it through several levels of approval and state-wide teacher committees alerted him to the need to explore these possible conflicts that arise. Thus, the central theme of this task was *coping with conflicts, dilemmas, and problem situations* (Zaslavsky, 2008). The mathematics teacher educator expanded his MTEPCK in determining effective ways to introduce these conflicts and allow practising teachers to resolve them. This also required that he explore the conflicts and be open to unexpected new questions that may arise.

A second theme of this task was *fostering awareness to similarities and differences* (Zaslavsky, 2008). The challenge for the mathematics teacher educator rests in sequencing the learning events that lead to raising awareness of similarities and differences. The mathematics teacher educator's MTEPCK grew via his need to enact Smith and Stein's (2011) *five practices*. The importance of anticipating the classroom interactions and scaffolding questions to address anticipated issues, listening to practising teachers work in pairs, and determining which pairs should report out and in what order, not only underscored the need for the latter, but also helped the mathematics teacher educator understand how to model these practices for the students.

Consistent with Aspinwall et al.'s (1997) research on the effects of uncontrollable visual imagery on student learning, Zaslavsky's (2008) theme *identifying and overcoming barriers to student learning* became critical in designing an inquiry-based experience for practising teachers that attended to conflicts and fostering awareness of similarities and differences. The use of the task deepened the mathematics teacher educator's MTEPCK regarding instances of uncontrollable visual imagery; the importance of understanding the practising teachers as learners and of seeing the dynamic sketches from their perspective became a critical aspect of the work to overcome possible barriers to student learning related to the mathematics or the representations used.

Facilitating the experimentation and discovery in this course helped the mathematics teacher educator to model the theme of *developing adaptability* (Zaslavsky, 2008). Embracing new or unexpected questions and observations from the practising teachers not only became a critical component in the learning experience for the practising teachers but also in enhancing mathematics teacher educator pedagogical content knowledge regarding effective ways to model flexibility and adaptability to unexpected situations for practising teachers.

Building Functions in an Undergraduate Course

Evolution of the task. In the course, *Functions and Modeling* as described above, an exploration of function patterns (Armendariz & Daniels, 2011, p. 28) engages prospective teachers in analysing patterns in data (e.g., as the input values increase by c, is there a pattern – a constant multiplier or constant increase – in the associated output values) and using these patterns to identify a function from the common functions studied in secondary school mathematics courses (i.e., linear, quadratic, power, exponential, and logarithmic functions) from which a function model could be built for the data. Students investigate how patterns in the domain values may result in patterns in the range values such as noticing that multiplying subsequent domain values by c results in a pattern of multiplying the corresponding range values by a constant k (that depends on c). Other than using the patterns identified in this lesson to build function models later in the course, there is no further investigation of these patterns, their connections to function transformations, or visual representations.

The exploration culminates in prospective teachers' proving algebraically why, for example, a linear function can be used to model data for which choosing domain values at equal-sized intervals (that is, consecutive domain values used differ by a constant c) will have corresponding consecutive range-values that differ by a constant k that depends on c. The latter domain-range pattern was called an *add-add pattern.* Similarly, domain-range patterns for which other function models may be appropriate are *multiply-multiply* for power functions, *add-multiply* for exponential functions, and *multiply-add* for logarithmic functions. The mathematics teacher educator, when using the materials for the first time, wondered whether prospective teachers could use the meanings garnered from the patterning work in the original lesson to make connections to visual representations and effects of function transformations. He assigned the task shown in Figure 11.7, which he had used previously in a practising teacher professional development setting with the expectation that the patterning work would transfer seamlessly to reasoning needed to complete the task.

Part I

Consider the functions $f(x) = x, s(x) = x^2, h(x) = \sqrt{x}, w(x) = e^x, v(x) = \log x\,, r(x) = \frac{1}{x},$ and $z(x) = |x|$.

Which of these functions, if any, exhibit the following, for any positive real number α and function g, where g is not the identity function or a constant function. Assign each group member one of the following roles: visual method (use graphical methods with or without technology), analytical method (use algebraic methods), numerical method (use tables of values with or without technology), and verifier.

(1) $f(x+\alpha) = \alpha + f(x)$
(2) $f(\alpha x) = \alpha f(x)$
(3) $f(\alpha x) = \alpha + f(x)$
(4) $f(x+\alpha) = \alpha f(x)$
(5) $f(x+\alpha) = g(\alpha)f(x)$
(6) $f(\alpha x) = g(\alpha)f$
(7) $(x)f(\alpha x) = g(\alpha) + f(x)$
(8) $\alpha f(x) = f(g(\alpha)x)$
(9) $f(x+\alpha) = g(\alpha) + f(x)$

Part II:

For a function that satisfies (1) in part I, explain the effects on the graph, if the graph is replaced by $f(x+\alpha)$ versus $\alpha + f(x)$. Similarly, repeat this for items (2)-(9) in part I.

Figure 11.7. Connecting patterns to visual representations (from Epperson, 2010)

The prospective teachers neither made natural connections using the structure of the equations presented to the patterns they had previously observed, nor readily connected the patterns to visual representations or transformations of functions. The next time the mathematics teacher educator taught the course, he followed the patterns exploration with tasks such the sample given in Figure 11.8.

For a given function f, we define a new function $g(x) = f(x + c)$ where $c > 0$. The graph of the new function g is a horizontal translation (shift) of the graph of f, but it also appears to be a vertical dilation of the graph of f. In order to observe this, which function pattern must f have? Explain your reasoning. Identify the function type that has this pattern.

Figure 11.8. Sample revision of task connecting patterns to visual representations

For this task, prospective teachers seemed more readily to make connections. This contrasted with the mathematics teacher educator's previous experience using the task seen in Figure 11.7. The change in the task, besides reducing its length, targeted the links between the graphical observations and the function patterns prospective teachers had determined by removing the investigation of the equations that required prospective teachers to connect the algebraic equation to transformations resulting in identical graphical representations and patterns previously determined. Before, for example, prospective teachers struggled to connect the equation $f(x + \alpha) = g(\alpha) f(x)$ (from Figure 11.7, Item (5)) to the idea that this would imply that a horizontal translation of the graph of f would also appear to be a vertical dilation of the same graph and then link this to their patterns work.

With significant revisions in the prospective teacher course, more focus was placed on developing covariational reasoning (how two quantities vary together; see also Carlson, Jacobs, Coe, Larsen, & Hsu, 2002), distinguishing characteristics of functions, and attending to the assumptions made when creating function models for given data. In Fall 2017, prospective teachers spent approximately two hours (approximately one and a half 80-minute class periods) working on the lesson "Functions Arising from Patterns." For homework, prospective teachers were given the exploration, "Reconciling Visual Imagery with Algebraic Forms," which was part of a newly developed lesson created by the authors to address our learning from the previous iteration of the course. In addition to our own experiences in the course, we also learned from classroom video, expert and advisory board feedback, mathematics teacher educator interviews, student interviews, and student work. Prospective teachers were encouraged to use graphing technology of their choice to explore four scenarios like the one above in which known transformations of certain functions appear to be a different transformation. The second part of the lesson – completed in class after discussion of the homework – gave them access to a Desmos link set up for verifying their resolutions from the assigned exploration. The two explorations (the one assigned as homework for discussion in class and the one completed entirely in class) took 15 minutes to complete in class. The mathematics teacher educator opened the class by discussing the first scenario and prompting them to think about the effects on the graph of $f(x)$ when it is replaced by $f(x + c)$ and then asking them, "for which kind of function would this to also appear to be a vertical stretch? And, the answer to that relates to what we worked on last time." After this statement by the mathematics teacher educator, a prospective teacher said to her group, "It's exponential so if we add 2 to the $x + c$ [sic], that's the same as to the x [sic] times 2 to the c." Another student in the group said, "Ohhhhh, so it's going to stretch as well." The prospective teachers used the Desmos link to test their conjectures and wrapped up their discussion.

Although the mathematics teacher educator may have over-scaffolded the discussion, prospective teachers seemed to be making better connections between visual representations, algebraic representations, and the observed patterns than in previous semesters. On a subsequent exam, the mathematics teacher educator used a question similar to the one in Figure 11.8 (see Figure 11.9).

10. For a given function f, we define a new function $g(x) = c \cdot f(x)$ where $c > 1$. The graph of the new function g is a vertical stretch of the graph of f, but it also appears to be a horizontal translation (shift) of the graph of f. In order to observe this, which function pattern must f have? Explain your reasoning. Identify the function type that has this pattern.

Figure 11.9. Exam question 10 on patterns and visual representations

On the exam, 25% of the prospective teachers received full credit for their response (see, for example, Figure 11.10). Also, 33% received significant partial credit but reversed the function pattern so they identified the function as a logarithmic function. This was in contrast to our experiences in previous semesters in which interpretation for full or partial credit fell below 20%.

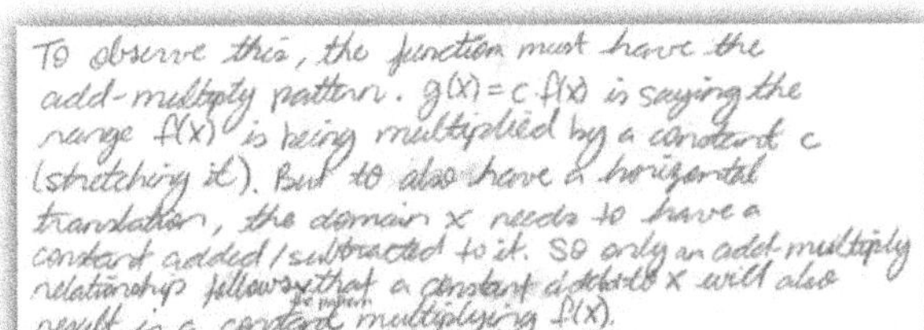
To observe this, the function must have the add-multiply pattern. g(x) = c·f(x) is saying the range f(x) is being multiplied by a constant c (stretching it). But to also have a horizontal translation, the domain x needs to have a constant added/subtracted to it. So only an add-multiply relationship follows that a constant added to x will also result in a constant multiplying f(x).
The exponential function follows this data.

Figure 11.10. Student response receiving full credit on exam question 10

The concern of the original mathematics teacher educator had been that prospective teachers were not making graphical and algebraic connections to their work on "Functions Arising from Patterns." We are moving closer to this goal, but continue to revise the task and explorations leading up to it to address, for example, aspects of the task that my influence why so many of the prospective teachers inadvertently reversed the pattern when interpreting the exam problem. Prospective teachers enjoy uncovering the patterns and applying them later in the course to building functions to model data. The development of these tasks to facilitate linking visual representations, algebraic forms, and the patterns highlights relationships among several representations that directly link to secondary school mathematics.

Mathematics teacher educator learning. The original task on function patterns asked students to connect the function pattern to its corresponding algebraic representation. One of the author's experiences using visualisation of transformations of functions in the graduate course provided the motivation for Zaslavsky's (2008) theme of *fostering awareness to similarities and differences* as this task evolved. As mathematics teacher educators, we saw that careful work in two representations (tabular data patterns, algebraic representations) did not transfer or connect to understanding similarities and differences visually. We developed mathematics teacher educator pedagogical content knowledge concerning ways to facilitate this transfer for prospective teachers in an inquiry-based setting.

Zaslavsky's (2008) theme of *selecting and using (appropriate) tools and resources for teaching* was also embedded in this task. Understanding how to best use the technology and which examples illustrate mathematical concepts most effectively arose as prospective teachers used technology to investigate the apparent contradictions to the "rules" they had learned about function transformations. Being aware that some examples of transformations may, when using technology, generate misconceptions better prepares prospective teachers to attend to this possibility. As mathematics teacher educators, our MTEPCK for understanding how and when to introduce the mathematics-specific technology (e.g., how to incorporate the technology in an exploration or when to make links to dynamic sketches available) was enhanced by our experiences in developing this task.

Collecting classroom video, student work, and studying the lesson with research colleagues relate to the theme of *learning from the study of practice* (Zaslavsky, 2008). Our MTEPCK grew through our study of classroom video and seeing the impact on group discussions arising from the fact that not all students had done the prep-work associated with the lesson and that some misconceptions persisted. This prompted changes in the next iteration of the lesson in fall 2018 in that prospective teachers were made accountable for submitting their prep-work before class via an online course management system so that they could be placed in groups that aligned with the quality of the prep-work submitted. Persistent misconceptions were addressed in group questioning and a summary class discussion.

The engaging and accessible nature of the function patterns tasks for prospective teachers and the attempts to incorporate and connect visual representations to the observations and conclusions from the task helped solidify our MTEPCK concerning how to describe or demonstrate a concept for prospective teachers.

DISCUSSION

As mathematics teacher educators teaching mathematics courses for teachers, we balance many goals, including promoting teachers' critical mathematical understandings while modelling evidence-based teaching practices. At the same time, we aim for the tasks and learning experiences in these courses to be grounded in secondary school mathematics so that teachers see meaningful connections between the course content and their practice. Through the process of fulfilling these multiple goals, we have learned about teachers' thinking and their existing mathematical understandings, ways in which to facilitate deeper mathematical learning, and ways in which to implement inquiry-based methods in our courses.

Zaslavsky's (2008) seven themes provided a frame for reporting our learning from the development of the tasks highlighted in this chapter. As seen in Table 11.1, mathematics teacher educator learning related to the tasks for the three topics described above involved each of the themes in our discussion at least once. However, even though a theme was not explicitly discussed (or was not a major area of focus) does not necessarily mean that the theme was absent in our planning, implementation, and reflection.

The theme, *fostering awareness to similarities and differences*, gave us insight into teacher thinking by revealing several instances in which teachers did not automatically make connections for themselves such as connecting different

Table 11.1. Mathematics teacher educator learning themes from Zaslavsky (2008) in graduate (practising) and undergraduate (prospective) courses for the topics of Functions and Equations, Visualising Complex-valued Zeros, and Building Functions

Theme	*Functions and equations*	*Visualising complex-valued zeros*	*Building functions*
Developing adaptability			practising
Fostering awareness …	prospective	prospective	prospective, practising
Coping with conflicts …	practising	prospective, practising	practising
Learning from … practice	prospective, practising	prospective	prospective
Selecting … tools …		prospective, practising	prospective
Identifying … barriers …	prospective	prospective	practising
Sharing … dispositions	prospective		

mathematical representations to explore similarities and differences between mathematical concepts or ideas. Similarly, the theme, *coping with conflicts, dilemmas, and problem situations*, revealed difficulties in teacher thinking by exposing the undue influence of memorized rules and algebraic representation on their approach to problems, the effects of limited exposure to mathematics content, or the lack of meaningful experiences in exploring dilemmas in their own understanding.

Insight into teacher thinking also emerged from the theme, *identifying and overcoming barriers to students' learning,* by requiring us to put ourselves in the place of the teachers to identify the situations in which they tended to rely heavily on rote knowledge or in which they felt uncomfortable questioning "obvious" mathematical statements. In addition, the theme, *learning from the study of practice*, helped underscore the need for scaffolding teacher thinking via preparatory assignments and allowing sufficient time for discovery and thinking in class.

Our learning about facilitating teacher learning links to several themes for the topics in this chapter. The theme, *fostering awareness to similarities and differences,* underscored the need for careful attention to structuring tasks, developing just-in-time questions to be used in facilitation, and sequencing mathematics teacher educator moves. Furthermore, the theme, *coping with conflicts, dilemmas, and problem situations*, revealed the need for developing strategies for helping teachers bridge ideas, question assumptions, and effectively facilitate spontaneous bursts of confusion, novel insight, or unique ideas.

The theme, *identifying and overcoming barriers to students' learning*, helped us understand the importance of incorporating opportunities for teachers to question their own understandings as a way to facilitate teacher learning. In addition, the theme, *learning from the study of practice*, provides opportunities for a systematic way to incorporate reflection on teacher thinking, sequencing of tasks, and classroom interactions to facilitate teacher learning.

Moreover, the design research we have conducted in the *Functions and Modeling* course has afforded us rich opportunities to learn from practice and improve tasks in a timely manner. With grant funding, we were able to hire a graduate student who provided assistance in all aspects of the research, including attending and video-recording all class meetings, keeping detailed records of written work from teachers in the courses, interviewing the mathematics teacher educator after each class meeting, conducting interviews with teachers after the completion of the course, and transcribing and analysing data. Our research also provided structure for the principal investigator team to observe each other's instruction and to meet weekly and reflect on current practices while revising future lessons. This data and time for discussion and structured reflection has been invaluable. As is evident in this chapter within the descriptions of tasks from the *Functions and Modeling* course, we were able to gain insight about prospective teachers' needs and how to modify tasks accordingly, and we have collected both quantitative and qualitative evidence to suggest that prospective teachers have positively benefitted from these changes.

In addition, our learning comes from reflective practice over 15 years of experience with the courses we teach. Positive change has come incrementally over this time.

Our learning as mathematics teacher educators derives from self-reflection on our practice, open interactions with and assessment of our students' learning, and research-based strategies facilitated by resources that enabled peer review and detailed analysis of classroom implementation, student work, and a team-based approach toward improving practice.

ACKNOWLEDGEMENTS

This research described in this chapter is based upon work partially supported by the National Science Foundation (NSF) under grant number DUE-1612380. Any opinions, findings, conclusions or recommendations are those of the authors and do not necessarily reflect the views of the NSF. We also thank Janessa M. Beach for research assistance.

REFERENCES

Academy of Inquiry-Based Learning. (n.d.). *Supporting instructors, empowering students, transforming mathematics learning*. Retrieved from http://www.inquirybasedlearning.org/

Álvarez, J. A. M., Jorgensen, T., & Rhoads, K. (2018). Identifying subtleties in preservice secondary mathematics teachers' distinctions between functions and equations. In A. Weinberg, C. Rasmussen, J. Rabin, M. Wawro, & S. Brown (Eds.), *Proceedings of the 21st Annual Conference on research in undergraduate mathematics education* (pp. 1562–1567). San Diego, California: RUME.

Armendariz, E. P., & Daniels, M. (2011). *Functions in mathematics: Introductory explorations for secondary school teachers.* Austin, TX: The UTeach Institute.

Aspinwall, L., Shaw, K. L., & Presmeg, N. C. (1997). Uncontrollable mental imagery: Graphical connections between a function and its derivative. *Educational Studies in Mathematics, 33*, 301–317.

Banchi, H., & Bell, R. (2008). The many levels of inquiry. *Science and Children, 46*(2), 26–29.

Boulter, D. R., & Kirby, J. R. (1994). Identification of strategies used in solving transformational geometry problems. *The Journal of Educational Research, 87*, 298–303.

Breidenbach, D., Dubinsky, E., Hawks, J., & Nichols, D. (1992). Development of the process conception of function. *Educational Studies in Mathematics, 23*, 247–85.

Carlson, M. P. (1998). A cross-sectional investigation of the development of the function concept. *Research in Collegiate Mathematics Education, 3*, 114–162.

Carlson, M. P., Jacobs, S., Coe, E., Larsen, S., & Hsu, E. (2002). Applying covariational reasoning while modeling dynamic events: A framework and a study. *Journal for Research in Mathematics Education, 33*, 352–378.

Chamberlin, M. T., & Candelaria, M. S. (2018). Learning from teaching teachers: A lesson experiment in area and volume with prospective teachers. *Mathematics Teacher Education and Development, 20*, 86–111.

Chapman, O. (2008). Mathematics teacher educators' learning from research on their instructional practices: A cognitive perspective. In B. Jaworski & T. Wood (Eds.), *The international handbook of mathematics teacher education: The mathematics teacher educator as a developing professional* (Vol. 4, pp. 115–134). Rotterdam, The Netherlands: Sense Publishers.

Chauvot, J. B. (2009). Grounding practice in scholarship, grounding scholarship in practice: Knowledge of a mathematics teacher educator–researcher. *Teaching and Teacher Education, 25*, 357–370.

Chazan, D., & Yerushalmy, M. (2003). On appreciating the cognitive complexity of school algebra: Research on algebra learning and directions of curricular change. In J. Kilpatrick, W. G. Martin, &

D. Schifter (Eds.), *A research companion to principles and standards for school mathematics* (pp. 123–135). Reston, VA: National Council of Teachers of Mathematics.

Chick, H., & Beswick, K. (2018). Teaching teachers to teach Boris: A framework for mathematics teacher educator pedagogical content knowledge. *Journal of Mathematics Teacher Education, 21*, 475–499.

Cobb, P., Confrey, J., di Sessa, A., Lehrer, R., & Schauble, L. (2003). Design experiments in educational research. *Educational Researcher, 32*(1), 9–13.

Conference Board of the Mathematical Sciences. (2012). *The mathematical education of teachers II.* Providence, RI & Washington, DC: American Mathematical Society and Mathematical Association of America.

Cooney, T. J., Beckmann, S., & Lloyd, G. M. (2010). *Developing essential understanding of functions for teaching mathematics in grades 9–12.* Reston, VA: National Council of Teachers of Mathematics.

Cruz, I., Febles, M., & Díaz, J. (2000). Kevin: A visualiser pupil. *For the Learning of Mathematics, 20*(2), 30–36.

Design-Based Research Collective. (2003). Design-based research: An emerging paradigm for educational inquiry. *Educational Researcher, 32*, 5–8.

Epperson, J. A. M. (2009). *Can inservice mathematics teachers answer mathematical questions that arise from classroom use of dynamic software?* Paper presented at the Mathematical Association of America's MathFest 2009, Portland, OR.

Epperson, J. A. M. (2010). *New 'rules' for transformations of functions.* Workshop presented at the 2010 Conference for the Advancement of Mathematics Teaching, San Antonio, TX.

Epperson, J. A. M., & Rhoads, K. (2015). Choosing high-yield tasks for the mathematical development of practising secondary teachers. *Journal of Mathematics Education at Teachers College, 6*, 37–44.

Even, R. (1993). Subject-matter knowledge and pedagogical content knowledge: Prospective secondary teachers and the function concept. *Journal for Research in Mathematics Education, 24*, 94–116. doi:10.2307/749215

Even, R. (2008). Facing the challenge of educating educators to work with practising mathematics teachers. In B. Jaworski & T. Wood (Eds.), *The international handbook of mathematics teacher education: The mathematics teacher educator as a developing professional* (Vol. 4, pp 57–73). Rotterdam, The Netherlands: Sense Publishers.

Hiebert, J. (2003). What research says about the NCTM standards. In J. Kilpatrick, W. G. Martin, & D. Schifter (Eds.), *A research companion to principles and standards for school mathematics* (pp. 5–23). Reston, VA: National Council of Teachers of Mathematics.

Hitt, F. (1998). Difficulties in the articulation of different representations linked to the concept of function. *Journal of Mathematical Behavior, 17*(1), 123–134.

Kieran, C. (2007). Learning and teaching algebra at the middle school through college levels: Building meaning for symbols and their manipulation. In F. Lester (Ed.), *Second handbook of research on mathematics teaching and learning* (pp. 707–762). Greenwich: Information Age Publishing.

Koichu, B., Zaslavsky, O., & Dolev, L. (2016). Effects of variations in task design on mathematics teachers' learning experiences: A case of a sorting task. *Journal of Mathematics Teacher Education, 19*, 349–370.

National Governors Association Center for Best Practices & Council of Chief State School Officers. (2010). *Common core state standards for mathematics.* Washington, DC: Authors.

Shulman, L. S. (1986). Those who understand: Knowledge growth in teaching. *Educational Researcher, 15*, 4–14.

Smith, M. S., & Stein, M. K. (2011). *Five practices for orchestrating productive mathematical discussions* (1st ed.). Reston VA: National Council of Teacher of Mathematics.

Stein, M. K., Engle, R. A., Smith, M. S., & Hughes, E. K. (2008). Orchestrating productive mathematical discussions: Five practices for helping teachers move beyond show and tell. *Mathematical Thinking and Learning, 10*(4), 313–340.

Strauss, A., & Corbin, J. (1990). *Basics of qualitative research: Grounded theory procedures and techniques.* Newbury Park, CA: Sage.

Usiskin, Z., Peressini, A., Marchisotto, E. A., & Stanley, D. (2003). *Mathematics for high school teachers: An advanced perspective*. Upper Saddle River, NJ: Pearson Education, Inc.

UTeach Institute. (n.d.). *The UTeach model.* Retrieved from https://institute.uteach.utexas.edu/Uteach-model

Villarreal, M. (2000). Mathematical thinking and intellectual technologies: The visual and the algebraic. *For the Learning of Mathematics, 20*(2), 2–7.

Watson, A., & Mason, J. (2007). Taken-as-shared: A review of common assumptions about mathematical tasks in teacher education. *Journal of Mathematics Teacher Education, 10*(4), 205–215.

Watson, A., & Sullivan, P. (2008). Teachers learning about tasks and lessons. In D. Tirosh & T. Wood (Eds.), *The international handbook of mathematics teacher education: Tools and processes in mathematics teacher education* (Vol. 2, pp. 107–134). Rotterdam, The Netherlands: Sense Publishers.

Zaslavsky, O. (2008). Meeting the challenges of mathematics teacher education through design and use of tasks that facilitate teacher learning. In B. Jaworski & T. Wood (Eds.), *The international handbook of mathematics teacher education: The mathematics teacher educator as a developing professional* (Vol. 4, pp. 93–114). Rotterdam, The Netherlands: Sense Publishers.

Zimmermann, W., & Cunningham, S. (1991). Editor's introduction: What is mathematical visualization? *Visualization in Teaching and Learning Mathematics MAA Notes, 19*, 1–7.

James A. Mendoza Álvarez
Department of Mathematics
The University of Texas at Arlington

Kathryn Rhoads
Department of Mathematics
The University of Texas at Arlington

Theresa Jorgensen
Department of Mathematics
The University of Texas at Arlington

SIMON GOODCHILD

12. MATHEMATICS TEACHING DEVELOPMENT IN HIGHER EDUCATION

This chapter focuses on mathematics teaching development in higher education contexts. It draws on my experience as a mathematics teacher educator in a setting within a Norwegian Centre for Excellence in (Higher) Education that focuses on the improvement of students' experiences of mathematics education, especially within higher education programmes in which mathematics is studied as a service subject. When compared with mathematics teaching development in schools, higher education mathematics teachers have different competencies and expertise in the knowledge to be taught. This chapter discusses some of the challenges encountered in attempts to motivate mathematics teaching development in higher education. It presents a rationale for introducing the notion of teaching paradigms, akin to Thomas Kuhn's research paradigms, that is then used to consider some of the technological resources, didactical approaches, and pedagogical arrangements that could ameliorate many of the problems experienced in higher education 'service' mathematics courses. It concludes with a discussion of what has been learned and what remains to be learned.

INTRODUCTION

This chapter is a personal reflection on the author's experience as a mathematics teacher educator on a segment of his developmental trajectory that has been driven by moving into a new arena of mathematics teacher education. The mathematics teachers in focus are working in higher education, teaching mathematics to prospective engineers, scientists, economists and mathematicians. The mathematics teachers could be contributing to mathematics content courses in teacher education programmes, but not necessarily as 'teacher-educators.' Thus, the chapter is a reflection on the author's developmental experiences encountered in the transition *from* engaging in school-mathematics-teacher education and professional development *to* higher education mathematics teaching development. In the author's national context (Norway) there is increasing awareness of a need to attend to the quality of teaching mathematics at higher education. It is likely that an increasing number of mathematics teacher educators will need to extend the scope of their activity to include professional development opportunities for higher education mathematics teachers. This chapter may contribute to raising the awareness of the challenges that are faced when mathematics teacher educators extend their professional reach.

© KONINKLIJKE BRILL NV, LEIDEN, 2020 | DOI: 10.1163/9789004424210_013

The chapter builds on the reflection written by the author for the first edition of this Handbook (Goodchild, 2008). That chapter described the trajectory of the author's professional development through mathematics teaching, mathematics teacher education, practitioner research, mathematics education research and eventually as a mathematics teaching developmental researcher. The chapter ended at a point in the author's developmental trajectory when he was embedded within a project in which university 'didacticians' collaborated in a co-learning agreement with school teachers. The earlier chapter also set out in some detail the teaching-developmental research methodology and its framing within a theory of communities of inquiry. This chapter will not repeat this ground, the interested reader is referred to the earlier chapter.

The mathematics teacher educator's professional development addressed in this chapter relates to the experience of translating mathematics teaching developmental *activity*, learned through working with school teachers, into the arena of higher education mathematics teachers. The development under consideration is rooted in the author's response to the question: How does mathematics teaching developmental research translate into the practice of higher education mathematics teachers? The reader's attention is drawn to the use of the word '*activity*' in the opening sentence of this paragraph rather than *research* because the chapter is based on experiential evidence rather than systematic inquiry that might count as research. The experience has been gained through the author's encounters and meetings with higher education mathematics teachers in both informal and formal settings in conferences, workshops and seminars. The mathematics teaching developmental *activity* both drives and is driven by research and scholarship into teaching and learning mathematics at higher education. The transition of the author as mathematics teacher educator to teaching development in higher education was precipitated by a successful proposal within a Norwegian programme to develop centres for excellence in (higher) education.

In summary, in this chapter the mathematics teacher educator is the author. The professional development under consideration is that of the author in the transition from working with school teachers to working with university mathematics teachers. The teachers referred to in this chapter are employed in higher education to teach, and often to research, mathematics. For the most part these teachers are not themselves 'mathematics teacher educators.'

A NEW DEVELOPMENT: MATRIC: CENTRE FOR RESEARCH, INNOVATION AND COORDINATION OF MATHEMATICS TEACHING

The developmental trajectory described in this chapter occurred because the author became leader of a project (MatRIC) within a national programme of quality enhancement in higher education. Personal and professional foundations in mathematics teaching developmental research projects with school teachers, described by Goodchild (2008) in the first edition of the Handbook, were fundamental to the development of MatRIC, which is expected to contribute to the development of mathematics education in higher education. As MatRIC is different from the school mathematics teaching development projects, so are the competencies needed to lead mathematics teaching development in the higher education context. The earlier projects, which focused on school teaching development, were based upon a relatively small number of schools local to the university, each with a small group of teachers, who had agreed to participate. The projects entailed workshops at the university, teachers implementing lesson ideas and approaches introduced in workshops. In addition, actions were focused on both teachers and university based didacticians researching (pupils) learning mathematics, (teachers) teaching mathematics and (didacticians) leading mathematics teaching development. By contrast, MatRIC, as a teaching development project, has a substantially different structure, goals and action plan, for this reason some space is devoted to describing MatRIC.

MatRIC is part of the Norwegian centre for excellence in higher education (CfE) programme, which began in 2012 with the establishment of a pilot centre, ProTed, that focused on the professional development of school teachers in initial teacher education programmes. At the time of writing ProTed remains the only CfE that focuses specifically on school teacher professional development. The CfE programme was informed by international experience, especially from the United Kingdom and Sweden. The programme's goals are to raise the quality of teaching and learning in higher education, and to raise the status of education in Norwegian higher education to match that of research. MatRIC, Centre for Research, Innovation and Coordination of Mathematics Teaching, came out of the first open competitive round for CfE proposals in 2014 with two other centres, bioCEED focusing on biology education and CEMPE focusing on music performance education. A second open round in 2016 resulted in a further four centres.[1]

There are three central pillars to the Norwegian CfE programme. First is the notion of research and development-based education. Teachers in higher education are also researchers and they are expected to infuse their teaching with methodologies and results from their own and others' research. The goal is that students experience learning at the cutting edge and experience the excitement of knowledge creation. Second, it is expected that classes, courses and programmes will be designed to ensure the greatest possible participation and engagement of students in the educational process. Within Norway, there is much to be explored, implemented and researched about students as partners in their education and the Norwegian CfEs are looking outside the country for inspiration, such as in Healey, Flint and Harrington (2014). Third, a major task laid upon the CfEs is to disseminate examples of good practice, innovation and results of research into teaching and learning in higher education. These central pillars are set out in the criteria for the award of CfE status, the centres are intended to:

- offer excellent research and development-based education
- develop innovative ways of working with research and development-based education

- encourage student engagement and ownership of learning
- contribute to the development and dissemination of knowledge and practices about the design of teaching and learning environments that are conducive to learning (Norwegian Agency for Quality Assurance in Education, 2016).

MatRIC's vision is about students enjoying transformed and improved learning experiences of mathematics in higher education.

> [The] status is awarded to academic communities that have already demonstrated excellent quality and innovative practices in education and that have plans in place for further development and innovation. (Ibid.)

In MatRIC's case the academic community is the mathematics education group and mathematics teachers in the Faculty of Engineering and Science at the University of Agder. From the outset it has been recognized that excellence in teaching mathematics in higher education is distributed throughout Norway, and an essential role of MatRIC is to network mathematics teachers, both within Norway and internationally to enable the sharing and dissemination of good practice in teaching and learning mathematics. MatRIC's role is to enhance the quality of learning mathematics in Norwegian higher education institutions, and this role is enacted through engagement with higher education mathematics teachers. The author is the Director of MatRIC, and in the context of this chapter and handbook he is the mathematics teacher educator. This chapter is a 'reflective case-study' of the development of the author's professional disposition, attitudes and priorities as a mathematics teacher educator.

Personal Position Statement

In leading MatRIC's engagement with higher education mathematics teachers I build on the experience of mathematics teaching developmental research in schools which I described in Goodchild (2008) and with colleagues in Goodchild, Fuglestad, and Jaworski (2013). The focus on teaching development rather than teacher development is an intentional ethical stance that draws attention to the goal of developing practice (the goal of the mathematics teacher educator) rather than people. As described in the chapter of the first edition of this Handbook, the methodology incorporates a synthesis of community of practice theory (Wenger, 1998) and inquiry, with the endeavour to develop inquiry communities (Jaworski, 2006a). Teaching and learning development occur as teachers implement within their practice an inquiry cycle, which is a six-stage cyclical process summarized as: plan, act, monitor, reflect, evaluate and feedback into plans for a new cycle. From these a theory of change is developed based upon community – enterprise, engagement and repertoire (Wenger, 1998), and *critical alignment* (Jaworski, 2006b) of practitioners as they collaborate and exert their agency to transform and improve teaching and learning.

The synthesis of community of practice theory and inquiry requires some theoretical contortions. Community of practice theory does not directly consider human agency (Wenger, 1998), and the notion of goal directed action is challenged by scholars that have contributed to the development of the theory (Lave, 1988). However, the exercise of agency lies at the heart of 'critical alignment,'

> in which it is possible for participants to align with aspects of practice while critically questioning roles and purposes as a part of their participation for ongoing regeneration of the practice. (Jaworski, 2006a, p. 190)

Moreover, goal directed action is fundamental to an inquiry approach to teaching and learning. These theoretical issues have led to the use of cultural-historical activity theory to frame the analysis of community of inquiry. In a later section of this chapter, the presentation of mathematics teaching development is considered from the cultural-historical activity theory perspective. Also, within this chapter, there are references to claims from cognitive and constructivist psychology when discussing teaching and learning at the level of the individual learner. The use of ontologically inconsistent and incommensurable theories to provide different insights into teaching and learning is problematic, but not new (Goodchild, 2001; Jaworski & Potari, 2009).

An Action Plan Based on International Experience

The award of centre for excellence status to a group that has been, and continues to be, deeply involved in school teachers' professional development and mathematics teaching developmental research was an opportunity to translate the knowledge gained from school to higher education mathematics teaching and learning. The context of this chapter is the transformation of thinking necessary when moving from school to higher education mathematics teaching development. As a centre for excellence in higher education, MatRIC was a new development within the Norwegian higher education mathematics community and it was a new development within the author's professional life. However, the idea of a centre for excellence in higher education mathematics was not a new development from an international perspective. For example, in the United Kingdom from 2005 to 2010 a national programme funding centres for excellence in teaching and learning was pursued by the now discontinued United Kingdom Higher Education Funding Council. One centre for excellence in teaching and learning, established jointly at the Universities of Coventry and Loughborough, named *sigma*, focused especially on providing mathematics support to students. This Centre proved to be one of the most successful centres for excellence in teaching and learning and developed into an international network[2]; its impact can be seen in the growth of mathematics support based on the *sigma* model at many universities. In Germany The Centre for Higher Mathematics Education (khdm: Kompetenzzentrums Hochschuldidaktik Mathematik) khdm[3] is based on a consortium of universities – Paderborn, Hannover and Kassel. Khdm received funding from the Volkswagon Foundation for a five-year period

2010–2015, and now continues from local resources. MatRIC has been inspired by both *sigma* and khdm, learning from their approaches and seeking advice from their leaders. The concentration in this chapter on experiences within MatRIC reflects the author's position and is the basis for the account of learning from mathematics teaching development in higher education.

MatRIC sets out to motivate higher education mathematics teachers to reflect upon their practice and consider alternative approaches and innovations that will transform and improve students' experiences of learning mathematics. Much effort is invested in organizing national events – conferences, workshops and seminars, in which leading international higher education mathematics educators contribute alongside Norwegian higher education mathematics teachers who are working innovatively in their own institutions. These events provide opportunities for dissemination, networking and community building, as well as stimulation for innovation and research by participants when they return to their home institutions. MatRIC also distributes some 'seed-corn' grants to facilitate small research and development projects. Locally, within MatRIC's host institution a prototype of higher education institution-wide mathematics teaching developmental research has been established. The recurrent challenges of teaching and learning mathematics at higher education, such as low performance, poor retention, motivation, student engagement and attendance, and very large groups are being explored and addressed through innovation and partnerships between mathematics teachers and mathematics education researchers.

CHALLENGES FACED IN MATHEMATICS TEACHING DEVELOPMENT AT HIGHER EDUCATION

The mathematics teacher educator's transition to higher education mathematics teaching development requires an understanding of the context and how teaching university students differs from school teaching. The experience gained from mathematics teaching developmental research with school teachers does not translate easily to higher education contexts for two major reasons:

- The teachers have a different background of professional preparation and development for and within their practice.
- The two arenas – school and higher education are very different contexts for teaching and learning mathematics.

The above reasons are complicated because higher education mathematics is passing through a period of rapid change. However, there is a widespread recognition of the need to transform and improve mathematics education at higher education, and there is a growing body of research that points the way to the desired improvements (see for example Rasmussen & Wawro, 2017). In the transition from the education (pre- or in-service) of school mathematics teachers to working with higher education mathematics teachers the mathematics teacher educator must understand and learn to accommodate to these differences.

The Professional Education and Work of the Higher Education Mathematics Teacher

Teachers appointed to Norwegian higher education institutions are now required to participate in a course in university pedagogy. The requirement to follow such courses is a fairly recent development. The courses are organized locally within the higher education institutions and there is no national framework for the courses and no attempt at standardization across institutions. The courses may require about 150 hours engagement in general higher education pedagogy. One of MatRIC's actions has been to set up a complementary mathematics teaching induction course for higher education teachers. This course focuses especially on teaching and learning mathematics, it requires about 100 hours engagement in face to face sessions and project work. It is open to higher education mathematics teachers throughout Norway, with MatRIC covering all costs – travel, accommodation and the course. It is essential to look beyond ourselves, therefore MatRIC imports international expertise to address topics for which local expertise is, so far, undeveloped. Experienced mathematics educators are invited from the United States of America and Europe to lead the classroom-based sessions covering a curriculum designed to include the most salient features of higher education mathematics teaching and learning. Recruitment to the MatRIC course has been low, within single figures for each cohort. Low recruitment could be due to several reasons: perhaps the course is incorrectly perceived as being in competition with, rather than complementary to, the general pedagogy course. Although MatRIC covers all costs, participants are required to make time for their engagement, and the course competes for time with participants' teaching duties and research activities. Furthermore, there is no compulsion to take the course, although the advent of a national scheme to recognize excellent teachers may make the course more attractive as an additional entry into a teacher's curriculum vitae. More generally, attracting participation in MatRIC events is a continual challenge and the mathematics teacher educator must learn and develop new skills in marketing.

Compare the above with the professional preparation of school teachers who normally participate in an extensive, subject focused teaching education before entering full-time service. Then during professional service, they will have regular opportunities for engaging in professional (teaching) development. The mathematics teacher educator must adapt to a different culture of professional development in which the contrasting approaches to professional preparation and development create different perceptions of the value of mathematics teaching development activity.

The higher education academic is employed to engage in both research and teaching. As noted above, the centre for excellence programme in Norway was introduced in part to raise the status of teaching in higher education, as also was a scheme to recognize excellent teachers. Nevertheless, most academics believe their research and publication profile will count most as they seek promotion through the employment ranks and strive to gain international recognition. It is true that linking teaching development to research, as in developmental research and the drive for research and development-based education, appears to bridge the divide between research and teaching. However, mathematics education research papers will not, in Norway, contribute to an academic's profile in mathematics. The point is, teaching developmental research with school teachers goes to the heart of their professional practice, for the higher education teacher it is a weak competitor vying for attention with their research activity, and the mathematics teacher educator must adapt to the different context in which their efforts may not be so highly valued.

The transition from a school focus to a higher education focus requires the mathematics teacher educator to recognize her or his limitations. Whereas, in most cases the school teacher has a much more thorough preparation for teaching, the higher education teacher will probably have a much deeper knowledge of the subject to be taught. Deep and extensive knowledge and understanding of mathematics is surely an essential attribute for a higher education teacher. Superior or deeper mathematical knowledge can also be an impediment to participation in teaching development activity led by a mathematics education scholar (mathematics teacher educator) rather than a mathematician. This is evident from the following quotations, the first of which comes from one of the mathematics teaching development projects with school teachers in which the author engaged.

> Mathematics Teacher: … I thought very much about you [referring to the mathematics education researcher] should come and tell us how we should run the mathematics teaching. That was how I thought, you are the great teachers. (Daland, 2007, p. 168)

This can be compared with the following which Elena Nardi wrote to characterize views expressed by mathematicians in her study of university mathematics teaching.

> The main bone of contention in the suspicion, even hostility often characterising the relationship between mathematicians and researchers in mathematics education is the substantially different epistemologies of the two communities. (Nardi, 2008, p. 264)

> … it is still the case that the image of a mathematics department that pays a lot of attention and contributes to research in mathematics education would be poor from other mathematicians' point of view: the mathematics community does not in its bulk look to this type of research as a source of knowledge or ideas about mathematics teaching. It just doesn't … whether it should or not is a different matter of course. (Mathematician's voice in Nardi, 2008, p. 268)

The Context of Higher Education Mathematics Teaching

The transition from school-focused-mathematics teacher educator to higher education-focused-mathematics teacher educator requires sensitivity to the possibilities of teaching and learning when the physical and structural conditions are very different. In the developed world, at least, school mathematics is taught in relatively small classes with 20 to 40 students. School teachers establish a fairly close relationship with their students and provide them with feedback that enables them to make progress in the subject. The students composing a single class will usually be engaged in the same educational programme working towards similar examination targets.

Higher education mathematics, especially when studied as a service subject, which is where MatRIC's attention is predominantly focused, is often taught to very large groups of students (200 to 400+) in large auditoriums. In their first encounter with mathematics in higher education, students are often brought together from different programmes of study such as different varieties of engineering (mechanical, electrical, building, etc.) or human sciences etc.[4] Consequently, the groups are rather heterogeneous, the teachers are fairly remote figures and the course is of a general nature with few, if any, illustrations and examples from the students' chosen programmes. It is true that in most universities the large lecture classes are also broken down into smaller tutorial groups or problem-solving classes. In these small groups the tutor responsible may be a student or post-doctoral teaching assistant who has little or no specialized education or preparation for the task. The higher education mathematics teacher's task is thus constrained to providing some exposition of the content, arranging appropriate exercises and problem sheets, and managing the formal summative assessment of the students.

Given the very large classes that limit opportunities for providing formative assessment and feedback, and the often weak prior knowledge on the part of the students and, in many cases, rather inexperienced teachers, it is unsurprising that many universities in Norway, Europe, Australia and the United States of America and are concerned with students' poor performance, slow progression and high drop-out (course retention) in mathematics (e.g., Chen, 2013). Structural factors within higher education create great challenges for mathematics teaching development activity. The context means the mathematics teacher educator may need to revise opinions about the place of technology to support students' independent learning, this issue is taken up in a later section of the chapter.

In Norway, MatRIC has been working during a time of great change in higher education at an institutional level. Universities and university colleges have been amalgamated to create large multi-campus institutions. The process has required academics and administrators at all levels to engage in discussions around organizational structures and managerial alignment. The energy consumed by the structural changes has left little in reserve for teaching development.

A PARADIGM FOR TEACHING?

The mathematics teacher educator has also to be sensitive to firmly held beliefs about teaching mathematics at higher education. In relation to this, I offer the following reflection. Teachers are often very articulate in explaining and justifying their teaching practice in ways that are very different from my own such that I have been led to wonder whether I am experiencing and adapting to a change in teaching paradigm. The notion of scientific 'paradigms,' was introduced by Thomas Kuhn who writes:

> [People] whose research is based on shared paradigms are committed to the same rules and standards for scientific practice. That commitment and the apparent consensus it produces are prerequisites for normal science, i.e., for the genesis and continuation of a particular research tradition. (Kuhn, 1996, p. 11)

I ask the reader to consider, in the above quotation replacing 'research' and 'science' with the word 'teaching': People whose *teaching* is based on shared paradigms are committed to the same rules and standards for *teaching* practice. That commitment and the apparent consensus it produces are prerequisites for normal *teaching*, that is, for the genesis and continuation of a particular *teaching* tradition.

It seems that a large proportion of mathematics teaching at higher education is committed to a set of standards that Artemeva and Fox (2011) have termed "chalk talk." Artemeva and Fox observed and interviewed 50 mathematics teachers of varying levels of experience, 34 of whom were experienced professors, working in 10 universities in seven countries. The national backgrounds of the teachers were more varied with 16 different first languages spoken by the sample.

> Across all the observed local contexts, mathematics teachers enacted the same teaching genre through speaking aloud while writing on the board, drawing, diagramming, moving, gesturing, and so on. As genre researchers, we identified this typified and recurring practice as chalk talk. (Artemeva & Fox, 2011, p. 355)

A more recent study by Olov Viirman in Sweden confirms the findings by Artemeva and Fox, who also included Sweden in their study. Viirman observes: "it is reasonable to claim that the overall form of the teaching practices used by all seven teachers is similar: content-driven lectures conducted using chalk talk" (Viirman, 2014, p. 69). However, Viirman's analysis of their teaching practices using commognition (Sfard, 2008) as an analytical framework enables him to make a more nuanced interpretation: "despite these outer similarities the discursive practices of the teachers are very different" (Viirman, 2014, p. 70).

The need to look more deeply into higher education practices, before drawing conclusions is also demonstrated in the research reported by Petropoulou, and colleagues, (Petropoulou, Jaworski, Potari, & Zachariades, 2015). They point to 12 different teaching actions that one lecturer, a research mathematician, uses to "draw [students] into mathematical production offering mathematical challenge" (p. 2226). These actions include, for example: directing discussion, drawing on students' experience, checking for consensus, and posing a problem. When required, mathematicians justify their "chalk talk" practice, as in the statement by the London

Mathematical Society (2010): "… one needs to see someone else, the lecturer working through and creating results. … The lecturer must be able to create, to write out, during the lecture itself, a large body of argument …" (p. 3).

The approach has served mathematicians for a long time, if only to allow the next generation of research mathematicians to emerge, because they demonstrate the validity of the claim:

> Yet, it is clear that one learns by example and precept, by sitting at the feet of the masters and imitating what they do; and it is equally clear that the masters are able to transmit something of their strategy and insight. (Davis & Hersh, 1981, p. 284)

It seems that there may be a paradigm for higher education mathematics teaching, in which teachers are "are committed to the same rules and standards for *teaching* practice." Mathematicians do not normally need to defend their approach and articulate strong arguments to support their practices. Moreover, as pointed out above, higher education mathematics teachers do not normally receive any subject specific teacher education, and thus their knowledge is based on what Lortie (1975) describes as the 'apprenticeship of observation.' Nevertheless, it would be wrong to imply that mathematics teachers in general follow unreflectively in the footsteps of their masters, expressing a rationale for their practice only when challenged. For example, in her doctoral study Stephanie Thomas followed the teaching practices of a mathematician who reflected deeply about the pedagogy of teaching linear algebra to a class of about 200 first year students at a United Kingdom university (Thomas, 2012).

"Chalk talk" may not be generally representative of all mathematics teaching in higher education. It may be that a paradigm shift is underway, provoked in part by the emergence of digital technologies and in part by the increasing number of students required to study mathematics in higher education. Many higher education teachers are exploring so called 'flipped classroom' approaches, which require the students to prepare for classes by watching a video, often recorded by their teacher, and then in class students engage in problem solving, discussion and other active learning approaches. Wes Maciejewski (2016) reports a quasi-experimental study in which a calculus course at a Canadian research-intensive university with nearly 700 students was divided into seven sections, four of these experienced a 'flipped classroom' approach, the other three conventional teaching approaches. Students on the 'flipped' approaches performed better, by about 8% on the traditional examinations. A similar study with first year higher education calculus classes in Sweden (Cronhjort, Filipsson, & Weurlander, 2017) involving nearly 250 students comparing 'flipped' and conventional lecture approaches also showed similar gains for the flipped classroom. These researchers also used a questionnaire to explore students' self-reported levels of engagement, revealing the students in the flipped approach to be more engaged in their mathematics than the other students.

Positive effects of active learning in higher education have been reported by many studies, a notable meta-analysis reported by Freeman, Eddy, McDonough, Smith, Okoroafor, Jordt, and Wenderoth (2014) included 225 studies that compared active approaches with conventional lecture approaches. Students in the active learning classes performed on average 6% better than those in conventional classes, and they were 1.5 times less likely to fail. These studies are noteworthy for their message about the beneficial effects of active learning, but the point to be emphasized here is that there is a growing interest among higher education mathematics teachers in alternative teaching approaches, they are not generally committed to the "chalk talk" paradigm.

It may then be asked, what is the point of introducing the notion of a teaching paradigm? Artemeva and Fox (2011) refer to "chalk talk" as a 'genre,' is that not sufficient? The value of considering a teaching paradigm lies in the rationale for the genre/approach that adherents do not normally need to explore or explain. Just as with theoretical arguments that can be unproductive when discussants occupy unstated but contrasting paradigmatic positions, so it is possible for discussions about teaching to be unproductive. As the passages quoted above from Elena Nardi's work imply, very often mathematicians and mathematics education researchers base their work on quite different teaching paradigms. The mathematics teacher educator transitioning to work with high education mathematics teachers needs to accommodate to an alternative set of rules and standards for teaching and develop ways to challenge and promote alternative approaches that will not be lost as a product of non-intersecting paradigms.

TECHNOLOGICAL RESOURCES, DIDACTICAL APPROACHES, PEDAGOGICAL ARRANGEMENTS

In addition to the professional and structural challenges outlined above, MatRIC's vision for transformation and improvement of students' learning experiences in mathematics is being pursued during a considerable period of technological change. Modern and emergent technologies and the development of social media have opened up new and exciting approaches for communicating, learning and discussing mathematics. The mathematics teacher educator may need to revise opinions about the relative value of technological learning resources and teaching/learning through human interaction. This, not because different models or theories of learning are applied, but because teacher-student ratios make technological solutions more attractive and students' maturity as independent learners make the tools more viable. The technology that can transform teaching and learning mathematics is also having an impact on the mathematics pursued by research mathematicians and used in industry and commerce. The impact of technology is having an effect on Norwegian education as, for example, a new school curriculum for mathematics being developed at the time of writing is required by ministerial decree to include programming at the expense of other mathematical content such as geometry and statistics. Similarly, at higher education, revised frameworks for engineering and economics including fresh demands on students' mathematical competencies are being introduced. Computational methods are being introduced in all branches of science; one of the most recent centres for excellence focuses on 'computing in science education' and aims to include an element of computing in a high proportion of science and mathematics courses at the University of Oslo.[5]

Technology is transforming the way that calculations and mathematical procedures can be carried out. A student with a suitable application installed on her or his smartphone can take a photograph of an equation and obtain the solution without further effort. Digital technology is also changing the nature of mathematics being pursued in the sciences where it is applied, and numerical approaches appear to be displacing analytical methods. Mathematics is also becoming increasingly significant in a broad range of disciplines. Often this is related to the existence and manipulation of 'big data' and computational methods. An internet search for "mathematics and <name of another discipline>" is likely to find academic papers or books on the use of mathematics in the other named discipline – to test this hypothesis I tried as the other discipline 'archaeology,' 'history,' 'sociology,' and 'divinity' – my hypothesis was not falsified. This accords with the assertion of the Soviet mathematician V. M. Glushkov who, in the 1980s, referred to the 'mathematization of knowledge.'[6] He explains: "Mathematization and computerization open new ways for studying complex social, engineering and natural systems, foreseeing remote consequences of made decisions" (Glushkov, 1983). In my engagement with pure mathematicians and mathematics teachers with an education in another discipline I have occasionally observed a tension arising from this 'mathematization.' It emerges when mathematicians apparently guard the 'pure truth' of their discipline in response to users of mathematics who may not be so precise in their definitions of mathematical objects, terms, representations and procedures. This tension can interfere with attempts to engage mathematics teachers with roots in different disciplines in coordinated teaching development activity.

The expansion of higher education is an issue that has been confronting university teachers in many countries. The quality of school leavers' competencies may not have changed but given that the proportion of school leavers that embark on university courses is growing, the spread of performance on entry is increasing. This appears to affect the overall performance of students required to study mathematics as a service subject (e.g., Hawkes & Savage, 2000). Increasing numbers of students without the necessary competencies in mathematics are joining programmes (e.g., Opstad, Bonesrønning & Fallan, 2017). Students struggle, give up, fail or perform badly. The growth of student mathematics support, especially in the United Kingdom and Republic of Ireland shows evidence of the seriousness with which Science, Technology, Engineering, Mathematics (STEM) faculties address the challenge (Pell & Croft, 2008). MatRIC invests a lot of effort and resource in providing student support and preparation of student mentors. However, it is likely that the main reason for poor performance is inadequate intake knowledge of mathematics. Mathematics teachers are aware of this and they may, therefore, resist arguments about the value of teaching developmental activity.

The mathematics teacher educator moving from mathematics teaching development in school to university must adapt to the structural, professional, and practice arena. However, there are principles of engagement that can be taken from one arena to the other. The ethical stance set out by the author in the first edition of the handbook (Goodchild, 2008), is one such transferable principle. Also, given the possibility of a paradigm clash three methodological principles for working with teachers set out by Cooper and McIntyre apply:

- Empathy: “with [teachers’] expressed views, however idiosyncratic these might be.”
- Unconditional positive regard: showing “an overt sense of liking and interest in informants as individuals.”
- Congruence: entailing honesty and authenticity in conversation with teachers (Cooper & McIntyre, 1996, p. 26).

Modern and emergent technologies might provide means of transforming and improving students’ learning experiences. The ‘flipped classroom’ approach, described above is one such application of technology.

Consider the following statements by the educational psychologist David Ausubel. First an aphorism:

> Knowledge is meaningful by definition. It is the meaningful product of a cognitive (“knowing”) psychological process involving the interaction between “logically” (culturally) meaningful ideas, relevant background (“anchoring”) ideas in the particular learner’s cognitive structure (or structure of his knowledge), and his mental “set” to learn meaningfully or to acquire and retain knowledge. (Ausubel, 2000, p. vi)

And an epigram:

> If I had to reduce all of educational psychology to just one principle, I would say this: The most important single factor influencing learning is what the learner already knows. Ascertain this and teach him accordingly. (Ausubel, 1968, p. vi)

As described above, in many higher education institutions, students without the necessary starting competencies in mathematics are recruited to STEM and other programmes which invariably include courses in higher mathematics. In the first year of study mathematics is taught to very large heterogeneous groups in which contact between the teacher and student is limited, and the curriculum forces a rapid pace through the content to be covered. It is hypothesized that students have registered to study programmes in which they often fail to relate the mathematics taught in the large general lectures, and they lack motivation to study mathematics. Furthermore, they are unprepared for the independence they have to direct their own learning. It is asserted that in mathematics courses, overall students' performance, progression, and retention will only be significantly and substantially improved if there are fundamental changes (with financial consequences) in institutional structures – recruitment, induction, organization/grouping and curriculum design. The structures of higher education institutions exist in stark contradiction to the research into teaching and learning carried out in those same institutions. In addition, higher education mathematics teachers who are committed to a "chalk talk" paradigm may see little if any reason to make changes in their well-established practice. These hypotheses and beliefs are extrapolated from research in mathematics education at school level. Research in undergraduate mathematics education is still in its infancy and there remains much to be done, especially within large scale studies to validate the assumptions. To be fair, it should be noted that some universities are making structural changes to the design of auditoriums, (large auditoriums are being remodelled with circular/polygonal tables or grouped settings) to facilitate group work and active learning.

It is well known that formative assessment and feedback play a significant role in learning. The evidence from school level research is compelling (Black & Wiliam, 1998) and it is highly likely that higher education learners benefit in the same way from well-directed feedback. Individually targeted feedback is one of the casualties of very large classes, but the development of computer aided assessment programs, with in-built computer algebra systems can help. Programs such as STACK created by Chris Sangwin (2013) or Numbas developed by the e-Learning Unit of Newcastle University's School of Mathematics and Statistics are two of many such systems. These programs have developed a long way from the early attempts at computer aided assessment that depended upon multi-choice responses and were thus open to students working backwards from the possible answer, for example it is easier to try out, by differentiating, given solutions to an indefinite integral question to find which solution belongs to the question. The modern systems allow symbolic answers to be given, which can be tested through the in-built algebra system. Questions can be set that will test intermediate stages in an answer and provide feedback if an incorrect answer is 'trapped' on the way. However, creating challenging tasks that are structured to enable constructive feedback is far from easy and the lone mathematics teacher may not feel the time expended is well-used. Sharing questions is an approach being used by many mathematics teachers, such as the Finnish based Abacus network[7] that (at the time of writing) includes mathematics teachers in 26 universities in eight countries. It is thus possible to create assessment and feedback tasks that are manageable with very large groups of students.

Another issue mentioned above was the pace of 'delivery' that is dictated by demanding curriculum content and fairly short, semester length courses. In many higher education institutions lectures are streamed and available to students on-line. The initial reason for streaming may be for other reasons, such as at the author's university where auditoriums are too small to accommodate classes and half the class watches the streamed video of the lecture in a remote auditorium. The streamed lectures also make it possible for students to watch a recording and spend more time with parts of the lecture, if they want. Many students may not be able to attend the lectures for a variety of reasons. In Norway attendance is not compulsory in many courses, absentees can watch the lectures at a time suitable to them. There is also a very large volume of video resource available freely on-line, much of this has been produced by individual teachers for their own classes. These videos can be used by students in much the same way as they would use text books to complement their teachers' presentations.

Interactivity in lectures is made possible by the development of audience response systems. Initially these were small hand-held devices (clickers) that could be linked to a teacher's presentation. At points within the lecture the teacher can pause and ask a question, asking students to choose between several different possible answers. Interactivity is increased if students are asked to discuss their answer and agree with their neighbour before making their choice. When this technology was first introduced it was necessary to distribute the clickers at the beginning of a lecture, synchronize them with the presentation, and then collect them at the end of the lecture, the management of the technology interfered with their usability. Now most students have smartphones and it is possible to use internet-based software[8] without the management issues, students merely need to log on to the program.

The best advocates for the use of technology in teaching mathematics are those teachers who are using it effectively with their own classes. The author's role in mathematics teaching development is therefore to make opportunities for teachers who are introducing active, inquiry based learning and new technologies to present their approaches to the wider community. At a recent meeting, the head of one university mathematics department commented along the lines, 'it is possible to have too much technology in teaching and learning, I have never heard an advocate of technology in education saying this.' I hastened to say it! Also, I responded to the underlying message, learning mathematics entails doing mathematics. It is necessary to engage in challenges and problems solving, to be active in doing and thinking rather than passive in watching and listening. The teaching and learning technologies described above are a means of overcoming some of the structural and institutional issues that interfere with teaching and learning. Furthermore, digital technologies have opened up ways to represent mathematics through dynamic and interactive visualizations. Computational approaches make possible new ways of working mathematically and exploring mathematical objects. Technology is not just a partial remedy for an illness in the system, it opens up new opportunities to engage with mathematics. One thing appears to be clear and agreed by all regarding the use of technology and teaching mathematics – approaches based on the use of prepared power point slides are not effective – the ready-made presentation of the material interferes with the pace of delivery, it becomes too fast, and lacks spontaneity in developing mathematical arguments (Artemeva & Fox, 2011).

REACHING OUT WITH DEVELOPMENTAL ACTIVITY

It is well recognized that mathematics teaching development in schools is a slow, evolutionary process (Jaworski, 1998). The risks entailed in adopting novel approaches, additional time required for preparation, routine curriculum demands and the influence of, perhaps, unenthusiastic or even reactionary colleagues are experienced at both school and higher education levels, and all contribute to impeding the pace of development. There are additional forces at work in higher education, a prevailing and pervasive teaching paradigm and the institutional structures set out above. Success in mathematics teaching development is measured in small steps.

Consistent with the developmental research methodology outlined above and described in more detail in the previous edition of the Handbook (Goodchild, 2008), teachers engaging in their own research and inquiry forms a significant part of MatRIC's agenda. In addition to eight PhD fellows and a post-doctoral researcher attached to the Centre, small grants are made available to higher edition mathematics teachers as seed-corn funding to facilitate their own research projects. After four years 18 projects have been funded (receiving an equivalent of about 5000 Euro each). Around half of the projects funded have been awarded to established mathematics education researchers focusing on mathematics in teacher education programmes, the remainder has supported mathematicians, all of these including collaboration with a mathematics education researcher. On a national scale, even a country with a small population like Norway, the numbers are very small, but they are significant for several reasons. Research into university mathematics education is under-developed in Norway and so any activity is an improvement. Second, the partnerships between mathematicians and mathematics education researchers crosses the intellectual divide between the fields of scholarship described by Nardi (2008) and teaching paradigms. Third, higher education mathematics teachers are being brought into the research discourse of mathematics education and presenting their research in mathematics education conferences such as RUME[9] in the United States of America and INDRUM[10] in Europe.

The projects funded by the small grants are required to fit with a broad interpretation of MatRIC's goals. They have thus focused on teaching and/or learning mathematics, especially with the use of technology and the development of active learning approaches using mathematical modelling, problem solving and inquiry-based learning. As a consequence of interests upon which the original proposal for MatRIC was based there has been a concentration of attention on the use of video, digital visualization, mathematical modelling and computer aided assessment in teaching and learning mathematics. The strongest growth of a collaborative network has been in the use of computer aided assessment, and yet there have been no proposals to research how computer aided assessment may be an effective means of formative assessment, or how computer aided assessment used as a tool to provide feedback to students might influence their understanding of mathematics or performance in summative assessment.

In 2017 MatRIC was subject to a mid-term evaluation. This rigorous exercise took about nine months to complete and involved an international panel of experts. An observation made by the evaluation panel was that MatRIC had been successful in reaching the 'enthusiasts' and it was now necessary to draw others (I suppose they mean the unenthusiastic) into MatRIC's activity. MatRIC has invested a lot of resource, financial and human into building networks around the use of digital technologies (computer aided assessment, video, visualization and simulation) to address the problems of teaching and learning mathematics in higher education. The challenge from the experts could be restated, 'it is necessary to attract practitioners from one teaching paradigm to consider and apply possibilities developed within another.' One response to attract practitioners across the paradigmatic divide is to engage them in comparative studies, using the enthusiasts and the sceptics, to demonstrate the learning gains claimed for the alternative approaches. Unfortunately, empirical studies with samples that ensure valid and reliable results are not facilitated by strong and ethically defensible data protection regulations and students' apparent reluctance to give their consent for their personal and performance data to be used in research.

WHAT HAS BEEN LEARNED? WHAT REMAINS TO BE LEARNED?

Earlier in this chapter, the use of cultural-historical activity theory as an analytical framework in community of inquiry research was briefly mentioned. The use of cultural-historical activity theory especially sensitizes the research to contradictions and tensions within the practice arena (Engeström, 2001; Roth & Radford, 2011). Engeström (2001) refers to contradictions as a source, and a driving force for change. The mathematics teacher educator entering the domain of higher education mathematics teaching development needs to be conscious of the contradictions and reflect on how these might be harnessed. In this section I draw attention to the many contradictions and few tensions experienced through trying to stimulate mathematics teaching development while leading MatRIC. This section brings together the contradictions that have been introduced earlier in this Chapter.

In the foregoing there have been some brief references to many students' inadequate knowledge of mathematics on entry to their programme of study. This appears to be an issue especially where mathematics is a service subject and delegated to a department outside the programme to which the student has been recruited. For example, one research fellow attached to MatRIC has explored the differences across a variety of mathematics qualifications and years of study offered at Norwegian upper secondary schools, and the demands made by the mathematics course, which students follow in their first year in the bachelor level economics programme. Even presuming that students have learned the curriculum to which they have been exposed in school, there are many significant gaps between students' prior learning opportunities and the expectations of the mathematics course at the university. The course is required to cover given content to meet the demands of the economics programme. Strategies have been implemented to make up the differences, for example to split the cohort and create a 'slow' route which includes more classes – in fact it is denser rather than slower because the examination target is the same after the same course length. Another approach is to provide a preliminary mathematics course for those students with less mathematics in the background, facilitated by situating the main mathematics course in the second semester. Also, drop-in mathematics support is available and additional tutorial help by student teaching assistants. Several contradictions are evident in the above, beyond the basic mismatch of recruitment to course demands (Opstad, Bonesrønning, & Fallan, 2017). Students, who by choosing a mathematics-light course at school have already demonstrated that they may not have a positive attitude towards mathematics, are 'invited' to engage in additional classes of mathematics and engage in a course which is very demanding. These contradictions are tightly entwined with other moral issues: whether it is right to recruit students who can be expected to struggle with mathematics and with a high probability of dropping out or failing, or to refuse entry to all such students thus denying access to a programme in which, otherwise, the student will flourish. Contradictions of student engagement will be considered below.

When mathematics is taught as a so called 'service' subject, the intention is to facilitate the student learning the mathematics needed within a non-mathematics programme, such as engineering, economics, natural or life sciences. As pointed out above, the reality is that many mathematics service courses are combined, in one university in Norway (not that of the author) for example, all students from over 20 different programmes within life sciences are taught as one heterogeneous group. In other universities, engineering students (mechanical, electrical, electronic, building, etc.) are combined for the lower level mathematics courses. Mathematics is then taught as a rather general discipline, decontextualized from any single programme, with a curriculum that has been abstracted, not for the coherence or specific relevance of the subject studied but rather to ensure the coverage of the range of procedures needed by the programmes 'served.' The service subject in fact ceases to 'serve' through any relevance to the supposedly 'served' programme (perhaps the course does serve the Mathematics Department's finances).

Who are the best people to teach mathematics as a service subject? Research mathematicians, specialized mathematics teachers, or mathematical competent scholars from the programmes served? Of course, there is no single answer to this, there will be examples in each category that demonstrate excellent teaching, and the opposite. However, as a mathematics educator, I assert that teaching quality cannot be assumed, the competencies required for teaching mathematics need to be learned and developed. Additionally, the learning goals need to be understood by the teacher – in terms of procedural and conceptual knowledge, and the consequences if only one of these is developed by students. This is related to the contradiction of requiring higher education teachers to undertake general pedagogy courses, that do not take account of specific teaching and learning issues within mathematics, and yet teachers are required to teach subject disciplines.

It appears contradictory that at the point where the teacher can get close to students, that is in tutorials and problem-solving classes, the mentoring is transferred to student teaching/learning assistants. The student teaching/learning assistants may be good at explaining things and they will probably have been appointed on the back of their high performance in mathematics. It is also argued that student teaching/learning assistants will have temporally close experience of their own engagement with the material. However, it does seem strange to leave this crucial part of education in the hands of novices; and the approach is reminiscent of the 'monitorial system' that was popular in elementary education in the early nineteenth century. Some universities offer some preparatory education for student teaching or learning assistants, and the extent of the partnership between the student teaching/learning assistants and the main course teacher can vary from almost nil to regular meetings. It is difficult to believe that decisions for employing student teaching/learning assistants are made on grounds other than available resources, such as sufficient teachers or cost. However, it is possible that undertaking the role of student teaching/learning assistants can be hugely beneficial to the student assistant and may also be of value to the assisted learner.

In the foregoing the possibility of a 'teaching paradigm' characterized by the "chalk talk" approach described by Artemeva and Fox (2011) was introduced. The approach was justified by mathematicians because, they argue, it is necessary to model how mathematics is developed. This may be true, but it will be an 'institutionalized' mathematics that is presented, in other words, the

> royal way … the made track by which [another] may now reach the same heights without difficulty. (Hermann von Helmholtz 1821–1894, in FLM, 1985, p. 28)

"Chalk talk" does not reflect the approach of mathematicians working on actual problems and engaging with difficulties, following wrong leads, and making unsuccessful attempts. "Chalk talk" expounds a textbook like production of mathematics, it does not offer insight into the actions of a professional mathematician, it is not an apprenticeship to be a mathematician. It could be argued that higher education mathematical studies at undergraduate level, and especially service courses, are about students developing mathematical content knowledge and not an apprenticeship to become a mathematician. However, most mathematics teachers will agree that mathematics is learned through doing mathematics and engaging with mathematical problems. It can then be argued that students are not well-served if they do not get any insight into or experience of the way expert mathematicians work.

In many higher education institutions, another contradiction is the strange belief that a short 4 to 10-week course can make a fundamental difference to students' mathematical understanding and competencies that have developed over 13 years in school. There seems to be a message that in higher education there lies an expertise in teaching that can ameliorate the deficient performance from school teaching.

Students choose to enroll on a programme in higher education, but for many their attendance at scheduled classes appears to contradict their motivation to study. Their independence and freedom to choose to attend are carefully guarded by students. Furthermore, it is recognized that many students appear to encounter difficulties with mathematics, and support systems are created for them, but often the students at most risk of failure do not take up the opportunity. It is easy to place the blame on students' lack of motivation or commitment to study and succeed. The reasons for absence and not taking up opportunities for support are many: practical, such as limited time or domestic circumstances; emotional, such as fear of appearing ignorant or panic at being confronted with a mathematics problem; or motivational, such as when classes are unconnected to the chosen programme.

Theoretical contradictions exposed in this chapter should not be ignored. It was explained at the outset that MatRIC was framed within a developmental research methodology based on community of practice theory and inquiry, and the development of communities of inquiry. The ontological and epistemological inconsistencies lead to the adoption of cultural-historical activity theory as an analytical framework. Then half-way through the chapter the educational psychologist David Ausubel was quoted with references to cognitive and constructivist psychology. There is sufficient here to entertain mathematics education researchers in heated debate for a long time. What will mathematics teachers make of the debate or the paradigmatic contradictions between theories? It is my guess that mathematics teachers find it much easier to comprehend and use effectively a theory of learning that treats the individual student constructing their own understanding. This will make more sense to them than theories in which, for example, cognition is argued to be "stretched over, not divided among – mind, body, activity and culturally organized settings" (Lave, 1988, p. 1).

There are, in addition, some *tensions* experienced in mathematics teaching development in higher education as a result of differing opinions and teaching goals. It is possible to identify at least four different groups of scholars engaged in mathematics teaching development. First research mathematicians, second mathematics teachers who have been employed on the basis of qualifications in a subject with substantial mathematical content such as physics, and astronomy, third sometimes mathematics is taught by specialists within the programmes served by mathematics such as engineers or economists, fourth there are mathematics education researchers. Each of these groups is likely to have a different relationship with mathematics, and their members vary in their beliefs about mathematical truth and the understanding students need to acquire, and possibly their teaching paradigm. The differences can emerge as tensions as members of each group assert their position based on knowledge and experience as being paramount. I have experienced within MatRIC how these tensions can emerge and interfere with concerted, collaborative action aimed at teaching development.

Another tension I have experienced recently is relevant to countries such as Norway where second or third languages are essential for engaging with the international community. Considering the needs of students in later years of their studies, the programme teachers want mathematics taught in English. The mathematics teachers and mathematics education researchers argue that students straight from school, where they have learned in Norwegian, experience mathematics as difficult in their first language. To additionally require these students to learn in a different language just adds to the difficulty. The programme teachers' requirement is met with the argument that students should be able to develop their understanding of mathematical concepts in the language of mathematics most familiar to them and it will not be so problematic later to learn new names for established concepts.

The third tension experienced by all in higher education is felt in trying to balance the demands between research, education and an increasing administrative load. Promotion and professional status within universities depend largely upon a scholar's research and publication, also upon her or his ability to attract external funding, which is also related to reputation gained through research. Regular teaching commitments are taken very seriously, but teaching developmental activity, as in the case of the teaching course mentioned in a previous section of the chapter, often comes quite low in priority for attention.

CONCLUSION

Mathematics teaching development is situated within a practice that is constrained by structural and cultural contradictions and tensions. Attaining MatRIC's vision of *'students enjoying transformed and improved learning experiences of mathematics in higher education'* requires structural and cultural change. One of MatRIC's roles, and the role of mathematics teaching development in higher education, is to expose these contradictions and tensions so that they may be recognized by the practitioner and policy making communities and develop a movement for change. Mathematics teaching developers in higher education must be campaigners and champions for real change. However, projects such as MatRIC are constrained by human, financial and temporal resources, and their remit, like MatRIC's, is to work with teachers and teaching. It is likely that teaching and learning in higher education would change without any intervention; just as the context has been changing as described in an earlier section of this chapter. Mathematics teaching development projects such as MatRIC can facilitate teachers and students to adapt more readily to the changes and the contradictions and tensions described above through a variety of actions:

- Working with enthusiastic mathematics teachers to develop active, blended and inquiry-based learning approaches and research and development-based education within all forms of teaching such as lecture, small group tutorials and individual support.
- Applying modern and emergent technologies to address challenges of communicating and conceptualizing mathematical ideas and providing effective information to the learner about the quality of their learning.

- Developing approaches in which students become genuine partners in their education, sharing responsibility for their own learning and that of their peers.
- Demonstrating possibilities for relevant teaching development for mathematics teachers and student teaching assistants.
- Creating opportunities for mathematics teachers to share experiences, to learn from each other's successes (and negative/neutral results), and to learn from recognized international expertise in teaching mathematics.
- Developing research in university mathematics education.
- Identifying local, that is within the host institution, challenges in teaching and learning mathematics and using these as opportunities for testing innovation, transformation and improvement.

Mathematics teachers in school and in higher education work within different cultures. For the school teacher, professional development very often means teaching development, also it could be the development of educational management or leadership competencies. For the higher education academic, professional development and advancement has traditionally been linked to research, knowledge creation and publication. Teaching has been perceived as a 'sideline' activity. However, there are signs of cultural change in higher education, some of these as adaptation to structural changes arising from political and policy decisions. Mathematics teaching development through centres such as MatRIC are clearly born from policy and political decisions because they depend upon resources, but they open the opportunity to motivate change from within professional practice. The mathematics teaching development researcher and activist, who has come through school teaching developmental activity has to recognize and adapt to the higher education context and be ready to push against cultural resistances to research and innovation, which has the potential to transform and improve students' experiences of learning mathematics.

NOTES

1. Information about all the centres and the programme is available at https://www.nokut.no/en/services/the-centres-for-excellence-in-education-initiative-sfu/
2. http://www.sigma-network.ac.uk/about/the-sigma-network/
3. https://www.khdm.de/en/
4. From this point the use of the words programme or programmes will be restricted to 'programme(s) of study' meaning the areas and fields of study that depend upon mathematics as a service subject.
5. See http://www.mn.uio.no/ccse/english/
6. I am not sure if the expression 'mathematization of knowledge' is correctly attributed to Glushkov.
7. https://abacus.aalto.fi/
8. For example, Kahoot: https://kahoot.com/
9. Special Interest Group of the MAA on Research in Undergraduate Mathematics Education. http://sigmaa.maa.org/rume/Site/News.html.
10. International Network for Didactic Research in University Mathematics, a 'Topic Conference' of the European Society for Research in Mathematics Education. http://www.mathematik.uni-dortmund.de/~erme/index.php?slab=erme-topic-conferences

REFERENCES

Artemeva, N., & Fox, J. (2011). The writing's on the board: The global and the local in teaching undergraduate mathematics through chalk talk. *Written Communication, 28*(4) 345–379.

Ausubel, D. P. (1968). *Educational psychology: A cognitive view.* New York, NY: Holt, Rinehart and Winston.

Ausubel, D. P. (2000). *The acquisition and retention of knowledge: A cognitive view.* Dordrecht: Kluwer Academic Publishers.

Black, P., & Wiliam, D. (1998). Assessment and classroom learning. *Assessment in Education: Principles, Policy & Practice, 5,* 7–74.

Chen, X. (2013). *STEM attrition: College students' paths into and out of STEM fields* (NCES 2014-001). Washington, DC: National Center for Education Statistics, Institute of Education Sciences, U.S. Department of Education. Retrieved from https://nces.ed.gov/pubs2014/2014001rev.pdf

Cooper, P., & McIntyre, D. (1996). *Effective teaching and learning: Teachers' and students' perspectives.* Buckingham: Open University Press.

Cronhjort, M., Filipsson, L., & Weurlander, M. (2017). Improved engagement and learning in flipped-classroom calculus. *Teaching Mathematics and Its Applications. hrx007.* https://doi.org/10.1093/teamat/hrx007

Daland, E. (2007). School-teams in mathematics, what are they good for? In B. Jaworski et al. (Eds.), *Læringsfellesskap I matematikk – Learning communities in mathematics* (pp. 161–174). Bergen, Norway: Caspar forlag.

Davis, P. J., & Hersh, R. (1981). *The mathematical experience.* Brighton: The Harvester Press.

Engeström, Y. (2001). Expansive learning at work: Toward an activity theoretical reconceptualization. *Journal of Education and Work, 14*(1), 133–156.

FLM. (1985). Hermann von Helmholtz. *For the Learning of Mathematics, 5*(1), 28.

Freeman, S., Eddy, S. L., McDonough, M., Smith, M. K., Okoroafor, N., Jordt, H., & Wenderoth, M. P. (2014). Active learning increases student performance in science, engineering, and mathematics [On-line]. *Proceedings of the National Academy of Sciences of the United States of America, 111*(23), 8410–8415. Retrieved May 14, 2018, from http://www.pnas.org/content/pnas/111/23/8410.full.pdf

Glushkov, V. M. (1983). What the mathematization of knowledge can yield in the field of human decisions. *der 16. Weltkongress für Philosophie, 2,* 546–553.

Goodchild, S. (2001). *Students' goals: A case study of activity in a mathematics classroom.* Bergen: Caspar Forlag.

Goodchild, S. (2008). A quest for 'good' research. In B. Jaworski & T. Wood (Eds.), *The international handbook of mathematics teacher education: The mathematics teacher educator as a developing professional* (Vol. 4, pp. 201–220). Rotterdam, The Netherlands: Sense publishers.

Goodchild, S., Fuglestad, A. B., & Jaworski, B. (2013). Critical alignment in inquiry-based practice in developing mathematics teaching. *Educational Studies in Mathematics, 84*(3), 393–412.

Hawkes, T., & Savage, M. (2000). *Measuring the mathematics problem* [On-line]. London: Engineering Council. Retrieved from https://www.engc.org.uk/EngCDocuments/Internet/Website/Measuring%20the%20Mathematic%20Problems.pdf

Healey, M., Flint, A., & Harrington, K. (2014). *Engagement through partnership: Students as partners in learning and teaching in higher education.* York: The Higher Education Academy. Retrieved from https://www.heacademy.ac.uk/system/files/resources/engagement_through_partnership.pdf

Jaworski, B. (1998). Mathematics teacher research: Process, practice and the development of teaching. *Journal of Mathematics Teacher Education, 1,* 3–31.

Jaworski, B. (2006a). Theory and practice in mathematics teaching development: Critical inquiry as a mode of learning in teaching. *Journal of Mathematics Teacher Education, 9,* 187–211.

Jaworski, B. (2006b). Developmental research in mathematics teaching and learning: Developing learning communities based on inquiry and design. In P. Liljedahl (Ed.), *Proceedings of the 2006 Annual Meeting of the Canadian Mathematics Education Study Group* (pp. 3–16). Calgary, Canada: University of Calgary.

Jaworski, B., & Potari, D. (2009). Bridging the macro-micro divide: Using an activity theory model to capture socio-cultural complexity in mathematics teaching and its development. *Educational Studies in Mathematics, 72*(2), 219–236.

Kuhn, T. S. (1996). *The structure of scientific revolutions* (3rd ed.). Chicago, IL: University of Chicago Press.

Lave, J. (1988). *Cognition in practice*. Cambridge: Cambridge University Press.

London Mathematical Society. (2010). *Mathematics degrees, their teaching and assessment* [On-line]. Retrieved May 14, 2018, from https://www.lms.ac.uk/sites/lms.ac.uk/files/Mathematics/Policy_repors/2010%20teaching_position_statement.pdf

Lortie, D. (1975). *Schoolteacher: A sociological study*. London: University of Chicago Press.

Maciejewski, W. (2016). Flipping the calculus classroom: An evaluative study. *Teaching Mathematics and Its Applications, 35*, 187–201.

Nardi, E. (2008). *Amongst mathematicians: Teaching and learning mathematics at university level*. New York, NY: Springer.

Norwegian Agency for Quality Assurance in Education. (2016). *Awarding status as centre for excellence in education (SFU)*. Retrieved from https://www.nokut.no/siteassets/sfu/awarding_status_as_centre_for_excellence_in_education_2016.pdf

Opstad, L., Bonesrønning, H., & Fallan, L. (2017). Tar vi opp de rette studentene ved økonomisk-administrative studier? En analyse av matematikkbakgrunn og resultater ved NTNU Handelshøyskolen [Do we enrol the right students on economics-administration studies? An analysis of mathematics background and results at NTNU Business School]. *Samfunnsøkonomen, 1*, 21–29.

Pell, G., & Croft, T. (2008). Mathematics support_support for all? *Teaching Mathematics and Its Applications, 7*(4), 167–173.

Petropoulou, G., Jaworski, B., Potari, D., & Zachariades, T. (2015). How do research mathematicians teach Calculus? In K. Krainer & N. Vondrová (Eds.), *Proceedings of the Ninth Congress of the European Society for Research in Mathematics* (pp. 2221–2227). Prague, Czech Republic: Charles University in Prague, Faculty of Education and ERME.

Rasmussen, C., & Wawro, M. (2017). Post-calculus research in undergraduate mathematics education. In J. Cai (Ed.), *Compendium for research in mathematics education* (pp. 551–579). Reston, VA: National Council of Teachers of Mathematics.

Roth, W.-M., & Radford, L. (2011). *A cultural-historical perspective on mathematics teaching and learning*. Rotterdam, The Netherlands: Sense Publishers.

Sangwin, C. (2013). *Computer aided assessment of mathematics*. Oxford: Oxford.

Sfard, A. (2008). *Thinking as communicating: Human development, the growth of discourses, and mathematizing*. Cambridge: Cambridge University Press.

Thomas, S. (2012). *An activity theory analysis of linear algebra teaching within university mathematics* (Unpublished doctoral dissertation). Loughborough University, UK.

Viirman, O. (2014). *The function concept and university mathematics teaching* (Unpublished doctoral dissertation). Karlstad University, Sweden.

Wenger, E. (1998). *Communities of practice: Learning, meaning, and identity*. Cambridge: Cambridge University Press.

Simon Goodchild
Department of Mathematical Sciences
University of Agder

ROSEMARY CALLINGHAM, YERSHAT SAPAZHANOV AND ALIBEK ORYNBASSAR

13. BECOMING A MATHEMATICS TEACHER EDUCATOR

Perspectives from Kazakhstan and Australia

Mathematics education is well established or growing in many countries around the world, but less is known about it and those who are involved in it from countries, such as Kazakhstan, that were part of the United Soviet Socialist Republics. In this chapter, we contribute to this gap in the field by addressing mathematics education and development of mathematics teacher educators in Kazakhstan. Specifically, we focus on the career paths of some Kazakhstan mathematics teacher educators and contrast them with career pathways identified in Australia. We use Bronfenbrenner's ecological systems model to compare how influences at the macro-, meso- and micro-systems levels play out differently in these two countries and affect the experience of becoming a mathematics teacher educator.

INTRODUCTION

Mathematics education is well established in many western countries, such as Australia and United Kingdom, and is growing in South East Asian countries. Less is known, however, about mathematics education and those who are involved in it from countries that were, until recently, part of the United Soviet Socialist Republics. Kazakhstan is one of a group of countries known collectively as the Commonwealth of Independent States that achieved independence from Russia in the early 1990s. Since independence the country has undertaken reform of its education system, including the creation of new universities. Mathematics education usually occurs within Faculties of Mathematics or Science.

In this chapter, the career paths of some Kazakhstan mathematics teacher educators are described and contrasted with career pathways identified in Australia. We then use Bronfenbrenner's ecological systems model to compare how influences at the macro-, meso- and micro-systems levels play out differently in the two countries and affect the experience of becoming a mathematics teacher educator.

© KONINKLIJKE BRILL NV, LEIDEN, 2020 | DOI: 10.1163/9789004424210_014

BACKGROUND

The Republic of Kazakhstan is a land-locked country located in central Asia, sharing borders with Russia, China and the smaller countries of Kyrgyzstan, Uzbekistan and Turkmenistan. Although geographically it is very large, it has a population of about 18 million people, making the population density very low. Many of these people are located in the two main cities, Almaty, the traditional capital in the south of the country, and Astana (called Nur-Sultan from 2019), the new capital created in the north in 1997. Kazakhstan has rich natural resources, including oil and gas, as well as minerals.

Over the course of centuries, Kazakhstan has become a melting pot of cultures. Waves of invaders and later 'silk road' traders brought new customs and ideas. From 1920, Kazakhstan was part of the Soviet Union. After 1949, the Soviet Union established a nuclear test site that was used up to 1989. The Baikonur Cosmodrome, located in a remote area in southern Kazakhstan, is still used to launch manned Soyuz rockets to the international space station. In 1991, Kazakhstan became an independent republic recognised by the United Nations. Less than thirty years old as a nation, Kazakhstan is developing its own identity (Burkhanov, 2017). The country is now in transition to a modern market-based economy focussed on natural resources and agriculture (World Bank, 2017).

Australia also has a low population density overall but is a highly urbanised country. Although inhabited for over 40,000 years, its modern history dates back about 230 years. During that period there have been many waves of migration leading to a highly multicultural society (Australian Government, 2018). Its economy is generally strong, based on vast natural resources. It operates as a federated commonwealth, operating for the common good, with eight independent states and territories, each with its own government, and is recognised internationally as an entity (Constitution Education Fund of Australia (CEFA), 2018)

Education in Kazakhstan

Education is recognised in Kazakhstan as critical to the future development of the country. The education system has been reformed to develop citizens to meet the new economic goals and challenges. There are now four levels of education: pre-school, school – split into primary (Grades 1 to 4), secondary (Grades 5 to 9) and senior (Grades 10 to 11) – undergraduate, and post-graduate. There are moves to extend the school system to 12 years, in line with most developed countries.

In 2010, Kazakhstan became part of the Bologna process (Pons et al., 2015). Tertiary education has a three-tier system leading to the degrees of bachelor, masters and doctor as appropriate. Entry to the next level depends on successfully completing the previous level. Bachelor degrees are four years in duration, masters take two years and a doctorate usually requires three years.

Teacher education in Kazakhstan differs for prospective primary or secondary and senior school teachers. Primary school teachers are educated in pedagogical colleges or universities, with the length of the college course depending on the level of prior education (Mullis, Martin, Goh, & Cotter, 2016). The Government has recognised that teachers are an important element in improving educational standards and is developing an independent professional certification process as well as providing extensive professional learning opportunities. Entrance to a university course is by examination.

The technological focus of its economy has led to a recognition of the importance of mathematics and science. The nature of the mathematics content taught in schools is generally more difficult than expected for similar year groups in Australia. Students attending international schools who wish to enter a Kazakhstan university often require additional coaching in mathematics in order to meet the standards required by the university entrance examination (A. Almas, personal communication, November 4th, 2018).

The challenges of the shifts in education thinking impact on the nature of mathematics teacher educators and their work. In particular, the change to a modern economy calls for independence of thought. Soviet style pedagogy relied heavily on memorisation and instruction was teacher-centred and authoritarian (Burkhalter & Shegebayev, 2012). The Soviet system was designed to prepare people for a collective existence fulfilling those tasks required by the state (Long & Long, 1999) whereas it is clear from the current policies in Kazakhstan that a more open and critical society is the long-term aim (Burkhalter & Shegebayev, 2010).

Education in Australia

In Australia, education is a responsibility of the states and territories. Although there are some differences, in general the organisation of schooling is similar across the country, with children undertaking one year of funded kindergarten, seven years of primary education, and six years of secondary education (Wernert & Thomson, 2016). Teacher education, however, is subject to standards imposed by a national agency, the Australian Institute for Teaching and School Leadership (AITSL). AITSL prescribes the courses that are needed to qualify as a teacher. Two routes are possible: a 4-year Bachelor of Education or a 2-year Master of Teaching following a Bachelor degree. All teacher education programs are required to be certified through AITSL, meeting established standards (AITSL, 2016). No specific content is prescribed, but mathematics is recognised as a core subject.

Students enter university on the basis of Grade 12 results or, sometimes, through alternative pathways for mature-age students who do not have the necessary formal qualifications. Over the past 30 years or so, tertiary education has expanded in Australia (McKenzie & Schweitzer, 2001) with nearly 37% of 20-year-olds attending university or other tertiary institutions in 2011 (Parr, 2015). Students undertaking teacher education qualifications are required to pass a test of literacy and numeracy

(Literacy and Numeracy Test for Initial Teacher Education students (LANTITE)) at a standard deemed equivalent to that which 30% of the Australian population could achieve (Australian Council for Educational Research, 2019).

These contextual differences impact on mathematics teacher educator in diverse ways.

The next section provides narratives about how the participants in this study entered mathematics teacher education.

BECOMING A MATHEMATICS TEACHER EDUCATOR

Against the background of a society still accommodating a fundamental political shift, what are the drivers and pathways into becoming a mathematics educator in Kazakhstan? Mathematics educators in Kazakhstan still come largely from a background of Soviet era thought. To better understand the processes involved, the experiences of two mathematics educators in Kazakhstan are contrasted with the stories of several Australian mathematics educators. Information was collected through informal conversations with all parties, and further email correspondence to provide clarification.

The two mathematics teacher educators teaching in tertiary institutions in Kazakhstan were male and in their late twenties. Both were wholly educated in Kazakhstan but were of an age where their mathematics classrooms were still influenced by Soviet style teaching. To preserve anonymity, they have been given pseudonyms of Abay and Berik. Both were married with young children and were studying for a PhD while also teaching at a university. They were teaching in the pedagogical stream of a mathematics department, and their students were intending to become high school mathematics teachers. The content taught included high-level mathematics such as calculus.

The Australian information was provided by three female mathematics teacher educators (given the pseudonyms of Kate, Mary and Susie) and two male mathematics teacher educators (given pseudonyms of James and John). In contrast to the Kazakhstan mathematics teacher educators, the Australian mathematics educators were older – all but one of the respondents, John, were over 50 years old. Of the five who participated, only one had been fully educated in Australia (Susie), and one other person (Kate) had started her education overseas but completed it in Australia. The educational backgrounds and experiences were very varied including England, New Zealand, the West Indies and North Africa. All held a PhD qualification and had prior classroom teaching experience. Two (Mary and Susie) were experienced primary school teachers, and Kate, James and John had all taught in high schools. John had also taught prospective high school teachers in a teacher training institution in his home country. Only one Australian mathematics teacher educator (John) had young children. All of the female Australian mathematics teacher educators had independent children. All of the Australian mathematics teacher educators were teaching potential primary school teachers, but two (Kate and James) also had some involvement with middle school (Grades 7 to 10) teachers. None of the Australian mathematics teacher educators were teaching high-level mathematics.

School Experiences

Both Kazakhstan mathematics teacher educators described how they had struggled with mathematics in primary school. Abay described his primary school years: “Primary school wasn’t so easy for me, telling the truth I’ve struggled a lot.” Berik came from a village in south-eastern Kazakhstan, and also struggled with mathematics:

> Usually villages don’t give good education in my country, so I wasn’t good at mathematics. I always tried to solve any kind of questions by using application of this topic in daily life, but it was not always possible. Maybe that is a reason why I didn’t like mathematics.

Interestingly, both had entered a specialist school for mathematics and science following an examination at the end of primary education. For Berik this meant living away from home and he met parental resistance:

> After 6 years of village education I decided to change my school. Of course, my parents didn’t like that idea, anyway they let me study in specialized school. I stayed in a school dormitory, that gave me great opportunity to spent [sic] all my free time for studying. I was able to come home at weekends, unless my parents visited me.

Abay described his joy when he passed the examination to enter the specialist school:

> After 6 years of hell in school I had, I tried myself in examination for specialized school. Luckily somehow, I [was] admitted there and there was no limit for my happiness unless [once] lessons have started.

The decision to attempt the examination and enter a specialist school indicates strong aspirations for success. For both persons this was not a trivial decision and they were well behind their peers in the new school. It took them about two years of very hard work to catch up. Abay had the support of a good mathematics teacher and this motivated him to choose mathematics teaching as a career:

> … a math teacher, whom I liked from the first lesson, saw my potential in math and helped me to became smarter in all areas, especially in math. In two years being his student I learnt a lot and been [was] so motivated with his willing[ness] to help others, that I decided to become a math teacher too.

For Berik, life at the specialist school required considerable work. He worked with friends to improve his grades and stated that he came to “understand that my education is direct dependent on me. Nobody will help me until I ask.” In this instance, possibly because he was living in a dormitory, he turned to his friends rather than to a supportive teacher. Both individuals demonstrated very high levels of motivation and they studied extremely hard to achieve their goals. Berik described how he developed a strategy with an older friend:

> First it was necessary to fill all blanks in my education and second get better grades. First goal was reached by studying weekdays, I used free time to complete my knowledge and second goal reached by preparing any lesson topics in advance, it was possible because teacher give books with all topics that will be studied in given term. As I was more active in the lessons teacher could see progress and my grades started to increase.

What is particularly noteworthy about both these accounts are the challenges that these two people met and the work they put into overcoming disadvantage. In Australia accounts such as these would be rare, and the opportunity to attend specialist schools is very limited.

Most of the Australian mathematics teacher educators had positive early school experiences overall, and mostly in countries outside of Australia. All said that they found mathematics easy, but James reported one very negative experience in Grade 3 with a teacher who referred to him as "Mr Smartypants" for asking the question, "Are there more odd numbers than even numbers?" James moved into a middle school at Grade 6 and was placed in an accelerated class. This he enjoyed immensely – it involved problem solving and considerable opportunities for students to compete against themselves and others. He described the challenges posed by solving the "puzzles of trig identities" and the game of being the first to get one out. Kate also started her education overseas in a country where arithmetic, geometry and algebra were taught as separate subjects. When she came to Australia, she moved into a high school in a small town where she excelled at mathematics, completing a Grade 11 course in Grade 10. It was shock to her when she moved into a large city and found herself behind other students: a situation that required considerable hard work before she caught up. Mary hated her very early years of schooling, but found the primary years, from Grade 3 onwards, rewarding. Grade 6 was especially so: she was taught by a teacher who enjoyed music and dance but was also a high school mathematics teacher. She described "discovering π" and other similar adventures in mathematics and felt that the teacher's strong subject knowledge helped him take the problem-based approach. Susie also had positive classroom experiences up to Grade 11. At the end of that year she was moved from her tertiary entrance mathematics course to home economics because her teacher "felt I didn't have the capacity for Maths C [the most advanced course] based on my gender." This forced decision impacted on her future career choices.

University Study

Both Kazakhstan mathematics teacher educators were successful in the entrance examination into university. Abay chose to study mathematics education with a view to becoming a teacher. Berik elected to study "scientific mathematics" but during his senior year selected a course called "The 12-year System of Education." He wrote:

> We had different challenges and saw interesting teaching methods in that course. Our teacher told that everyone in this course should become a teacher regardless that half of group was from scientific math. She also said that most teachers weren't specialists in their fields, but students as we are, will become good teachers that can give good education and motivation. As I was from [a] village and education in some regions is far from good, her words hooked me and I realised that I must become a teacher.

Both these mathematics educators were focused on becoming teachers from that point on but took different career pathways before returning to university as mathematics teacher educators. One respondent (Abay) took a degree focussed on becoming a teacher, the other by chance took a single course in his final year and from that decided to focus on teaching. Berik, who took the scientific route, decided to undertake a program that aims to place teachers in more remote schools. He said:

> This program provides good salary in the village you go, but importance is that village must be different from your homeland. I chose the small city in the opposite side of my country. I was interested if I can prove that I am good teacher in different city where I am just a young man with big ambitions.

On entry to that program he undertook a one-month course for young teachers. During this course:

> Skilled teachers gave lessons and showed different tips and tricks that young teachers should know …. Experienced teachers showed how to teach the most difficult topics or topics that students usually have problems with.

Following this preparation, he taught for two years in a private school. During that time, he experienced similar seminars twice a year, and had to undertake tests every month. He was strongly motivated by these tests, striving always to be in the top ten. He stated:

> These years were the most productive for learning to teach. After I became master teacher and extra young teachers' preparation duties were added.

Abay began teaching in a local college during his final year at university. Colleges in Kazakhstan are specialised institutions enrolling students from Grade 9 onwards. Students who successfully complete a college course enter university at the second-year level, unless they choose to change specialisation when they begin university in Year 1. He then went on to teach mathematics in a high school for four years.

The Australian mathematics teacher educators had varied pathways. John was allocated a place to study teaching by his government. His friends were studying engineering and medicine and "teaching was not my interest, but I had no choice." He did choose, however, to make mathematics his major subject, with physics as a minor because, "I was strong in mathematics." In contrast, Mary liked the idea of teaching but had two young children so could not pursue this career immediately. She started tertiary study by distance in science and mathematics and then moved into a primary Bachelor of Education (BEd) course. At the time all prospective teachers were required to undertake a specialisation, so she chose mathematics, but also liked the idea of teaching "all the other stuff." Kate wanted to study medicine and went into a medical science degree as step in this direction. She worked in this field but found that technology was rendering the work "boring." At about the same time she moved to another Australian state and this move provided the opportunity to study secondary teaching through a mathematics and science Bachelor of Education. Susie undertook a variety of jobs, completed a Bachelor of Business before moving overseas and then back to Australia following her husband's work. Finding herself in a new town with two small children she decided to enrol in a degree in primary education while simultaneously undertaking a Graduate Diploma in Psychology at a different institution. She began teaching a Grade 6 class on completion of her degrees but was aware that her students did not like mathematics and started trying to find ways to make mathematics more enjoyable and engaging. At this point her husband's job again took her overseas and she took leave from her teaching position. Only James had wanted to be a teacher during his school years, and he rejected possibilities such as law and medicine to take a student scholarship in teaching through a traditional model of Bachelor of Science followed by a one-year Diploma of Education (DipEd). He described his DipEd year as "appalling." The focus was on procedural aspects of teaching, such as blackboard skills, taught at the local Teachers' College. All of these colleges have now been absorbed into universities.

Further Study and Entry into Tertiary Teaching

All of the mathematics teacher educators in this small-scale study were very highly qualified. Several of the Australian participants had more than one undergraduate degree, often in diverse fields. All participants had a masters level qualification, and some had more than one. In Kazakhstan, Berik and Abay both undertook a masters degree while they were teaching. This is a recognised pathway and encouraged by the Government. Several participants from both countries mentioned an influential course undertaken as part of professional development that prompted them to either undertake further study or to move into mathematics education. John chose to undertake a Masters degree in his home country, to "learn more about maths teaching." He later completed a second, research-focussed masters, gaining a scholarship to study in Europe. James undertook a masters by coursework after eight years in the classroom. During this degree he took several education-focussed courses which, in contrast to his earlier diploma, he described as eye-opening. Taking this degree was a "watershed" in his career, and he wanted other teachers to "know the stuff I didn't know" when he entered teaching. Susie had started a second masters level course during her period overseas and returned to teaching in Australia needing only a small research study to complete her degree.

Clearly this group, from both Kazakhstan and Australia, were high achievers, motivated to take on higher education and to work hard to achieve their goals. Of interest is the pathway into mathematics education specifically. In Kazakhstan, both Abay and Berik had successful experiences as classroom teachers in high schools or colleges. They taught mathematics in senior secondary schools and Berik also had several years of teaching mathematics in specialist university level institutions that had a strong focus on mathematics and science, before entering the university. Clearly their potential was noticed, because their tertiary positions resulted from invitations by senior figures in the various institutions. By Australian standards they were very young but in Kazakhstan their youth was not unusual.

In contrast, entry into tertiary teaching for the Australians appeared unplanned other than for John who intentionally applied for a position in the teacher education institution in his home country. Mary had had a very successful career in teaching. She was known as a good primary teacher by her colleagues. Her mathematics teaching expertise was also recognised by the education authority that paid for a 20-day specialist course in mathematics teaching. This "brilliant" course became a springboard into a Master's degree. Despite this background, during a regular school inspection, Mary was criticised for teaching mathematics in an integrated way, along with science. She felt that she was being told how to teach mathematics, and that the approach being taken was highly constrained. At this point she resigned from teaching. By chance she noticed an advertisement in a local paper for a mathematics teacher educator at a local teacher training college and decided to apply.

Kate also left classroom teaching, having decided that she "struggled in the [high school] classroom" with the growing behavioural issues not the mathematical content. She returned to university and studied a Graduate Certificate in Statistics, that included a research component. During this period, she tutored in university statistics and also taught laboratory science students at the local senior secondary school. She discovered that she enjoyed teaching adults and enrolled in a PhD in statistics education and ultimately moved into being a mathematics teacher educator.

Susie also had not planned to become a mathematics teacher educator. The school in which she was teaching became involved in a research study with the local university. Susie's class was nominated, somewhat against her will, as one of the classes for the study. The topic was teaching mathematics through an inquiry approach and over the course of the first year of the study, Susie noticed that her class were more engaged and enjoying mathematics more. This topic became the basis for her research study that she needed to complete her Master's degree. A scholarship from the local education authority allowed her to participate in an International Congress on Mathematical Education (ICME), where she presented her results. It was not Susie's intention, however, to leave teaching but when a serious flood destroyed her classroom and many years of resources she was prompted to rethink and decided to continue her studies towards a PhD. She had embarked on the PhD part-time while teaching but changed to full-time study, supplementing her scholarship with tutoring in mathematics education.

James' pathway was similar in some respects to those of Abay and Berik in Kazakhstan in that he studied mathematics at tertiary level on a teaching pathway, and then taught secondary school students. As he completed his masters, he was asked to teach a bridging mathematics course, for students who had not achieved the grades needed for tertiary entrance. He spent the next thirteen years teaching bridging mathematics or first year mathematics in university, before applying for a position in mathematics education in an Australian university. Following his European Masters course, John gained a scholarship to study for a PhD in mathematics education at a university in Australia. Towards the end of his PhD he secured a temporary appointment at another Australian university and this became a continuing position.

Being a Mathematics Teacher Educator

All the participants, from both countries, were highly qualified and had considerable school teaching experience. In Australia, however, regardless of background, all mathematics teacher educators were teaching prospective primary school teachers. All the Australians had studied some level of tertiary mathematics, although not necessarily pure mathematics. In contrast, the Kazakhstan mathematics teacher educators were teaching high level tertiary mathematics to students intending to become high school and senior college teachers. They were not involved with prospective primary school teachers as these are taught in specialist colleges in Kazakhstan.

The foci of the courses the Australian and Kazakhstan mathematics teacher educators taught were also different. In Kazakhstan, the emphasis was on learning mathematics in a pedagogical context, whereas in Australia the courses taught generally emphasised mathematics curriculum knowledge rather than the subject itself. These differences raise the issue of horizon mathematical knowledge (Ball & Bass, 2009; Ball, Thames, & Phelps, 2008); that mathematical knowledge that can inform teachers' decision making. Ball and Bass recognise that this is more than understanding or knowing specific aspects of mathematics but is "an awareness – more as an experienced and appreciative tourist than as a tour guide – of the large mathematical landscape." All of the Australian mathematics teacher educators mentioned the lack of mathematical understanding and the often negative dispositions their students had towards the subject. John, whose background was similar to that of the Kazakhstan participants in that in his home country there is a strong focus on mathematical content for both primary and secondary school teachers, said that he was "shocked" when he started teaching prospective teachers in Australia.

The two Kazakhstan mathematics teacher educators did not comment on content but on pedagogical approaches. Berik stated:

> As I am teacher of education department, it is important to give skill and knowledge that can be used while teaching. Mostly I try to show problems that

> young teachers meet. Also try to explain methods than can be used during the lessons from books, and explain methods that were more useful in education … through my approach [experience].

Abay also emphasised the teaching of content knowledge:

> Our faculty is Education and Humanities, and our math students are future maths teachers. Five-year experience is not a lot, but I think that it will be useful, as I will tell them about real experiences. But when I am preparing for lectures I try to also give them new information, and it is also useful for me to upgrade myself.

For the two Kazakhstan mathematics teacher educators their teaching background was critical – they drew on their pedagogical experience in a mathematics context. In contrast, the Australian mathematics teacher educators drew on their subject knowledge in a pedagogical context. This finding is interesting because it may reflect the different contexts in which the two groups were operating. It also suggests that these two groups drew on their professional knowledge in different ways. In some respects, the ways in which mathematics teacher educators in Kazakhstan and Australia talked about their current teaching reflects Chick, Pham and Baker's (2006) characterisation of classroom teachers' pedagogical content knowledge (PCK) (Shulman, 1987). Chick et al. identified teachers using mathematical knowledge in a pedagogical context when they, for example, demonstrated different ways of solving mathematics problems, showed deep understanding of mathematical ideas, or made connections within and between mathematical concepts. Pedagogical knowledge in a mathematics context was shown by teachers discussing classroom approaches and techniques or talking about strategies for engaging students. Aspects of teaching that Chick et al. indicated clearly showed pedagogical content knowledge included recognising the cognitive demands of a task, providing explanations of a mathematical process or identifying students' misconceptions. The two Kazakhstan mathematics teacher educators explicitly talked about engaging students, both those they were teaching and how their students might also interest school students in their own classrooms, through appropriate pedagogical approaches such as active engagement in learning. In contrast the Australian mathematics teacher educators frequently referred to the "struggle" to develop their students' mathematical knowledge and made comments such as "more content is needed in primary [teacher] education" (John). Mary also commented that in mathematics education "you had to know the mathematics." This finding will be further explored in the discussion.

Although these stories about becoming mathematics teacher educators are interesting, how can they be interpreted? What pointers are there for systems, institutions and individuals, and what influenced the decisions made to become mathematics educators? Bronfenbrenner's (1977) Ecological Systems Theory provides a framework for considering these narratives. The findings are discussed in relation to this theory in the next section.

ECOLOGICAL SYSTEMS THEORY

Bronfenbrenner (1977) contended that the development of an individual could not be disassociated from his or her environment, and that development should always be examined in that context. He posited several nested levels of influence. Closest to the individual is the microsystem that includes those experiences and circumstances that are unique to the person. The meso-system is at the next level, and includes less direct influences, such as school policies and programs. Beyond this level lies the exo-system, those external influences over which an individual has no control but which, nevertheless, impact on opportunities and decisions, such as country or state structures and policies. All of these levels operate with a macro-system comprising the overarching societal values. These systems are often shown as a series of nested circles as shown in Figure 13.1.

The systems are not discrete – they interact in complex and sometimes unpredictable ways. The do, however, provide a useful delineation of the influences on individuals, in this instance as they become mathematics teacher educators. Each of these systems will be examined separately, drawing on the diverse narratives of the participants in this study, but acknowledging that they are both nested and interactive. The context of Kazakhstan is qualitatively different from Australia and, because of this fact, similar micro-level experiences, for example, may play out quite differently in the two countries.

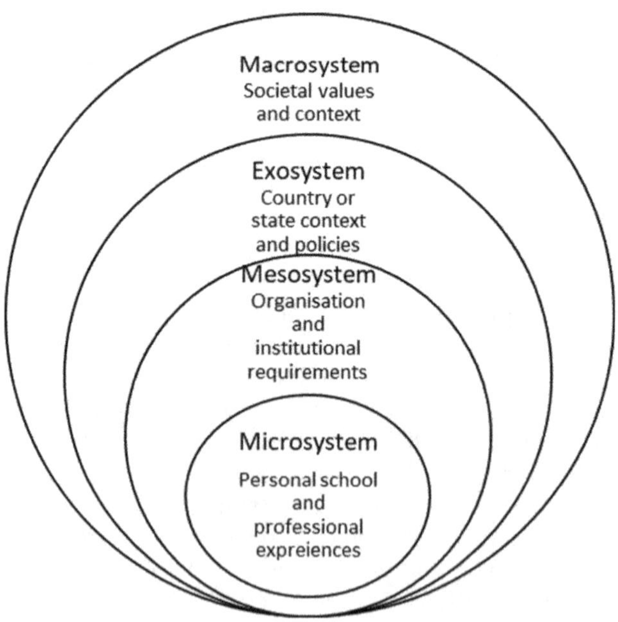

Figure 13.1. Ecological systems model (from Bronfenbrenner, 1977)

Microsystem Influences

These influences occur at the most personal level. For example, all participants talked about their early schooling although these experiences differed. Berik and Abay had negative elementary school experiences with poor teaching, possibly because of the influence of Soviet teaching styles. These styles have been characterised as didactic and rigid (Rourke, 1960), focussed on memorisation of facts (Wilson, Andrew, & Below, 2006) and controlled by fear (Burkhalter & Shegebayev, 2012). In contrast, the Australian mathematics teacher educators had largely positive elementary school experiences. They all stated that they were "good at mathematics" or "found maths easy." These positive experiences resonate with findings from the Trends in International Mathematics and Science Study (TIMSS) (e.g., Australian Bureau of Statistics, 2009) that Australian students had generally confident attitudes towards mathematics. The one negative early school experience, related by James, although unpleasant at the time, focussed on him being a "smartypants," reinforcing, albeit in a negative way, a self-perception of being good at mathematics.

Differing experiences were also evident at the secondary school level. Berik and Abay both went to specialist schools, and James also went to a school that extended its best students. The Kazakhstan mathematics teacher educators both described having to work very hard to catch up after their poor elementary school experience. In this period, they developed tenacity and a sense of being in control of their learning. James enjoyed mathematical problems and puzzles and recognised that he was good at mathematics. In contrast, Kate and Susie both had challenges in the later years of high school, leaving Susie with a feeling of being "no good at maths (thank you high school teacher)." Kate, like Abay and Berik, worked very hard to make up deficits in her mathematical knowledge but did not achieve the marks needed for her chosen career, and Susie was prevented from applying for her course of choice because she was unenrolled from a high-level mathematics course, despite good grades, and made to do home economics.

These early experiences impact on a person's self-efficacy (Bandura, 1994). Self-efficacy is an individual's belief about his or her capacity to undertake tasks and "to produce designated levels of performance that exercise influence over events that affect their lives" (p. 71). Berik, Abay and James developed a strong self-efficacy for mathematics and went on to study mathematics at tertiary level. In contrast, Kate and Susie both chose tertiary courses that required mathematics but at a lower level.

All the mathematics teacher educators had an image of themselves as good teachers. They commented that they had been successful. For example, Berik became a Master Teacher, and Mary was recognised as a quality mathematics teacher through the course she was able to complete as part of her professional learning. Kate enjoyed teaching adults in specialist fields. These micro-system influences helped them develop self-efficacy as teachers of mathematics. Susie's class was chosen to participate in a research study because the school Principal recognised her competence although Susie stated, "I was very aware that my students did not enjoy maths and I began looking for ways to make it more interesting." This search for improvement is typical of people who have high self-efficacy (Bandura, 1976). Rather than turning away from a challenge, they look for solutions to problems with Susie recognising in this instance that she could change her approach.

Meso-System Influences

The meso-system works at a level one step removed from the individual but, nevertheless, can have a direct impact on behaviour. For example, John had no choice about which university he went to, or which career to pursue. His choices were limited to the major he would study to become a teacher. He became increasingly interested in mathematics education and subsequently completed both a research masters and a PhD in the field. Although his choices were initially constrained, decisions that he made led to new opportunities.

Mary was also influenced by the meso-system through the negative evaluation of her teaching during a school inspection. Her self-efficacy for teaching was sufficiently robust at that stage in her career that she chose to resign from her position. In a second interaction at meso-system level, she noticed an advertisement for a mathematics teacher educator at a local teacher education college and ultimately began a new career. Mary provides an example of the interaction between micro-system and meso-system. The meso-system provided opportunities that the personal experiences of the micro-system, leading to her perception of herself as an effective teacher, enabled her to take advantage of.

Berik's choice to undertake the "to village with diploma" program was also at the meso-system level. This program, launched over 10 years ago to improve education in rural and regional areas of Kazakhstan, has been taken up by more than 1000 young professionals. As Berik pointed out the benefits are many:

> They get good bonuses: first, work experience, second, a good salary, and third, the housing issue. And, of course, the patriots of our country can thus repay a debt to the state. After all, now young specialists in villages are more in demand than ever. Today, there is an acute shortage of workers in education.

His motivation was to "test myself" in a new environment, and Berik's high school experiences in which he took charge of his own learning laid the foundation for this new challenge.

The careers of many of the individuals in this study were influenced at the meso-system level through the various opportunities that became available. James undertook a Masters course and was asked to teach bridging mathematics as a result. Susie moved into mathematics teacher education during her PhD studies when she worked as a tutor. John decided to apply to the local teacher training institution in his home country, and subsequently for study overseas. Both Berik and Abay were asked if they would like to teach mathematics to prospective teachers when they were studying at a local university.

For this group of people, the meso-system appeared to provide opportunities to change direction or make choices about their working lives. Decision making is affected by a range of factors, most of which occur within the micro-system level. Prior experiences (Jullisson, Karlsson, & Garling, 2005) play a large part. All of these mathematics teacher educators had some prior experience of teaching adults, albeit in different situations, or teaching in a tertiary setting before moving into mathematics teacher education. Personal characteristics also play a part, such as age and life stage (de Bruin, Parker, & Fischoff, 2007), and this was particularly evident in the narratives of the female mathematics teacher educators. They all came to mathematics teacher education after having children and at a point in their lives where there was little risk attached. This observation is in line with the work of Tversky and Kahneman (1981) who indicated that people changed their choices when a problem or situation was framed differently. Had the three women in this study been younger, or responsible for small children, their decisions might have been different.

For the women in this small study, family considerations also acted at the meso-system level. Decisions made by others impacted directly on the capacity of these people to undertake education, continue a career, or even to decide where they lived. There is no suggestion that any of these women resented or felt unduly constrained by this reality, and they used these challenges to their advantage. It would seem likely that similar considerations also affect male mathematics teacher educator, but this was less evident in the stories they told.

Exo-System Influences

The meso-system opportunities occurred within the conditions provided by the exo-system, that is the conditions set up by system or country policies. The Kazakhstan policy of "to village with diploma" provided an opportunity for Berik after he had decided to become a mathematics teacher. The prior schooling experiences of the study participants were also situated within particular systems' frameworks. Kate, for example, experienced a curriculum in which different fields of mathematics were taught as separate subjects, whereas Mary taught mathematics in an integrated way and left teaching when that approach was superseded.

The structure of education in Kazakhstan and Australia is very different. Both Berik and Abay went to specialist high schools following success in an examination. In Australia, few states have selective entry government high schools, and none of the Australian mathematics teacher educators had had this experience. The closest was James, in a country outside Australia, who was in a class that aimed to extend talented students.

Kazakhstan has a national curriculum in which the mathematics content is clearly prescribed. The approach to teaching is changing, with increased emphasis on child-centred approaches, and the texts and materials are approved annually by the Ministry of Education and Science (Mullis, Martin, Goh, & Cotter, 2016).

In contrast, in Australia, each state is responsible for implementing the national curriculum in ways that suit its context. There is no preferred approach to teaching mathematics and schools and teachers are free to choose and use whatever materials they deem appropriate (Wernert & Thomson, 2016). In summary, the Kazakhstan system is highly centralised, whereas the Australian approach is decentralised. These differences, although removed from a direct influence on individual mathematics educators, provide a background within which mathematics teachers and mathematics teacher educators act.

The nature of the policies and context of education are critical to the enactment of local policies and processes. For example, professional learning, whether leading to a formal qualification or not, appeared to impact on all the mathematics teacher educators in the study. The conditions under which the policies are enacted at the local level lead to meso-system opportunities for individuals. Mary, for example, received a very negative review following an implementation of a process of school inspections that took a particular stance in regard to the teaching of mathematics. This stance had changed following a policy shift by the education authority. Decisions made at the exo-system level trickle down to the local level and are felt as meso-system influences.

These influences may be quite subtle. In Australia, for example, the decision to introduce the LANTITE test for graduating teachers puts some pressure on mathematics teacher educators to ensure that their students acquire the relevant mathematical knowledge. In Kazakhstan there is pressure to change approaches to teaching, introducing active learning and moving away from memorisation. Institutions respond to these exo-system demands by implementing new structures and courses which also impact on mathematics teacher educators.

Macro-System Influences

The macro-system concerns the overarching societal values. These values change over time and may impact in various ways. Both Kazakhstan and Australia place high value on mathematics and education, but these values are enacted differently. The independence of the different states in Australia is highly valued and this leads to a system that, although nominally following the same curriculum, delivers this curriculum in a variety of ways to suit local conditions (Wernert & Thomson, 2016). In turn, mathematics education in the separate states develops differently, with mathematics being taught at diverse points within initial teacher education programs and emphasising local curriculum documents.

One of the biggest discussion points for education in Kazakhstan is the eradication of corruption (Rumantseva, 2004). Corruption was commonplace in the Soviet era, but the country has moved to a point where this is no longer accepted. Consequently, systems and processes are being developed to prevent such occurrences. In Kazakhstan universities, this issue was a central theme in discussions with academics.

Because these values are part of the overarching societal culture, the respondents did not comment explicitly about them, other than in very general terms. Community values are underpinned by taken-for-granted assumptions (Spencer-Oatey, 2012) but are likely to affect beliefs about education and mathematics education, which operate at the micro-system level.

DISCUSSION

Using Bronfenbrenner's (1977) model has provided some insights into the ways in which the context or environment shapes individuals. In turn this leads to some suggestions for institutions and policy makers about the future development of mathematics teacher educators.

The very high levels of mathematics achieved by this group in their own studies raises questions about horizon knowledge (Ball et al., 2008) for mathematics teacher educators. The notion of being a mathematical tourist (Ball & Bass, 2009) is one that could be usefully explored. A successful tourist samples the environment, tries the food, learns something of the history, explores the arts and crafts, and may learn a few words of the language, shares their experiences on social media but does not become an expert about the new place. Successful tourists have a disposition towards learning something about the places that they visit and sharing this knowledge with others on their return. Translated to mathematics, mathematical tourists want to engage with mathematical ideas, they read about new breakthroughs in mathematics, they explore the beauty of mathematics and communicate this interest to others. From this perspective the level or nature of the formal mathematics studied may be less important than the willingness to engage mathematically. In order to engage effectively, however, there needs to be some high-level mathematical knowledge as a platform, in the same way that experienced travellers become more engaged in the new context, travelling on local transport, for example, rather than in a tour bus.

Extending this metaphor to the intriguing finding that the two groups of mathematics teacher educators in this study took different perspectives on their work raises the question of whether mathematical tourism for mathematics teacher educators takes different forms. Are the Kazakhstan mathematics teacher educator being pedagogical tourists in a mathematical terrain and the Australian mathematics teacher educator acting as mathematical tourists in a pedagogical domain? Pedagogical content knowledge would seem to be a seamless blending of pedagogical and mathematical knowledge. Perhaps mathematics teacher educator are tourists in the landscape of pedagogical content knowledge with different dispositions towards the diverse aspects of teaching mathematics that they may experience during their travels. These dispositions at the personal, micro-level are likely to be influenced by the mathematics teacher educators' prior experiences at the micro- and meso-levels, reinforced or challenged by the exo- and macro-level environments. The Kazakhstan mathematics teacher educators appeared confident about their own and their students' content knowledge but were keen to explore pedagogical aspects of mathematics teaching. The Australian mathematics teacher educators, however, had a strong focus on developing their students' mathematical knowledge.

This study suggests there are potential dangers for the future of mathematics education in western countries. The lack of mathematical understanding among their students was of concern for all the Australian mathematics teacher educators, especially among prospective primary school teachers. In turn, this lack of mathematical understanding has the potential to impact on the next generation of mathematics teacher educators, who are generally recruited from schools. Understanding pedagogical approaches without having deep subject knowledge is likely to be less productive in that it can lead to some attractive activity in which there may be little learning taking place. For example, asking primary school students to design a house may engage children in an interesting activity that apparently draws on mathematical knowledge, but the mathematics may be missed by the students themselves (Moschkovich, 1998).

Such considerations are part of the broader discussion about the pedagogical content knowledge needed by mathematics teacher educators. It seems reasonable to expect mathematics teacher educators to have considerably higher levels of content knowledge than might be reasonable for teachers. Mary, particularly, explicitly commented that she believed this. Mathematics teacher educators need to have a broader horizon to take account not only of the mathematical knowledge required by the curriculum but also where that knowledge fits into the bigger mathematical picture. Mathematics teacher educators are the tour guides of mathematical tourism and need to be able not only to point out and explain the 'must-sees' of the mathematical pedagogical content knowledge terrain, but also to be able to respond to unexpected situations and change their itinerary if needed.

In contrast to Australia, in Kazakhstan there is a strong focus on teachers learning high level mathematics. This focus may restrict the breadth of experience implicit in the notion of mathematical tourism. When a curriculum is crowded with difficult mathematics there is less time for joyful exploration of mathematics of interest. There needs to be continuing debate about the level and nature of the mathematics content, and the horizon knowledge needed by mathematics teacher educators. At present, this discussion is taking place at the meso-system level among groups of mathematics teacher educators and has only impacted at the exo-system level in a limited way but there may need to be some policy input around appropriate experience and qualifications for mathematics teacher educators.

Strong content knowledge provides the platform on which to build innovative pedagogical approaches. Unless the mathematical focus of the activity, representation or explanation provided to students is understood deeply, important aspects of the mathematical understanding may be overlooked (Moschkovich, 1998; Shepard, 2000). Ball (1988) suggests that what prospective teachers experience during their teacher preparation courses has an impact on prospective teachers' mathematical knowledge and attitudes towards mathematics, and subsequently, on their classrooms. All the mathematics teacher educators had successful teaching experience prior to entering the field. This experience is recognised as important in Australia where university lecturers in education are usually required to hold qualifications that would allow them to teach in schools, and preferably also have experience in the classroom. As the Kazakhstan education system changes, it will be important to recruit mathematics educators from among the ranks of pioneering and innovative teachers.

The varied pathways into mathematics teacher education are also worthy of note. The male mathematics teacher educators generally set out to be teachers and moved into mathematics teacher education purposefully. The female mathematics teacher educators had less straightforward journeys. At the micro-system level their personal decisions were often constrained by family considerations, such as Susie moving overseas, Kate changing states, and Mary not able to take up teaching earlier because she had two small children.

The varied pathways are important because they lead to greater diversity among mathematics teacher educators. Diversity is important to avoid 'groupthink' (Fernandez, 2007). A growing body of research indicates that classrooms are becoming more diverse (e.g., Askew, 2015; Black & Atkin, 2005) and this fact alone suggests that mathematics teacher educators should also be diverse. Chick (2011) warns that we "God-like educators" risk privileging specific approaches to teaching mathematics and, more particularly, mathematics pedagogical content knowledge. There is a growing body of research about teacher knowledge demonstrated in varied cultural contexts (e.g., Chan et al., 2018), suggesting that different cultures emphasise different approaches.

One way of ensuring that there is ongoing debate is to encourage diversity—among mathematics teacher educators as well as among teachers. The mathematics teacher educators in this small study brought a wide variety of perspectives, developed at the micro-system level, to their work which impacts on their students in the meso-system.

Maintaining diversity of pathways may not be too challenging in Australia with its multicultural society and acceptance of different points of view part of the macro-system values. There the problem may be more related to establishing appropriate backgrounds and standards of mathematical knowledge for mathematics teacher educators. Countries such as Kazakhstan, however, have different macro-system influences. Soviet era thought was focussed on conformity (Webber, 2000) and embracing diversity and creating different routes into mathematics teacher education may be more difficult.

CONCLUSION

The outcomes of this small-scale study show that context makes a difference to the work and lives of mathematics teacher educators. At one level this is an obvious conclusion, but the exploration of context using Bronfenbrenner's (1977) ecological model has provided some insights into how differences arise, and the factors that may impact on an individual as he or she transitions into being a mathematics teacher educator.

If societies are serious about improving the quality of mathematics teaching, they must also take account of their mathematics teacher educators. There are rich possibilities for considering horizon knowledge, diversity, and approaches to teaching tertiary level prospective teachers. Research into these, and other aspects of mathematics educators' practices and decision making have the potential to explain how a culture of mathematics education can be developed and how beliefs, practices, and values with respect to mathematics education are transmitted to new generations of mathematics teacher educators and teachers. The development of a cross-cultural research agenda with mathematics teacher educators as the focus would be timely.

ACKNOWLEDGEMENT

The authors would like to thank all the mathematics teacher educators who provided their personal stories for this study.

REFERENCES

Askew, M. (2015). Diversity, inclusion and equity in mathematics classrooms: From individual problems to collective possibility. In A. Bishop, H. Tan, & T. Barkatsas (Eds.), *Diversity in mathematics education* (Mathematics Education Library). Cham: Springer.

Australian Bureau of Statistics. (2009). *Student achievement in maths and science.* (ABS Australian social trends 4102.0, June). Belconnen: Commonwealth of Australia.

Australian Council for Educational Research. (2019). *Literacy and numeracy test for initial teacher education students.* Melbourne: ACER.https://teacheredtest.acer.edu.au/about Retrieved from

Australian Government. (2018). *About Australia. Our people.* Canberra: Commonwealth of Australia. Retrieved from www.australia.gov.au/about-australia/our-country/our-people

Australian Institute for Teaching and School Leadership (AITSL). (2016). *Guidelines for the accreditation of initial teacher education programs in Australia.* Melbourne, Australia: AITSL.

Ball, D. L. (1988). Unlearning to teach mathematics. *For the Learning of Mathematics, 8*(1), 40–48.

Ball, D. L., & Bass, H. (2009). *With an eye on the mathematical horizon: Knowing mathematics for teaching to learners' mathematical futures.* Paper presented at the 43rd Jahrestagung der Gesellschaft für Didaktik der Mathematik, Oldenburg, Germany.

Ball, D. L., Thames, M. H., & Phelps, G. (2008). Content knowledge for teaching: What makes it special? *Journal of Teacher Education, 59*(5), 389–407.

Bandura, A. (1976). Self-reinforcement: Theoretical and methodological considerations. *Behaviorism, 4*(2), 135–155.

Bandura, A. (1994). Self-efficacy. In V. S. Ramachaudran (Ed.), *Encyclopedia of human behavior* (Vol 4, pp. 71–81). New York, NY: Academic Press.

Black, P., & Atkin, J. M. (Eds.). (2005). Introduction. *Changing the subject. Innovations in science, mathematics and technology education* (Ebook). London: Routledge.

Bronfenbrenner, U. (1977). Toward an experimental ecology of human development. *American Psychologist, 32*(7), 513–531.

Burkhalter, N., & Shegebayev, M. R. (2012). Critical thinking as culture: Teaching post-Soviet teachers in Kazakhstan. *International Review of Education, 58*, 55–72.

Burkhalter, N., & Shegebayev, M. R. (2010). The critical thinking movement in Kazakhstan: A progress report. *Research in Comparative and International Education, 5*(4), 434–445.

Burkhanov, A. (2017). Kazakhstan's national identity-building policy: Soviet legacy, state efforts, and societal reactions. *Cornell International Law Journal, 50*(1), 1–14.

Chan, M. C. E., Clarke, D. J., Clarke, D. M., Roche, A., Cao, Y., & Berik-Koop, A. (2018). Learning from lessons: Studying the structure and construction of mathematics teacher knowledge in Australia, China and Germany. *Mathematics Education Research Journal, 30*, 89–102.

Chick, H. L. (2011). God-like educators in a fallen world. In J. Wright (Ed.), *Proceedings of the Annual Conference of the Australian Association for Research in Education.* Retrieved from https://www.aare.edu.au/publications/aare-conference-papers/show/6143/god-like-educators-in-a-fallen-world

Chick, H. L., Pham, T. H., & Baker, M. (2006). Probing teachers' pedagogical content knowledge: lessons from the case of the subtraction algorithm. In P. Grootenboer, R. Zevenbergen, & M. Chinnappan (Eds.), *Identities cultures and learning spaces* (Proceedings of the 29th Annual Conference of the Mathematics Education Research Group of Australasia, pp. 139–146). Canberra, Australia: MERGA.

Constitution Education Fund of Australia (CEFA). (2018). *Why are we called the 'Commonwealth of Australia'?* Ultimo: Author. Retrieved from http://cefa.org.au/ccf/why-are-we-called-'commonwealth-australia'

de Bruin, W. B., Parker, A. M., & Fischhoff, B. (2007). Individual differences in adult decision-making competence. *Journal of Personality and Social Psychology, 92*(5), 938–956.

Fernandez, C. P. (2007). Creating thought diversity: The antidote to group think. *Journal of Public Health Management Practice, 13*(6), 670–671.

Jullisson, E. A., Karlsson, N., & Garling, T. (2005). Weighing the past and the future in decision making. *European Journal of Cognitive Psychology, 17*(4), 561–575.

Long, D., & Long, R. (1999). *Education of teachers in Russia.* Santa Barbara, CA: Greenwood Press.

McKenzie, K., & Schweitzer, R. D. (2001). Who succeeds at university? Factors predicting academic performance in first year Australian university students. *Higher Education Research & Development, 20*, 21–33.

Moschkovich, J. N. (1998). Rethinking authentic assessments of student mathematical activity. *Focus on Learning Problems in Mathematics, 20*, 4, 1–18.

Mullis, I. V. S., Martin, M. O., Goh, S., & Cotter, K. (Eds.). (2016). *TIMSS 2015 encyclopedia: Education policy and curriculum in mathematics and science.* Boston College, TIMSS & PIRLS International Study Center. Retrieved from http://timssandpirls.bc.edu/timss2015/encyclopedia/

Parr, N. (2015, May 25). *Who goes to university? The changing profile of our students* (The Conversation. Australia Edition). Retrieved from theconversation.com/who-goes-to-university-the-changing-profile-of-our-students-40373

Pons, A., Amoroso, J., Herczynski, J., Kheyfets, I., Lockheed, M., & Santiago, P. (2015). *OECD Reviews of school resources: Kazakhstan* (Pre-publication copy). Astana: OECD and the International Bank for Reconstruction and Development/The World Bank.

Rourke, R. (1960). Some observations on mathematics education in Russian secondary schools. *The Mathematics Teacher, 53*(4), 241–252.

Rumantseva, N. (2004). Kazakhstan: The issue of corruption. International *Higher Education, 37*, 24–25.

Shepard, L. A. (2000). *The role of classroom assessment in teaching and learning. CSE Technical Report 517.* Los Angeles, CA: National Center for Research on Evaluation, Standards, and Student Testing.

Shulman, L. S. (1987). Knowledge and teaching: Foundations of the new reform. *Harvard Educational Review, 57*(1), 1–21.

Spencer-Oatey, H. (2012) *What is culture? A compilation of quotations.* GlobalPAD Core Concepts. GlobalPAD Open House. Retrieved from http://www.warwick.ac.uk/globalpadintercultural

Tversky, A., & Kahneman, D. (1981). The framing of decisions and the psychology of choice. *Science, 211*, 453–458.

Webber, S. (2000). *School, reform and society in the new Russia.* New York, NY: Palgrave Macmillan.

Wernert, N., & Thomson, S. (2016). Australian chapter. In I. V. S. Mullis, M. O. Martin, S. Goh, & K. Cotter (Eds.), *TIMSS 2015 encyclopedia: Education policy and curriculum in mathematics and science.* Boston College, TIMSS & PIRLS International Study Center. Retrieved from http://timssandpirls.bc.edu/timss2015/encyclopedia/

Wilson, L., Andrew, C., & Below, J. (2006). A comparison of teacher/pupil interaction within mathematics lessons in St Petersberg, Russia and North-East of England. *British Educational Research Journal, 32*(3), 411–441.

World Bank. (2017). Kazakhstan's economy is rising – it is still all about oil. *Country Economic Update, Fall* [On-line]. World Bank. Retrieved from https://www.worldbank.org/en/country/kazakhstan/publication/economic-update-fall-2017

Rosemary Callingham
School of Education
University of Tasmania

Yershat Sapazhanov
Faculty of Education and Humanities
Suleyman Demirel University

Alibek Orynbassar
Faculty of Education and Humanities
Suleyman Demirel University

PART 4

RESEARCHING MATHEMATICS TEACHER EDUCATORS

MARGARET MARSHMAN

14. COMPETING PRESSURES ON MATHEMATICS TEACHER EDUCATORS

Mathematics teacher educators globally are under increasing regulatory and accountability pressures from governments and their agencies to prepare more and better mathematics teachers who will improve student outcomes. Mathematics teacher educators must also prepare teachers to respond to school mathematics curricula that not only include lists of the specific content knowledge to be taught but also that students can use mathematics in their daily lives, are able to solve mathematical problems, and that equity and social justice are considered. How mathematics is taught is influenced by teachers' beliefs about mathematics and mathematics teaching and learning. This chapter explores both the beliefs of mathematics teacher educators (mathematicians and mathematics educators) and the competing pressures that they experience and ends with a call for them both (mathematicians and mathematics educators) to engage in discussions with each other and with policy makers and governments so as to "find their voice" (Dinham, 2013, p. 91) and respond to these pressures.

INTRODUCTION

In Australia, as in many other countries including the United Kingdom (Hoyle, 2016) and Germany (Jackson, 2000), there has been a decline in the number of students studying mathematics (and particularly more demanding mathematics subjects) at secondary and tertiary levels (Kennedy, Lyons & Quinn, 2014; McGregor 2016). There is also a shortage of qualified mathematics teachers in schools, with one third of junior secondary school mathematics classes in Australia having a teacher without mathematics qualifications (Prince & O'Connor, 2018). The decline in student participation has been attributed in part to students finding mathematics 'boring' and 'irrelevant' (Kennedy, Lyons, & Quinn, 2014; Office of the Chief Scientist, 2012). Although the suituation is not the same in all countries (for example, in the United States numbers of secondary school students studying mathematics are increasing (National Centre for Education Statistics, n.d.)), these trends suggest a need to explore the preparation of secondary school mathematics teachers and consider some of the tensions that mathematics teacher educators experience in trying to address these issues.

© KONINKLIJKE BRILL NV, LEIDEN, 2020 | DOI: 10.1163/9789004424210_015

Students views of mathematics are shaped by those of their teachers (Beswick, 2005; Ernest, 1989; Philip, 2007) so the preparation of secondary school mathematics teachers is an important part of the solution to the problems outlined and one for which mathematics teacher educators are responsible. Prospective secondary school mathematics teachers need to explore different views of mathematics as part of learning how to engage students with mathematics and to teach mathematics effectively. They need "time and opportunity to think about, discuss, and explain mathematical ideas" and the chance to learn "to treat mathematics as a sense-making enterprise" (Conference Board of the Mathematical Sciences, 2012, p. 24). In Australia, the Office of the Chief Scientist (2014) recommended mathematics be taught as it is practised in order to engage students with mathematics, and so it is incumbent on mathematics teacher educators to know how mathematics is practised and to promote teaching in this way among prospective teachers.

Initial teacher education programs for secondary school mathematics teachers may be either undergraduate, where prospective teachers study mathematics and education courses in tandem or in parallel as a combined degree, or postgraduate for those who have previously completed mathematics qualifications. The structure of initial teacher education courses is an important factor in determining from whom prospective teachers learn about mathematics and how to teach it. The mathematics teacher educators involved may be mathematics academics who teach mathematics content courses and engage in research and scholarship in the field of mathematics (i.e., mathematicians) or mathematics educators who teach mathematics methods courses and are engaged in exploring the connection between mathematics content and pedagogy to support teachers entering secondary teaching (i.e., mathematics educators). Mathematicians may be teaching prospective teachers in initial teacher education courses or, in the case of postgraduate courses, may have taught them during their previous study of mathematics. In this latter context they may or may not know who the future teachers in their classes are. Mathematics educators teach in initial teacher education programs and typically focus on how mathematics is taught and learned rather than on mathematics content. These two groups of mathematics teacher educators – mathematicians and mathematics educators – may have differing epistemologies and perspectives on pedagogy and may communicate different visions of mathematics arising from their different beliefs.

Teacher educators, including mathematics teacher educators, are under increasing scrutiny from governments and the media, and subject to increasing government regulation. They are under increasing pressure to improve the quality of graduate teachers so that student outcomes are improved (Dinham, 2013; Lerman, 2014). There are also institutional and time pressures as teaching is usually only one part of their job descriptions. Most research on mathematics teacher educators has centred on individuals studying their own practice – learning through their research (see for example Brown, Helliwell, & Coles, 2018; Chick & Beswick, 2018). In relation to beliefs, there have been many studies of the beliefs of school mathematics teachers, but little research on mathematics teacher educators' beliefs, including about the nature of mathematics and mathematics teaching and learning (Rino, 2015).

This chapter begins by considering the nature and purpose of mathematics in school curricula as the context in which to consider the beliefs about mathematics and its teaching and learning held by two professional communities of mathematics teacher educators (mathematicians and mathematics educators). Their beliefs are then compared and discussed. Some other competing pressures on mathematics teacher educators are described and the chapter ends with a call for both mathematicians and mathematics educators to engage in discussions with each other as well as with policy makers and governments and "find their voice" (Dinham, 2013, p. 91) to respond to these pressures.

Throughout this chapter I draw on recent data from a study I am conducting with Merrilyn Goos where we sent Beswick's (2005) *Beliefs about mathematics, its teaching and its learning* survey to Australian mathematics teacher educators. The survey sought responses to items on 5-point Likert scales from Strongly disagree to Strongly agree. Of the 82 (of 120 invited) respondents who completed all items in the survey 60 (73%) taught mathematics (i.e., were mathematicians) and 22 (27%) taught mathematics education (i.e., were mathematic educators). The results were analysed using descriptive statistics, *t*-tests and ANOVAs with Bonferroni post-hoc tests, eliminating items that violated the Levene test for homogeneity of variance. Twenty-five of 39 prospective teachers from three universities also completed the same survey.

Seven of the mathematics teacher educators were interviewed. Five were mathematicians and two were mathematics educators. One of the mathematics educators also taught mathematics. The following questions were used as a basis for semi-structured interviews which were analysed thematically:

1. Will you please describe how you teach mathematics or statistics in a lecture and a tutorial?
2. How would you describe any perceived differences (if any) between the way mathematics is practised and the way mathematics is taught?
3. How would you describe any differences between how mathematics is taught in schools and university?

The indicative questions for the seven prospective teachers (6 were completing an undergraduate qualification, and 1 had returned to complete a postgraduate qualification after approximately 10 years in the workforce) were:

1. How would you describe the difference, if any, in the way you are taught mathematics and statistics and the way you are taught to teach mathematics?
2. Do you feel any tension between the ways you are taught mathematics and statistics and the way you think you learn it best?
3. How would you best describe the different ways your lecturers and tutors view mathematics? Do you ever find it confusing? Please explain.

I include data from the prospective secondary school mathematics teachers as a basis for considering the impact on them of the different messages about the nature of mathematics and about mathematics teaching and learning that they may be receiving from the different groups of mathematics teacher educators they encounter.

MATHEMATICS IN SCHOOLS

Curriculum summaries submitted to the *TIMSS 2015 Encyclopedia: Education Policy and Curriculum in Mathematics and Science* (Mullis, Martin, Goh, & Cotter, 2016) show that school mathematics curricula in many countries are designed with the intention that students will be encouraged to problem solve and reason. To this end, teachers should be encouraging students to: make sense of mathematics; look for patterns; reason; construct arguments; and solve problems (National Governors Association Center for Best Practices & Council of Chief State School Officers (CCSSO), 2010).

In their influential report on how students learn mathematics, Kilpatrick, Swafford, and Findell (2001) identified five interrelated strands that describe a mathematically proficient person. These were conceptual understanding; procedural fluency; strategic competence (ability to formulate, represent, and solve mathematical problems); adaptive reasoning; and productive disposition. In the United States these proficiencies form the basis for the Standards for Mathematical Practice found in the *Common Core State Standards* (Council of Chief State School Officers (CCSSO), 2010) and describe the knowledge, skills and capabilities that mathematics teachers are seeking to develop in their students. The mathematical proficiencies have also been a guiding force in the school mathematics reform agenda and are quoted in the mathematics curricula of several countries beyond the United States, including Finland (Andrews, 2013) and Ireland (Mullis et al., 2016), and are evident in many more (Mullis et al., 2016). Elsewhere they have informed the development of mathematics curricula. For example, the objectives of the new Chinese mathematics curriculum are to develop students': knowledge and skills; mathematical thinking; problem-solving; and emotions and attitudes (Guo, Silver, & Yang, 2018), while, in Australia, Kilpatrick et al.'s proficiencies were framed as four proficiency strands: understanding; fluency; problem-solving; and reasoning (Australian Curriculum, Assessment and Reporting Authority (ACARA), 2018). Other countries have foregrounded problem-solving. For example, the Singaporean mathematics curriculum is built around five inter-connected components of problem-solving: concepts; skills; processes; metacognition; and attitudes (Ministry of Education, Singapore, 2012) and countries that participate in the Trends in International Mathematics and Science Study (TIMSS) specifically mention problem-solving in their mathematics curriculum (Mullis et al., 2016). The Programme for International Student Assessment (PISA), measures mathematical literacy, defined as the ability to "formulate, employ and interpret mathematics in a variety of contexts" (Organisation for Economic Cooperation and Development (OECD), 2017) and therefore also emphasises skills in problem-solving.

Also important in school mathematics is developing students' understanding of the role of mathematics in their lives, not just its role as a discipline and within related fields such as science and engineering. In many countries this is interpreted as numeracy or mathematical literacy. The Organisation for Economic Cooperation and Development (OECD) (2017) defined mathematical literacy as:

> An individual's capacity to formulate, employ and interpret mathematics in a variety of contexts. It includes reasoning mathematically and using mathematical concepts, procedures, facts and tools to describe, explain and predict phenomena. It assists individuals to recognise the role that mathematics plays in the world and to make well founded judgements and decisions needed by constructive, engaged and reflective citizens. (p. 67)

School mathematics also has the potential to address social justice issues. In the case of South Africa, the focus of mathematics curriculum reform is on redressing inequality and repairing the damage done by apartheid (Adler & Davis, 2006). Sriramen, Roscoe and English (2010, p. 627) refer to the widely held view that it is imperative that social justice issues are addressed through school mathematics curriculum:

> Numerous scholars like Ubiratan D'Ambrosio, Ole Skovsmose, Bill Atweh, Alan Schoenfeld, Rico Gutstein, Brian Greer, Swapna Mukhopadhyay among others have argued that mathematics education has everything to do with today's socio-cultural political and economic scenario. In particular mathematics education has much more to do with politics, in its broad sense, than with mathematics, in its inner sense.

School mathematics curricula include lists of the specific content knowledge to be taught but there are also other demands including that teachers ensure that students can use mathematics in their daily lives, are able to solve mathematical problems, and that equity and social justice are considered. There is pressure on mathematics teacher educators to prepare school mathematics teachers to respond to these requirements.

BELIEFS

How mathematics is taught is influenced by teachers' beliefs about mathematics, and mathematics teaching and learning (Philip, 2007; Ernest, 1989). The way in which mathematics is taught influences students' beliefs about mathematics and about their capability to learn mathematics (Mosvold & Fauskanger, 2014; Grootenboer, 2008; McLeod, 1992; Pajares, 1992).

There have been many attempts to define beliefs (Philip, 2007; Pajares, 1992). Rokeach's (1968) original definition of beliefs is a personal acknowledgement of the truth which determines personal behaviour: "All beliefs are predispositions to action" (p. 113). This definition does not mean that an individual is necessarily aware of their beliefs or that the individual can articulate those beliefs (Leatham, 2006). Philip (2007, p. 259) considered beliefs to be:

> Psychologically held understandings, premises, or propositions about the world that are thought to be true. Beliefs are more cognitive, are felt less intensely, and are harder to change than attitudes. Beliefs might be thought of as lenses that affect one's view of some aspect of the world or as dispositions toward action. Beliefs, unlike knowledge, may be held with varying degrees of conviction and are not consensual.

Beswick (2005, p. 39) defined beliefs as "anything that an individual regards as true." Drawing on these definitions I take beliefs to be subjectively held truths that predispose an individual to action.

Individual teachers can hold seemingly contradictory beliefs about school mathematics, mathematics as a discipline, and how mathematics is learned in their classroom environment (Beswick, 2005, 2012; Jorgensen, Grootenboer, Niesche, & Lerman, 2010; Philipp, 2007). Apparent contradictions can be understood in terms of the clustered structure of belief systems (e.g., Green, 1971) and Liljedahl (2008) reported a range of reasons suggested by teachers, for which their espoused beliefs and practices may not align. Beliefs need to be inferred as they cannot be directly observed (Grootenboer & Marshman, 2016) and powerful personal and social influences can mean that individuals state beliefs that may be different from their actual beliefs. The beliefs an individual expresses may also change depending on the particular situation or context (Smith, Kim, & Mcintyre, 2016). It can take a variety of resources to surmise someone's beliefs (Leatham, 2006, p. 92). Similar classroom practices may arise from different though not contradictory beliefs (Beswick, 2007) as the same beliefs can lead to different practices in different contexts and in interaction with other beliefs (Beswick, 2012). Leatham (2006) described beliefs as constituting sensible systems, in which an individual's beliefs are internally consistent to them, make sense to them, and fit with their other beliefs. If an individual's beliefs therefore appear contradictory, we have not understood them (Leatham, 2006).

Understanding mathematics teacher educators' beliefs provides a window into how they may teach and the possible consequences of that teaching for the beliefs about mathematics of their students, including prospective teachers.

EARLY CATEGORISATIONS OF BELIEFS ABOUT MATHEMATICS

There have been several categorisations of beliefs about mathematics that can be useful as a starting point for investigating mathematics teacher educators' beliefs and have been used by a number of researchers. Here I focus on several that have been used by the literature describing mathematicians' beliefs as well as beliefs of mathematics teacher educators and teachers of mathematics. Among the most influential is that of Ernest (1989). He defined three views of mathematics: instrumentalist, Platonist, and problem-solving, and logically consistent characteristics of teaching by teachers holding each of these views. Individuals with an instrumentalist view regard mathematics as a collection of procedures, facts and skills. The teacher acting on instrumental beliefs would assist students to master the skills and procedures by following a textbook or other prescribed sequence (Ernest, 1989). According to the Platonist view mathematics is a structured, unchanging body of knowledge which is discovered not created (Ernest, 1989). To Hersh (1997) the Platonist view is one in which "mathematical entities exist outside space and time, outside thought and matter, in an abstract realm independent of any consciousness, individual or social" (p. 9). The Platonist teacher is an explainer who helps students build conceptual understanding. In the problem-solving view mathematics is seen as a creative, social, and cultural activity building a dynamic, increasing field of knowledge and the teacher is a facilitator assisting students to become confident problem posers and problem solvers (Ernest, 1989).

Elsewhere Ernest (1991) described views of mathematics in terms of a dichotomy. On the one hand is the absolutist view in which "mathematical truth is absolutely certain, that mathematics is the one and perhaps the only realm of certain, unquestionable and objective knowledge" (Ernest, 1991, p. 3) whilst on the other there is the fallibilist view that "mathematical truth is corrigible, and can never be regarded as being above revision and correction" (Ernest, 1991, p. 3). Others have considered that people's beliefs fit somewhere along a continuum between these two (Lerman, 1990). An alternate categorisation is formalist, traditionalist and constructivist (Mura, 1993). The formalist view is where mathematical symbols can be regarded as physical objects that are useful but the mathematical statement itself has no meaning. The traditionalist view of mathematics aligns with Ernest's instrumentalist view (Mura, 1993). According to the constructivist view, for a mathematical object to exist, a proof "demonstrates the existence of a mathematical object by outlining a method of finding ("constructing") such an object" (McKubre-Jordens, n.d.). Classifying beliefs is more complicated than these categorisations might suggest: Peoples' beliefs generally do not align with any one category. Nevertheless, Ernst's problem-solving view of mathematics aligns with the problem-solving emphasis that has been incorporated in many curriculum documents.

Discussions of teachers' beliefs about mathematics generally present some views as more desirable than others (Mura, 1993) and this is often reflected in school curricula. As discussed earlier, since the development of the mathematical proficiencies (Kilpatrick, Swafford, & Findell, 2001), and TIMSS and PISA testing, problem-solving appears in many countries' curriculum documents (Mullis et al., 2016). This suggests that a problem-solving view of school mathematics, has been widely considered desirable.

MATHEMATICIANS' BELIEFS ABOUT MATHEMATICS

Most studies of mathematicians' beliefs about mathematics do not specify whether they are mathematics teacher educators (e.g., Brandt, Lunt, & Meilstrup, 2016; Burton, 1998; Mura, 1993). Mathematicians may not be aware of which, if any, students in their classes are or will be prospective teachers. In addition, students taught by mathematicians may not decide to become teachers until sometime after they have completed their mathematics courses. Sometimes people (e.g., engineers, scientists) who have been working in industry are looking to change careers to mathematics teaching and complete postgraduate qualifications.

In an open-response survey Mura (1993) asked 116 Canadian mathematicians about their images of mathematics. The most common themes that emerged were: the design and analysis of models abstracted from reality; logic, rigour, accuracy and reasoning; the study of axiomatic systems; art, creativity; imagination, beauty and harmony; and a science or tool for other sciences (Mura, 1993). Mura classified many as formalists, compared with traditionalists and constructivists. These mathematicians were hesitant to respond to open philosophical and historical questions. Similarly, Grigutsch and Törner (1998) found that "mathematicians view mathematics as a discovery and understanding process" (p. 29) and agreed with Hersch (1997) that mathematicians were not interested in the philosophy of mathematics and that they had probably not previously considered it. For Brandt et al. (2016) the most important processes for professional mathematicians in the United States and Canada were identified as participating in original research, conjecturing/generalising/exploring, communicating, proving, and making connections.

Based on interviews of 60 United Kingdom and Irish mathematicians, Burton (1998) obtained an holistic picture of the views of mathematics held by mathematicians. These mathematicians described mathematics as: making sense of the world; seeing how mathematics connected with the 'real' world; and seeing the connectedness of the different parts of mathematics. They described their work with metaphors such as putting a piece into a jigsaw puzzle or climbing mountains and "you know if it works" (p. 134) or if you can make connections. In interviews, almost all the mathematicians described the collaborative or cooperative nature of their work in communities of practices and that mathematics epistemologically, "is personally- and culturally/socially-related" (Burton, 1998, p. 139). These mathematicians were expressing views that aligned with Ernst's (1989) problem-solving view of mathematics as a creative, social and cultural activity. This has resonances with school mathematics as presented in curriculum documents where students are encouraged to collaborate and make meaning of the mathematics to see connections between the mathematical concepts and use mathematical modelling (e.g., ACARA, 2018; Ministry of Education, Singapore, 2012).

Carlson and Bloom (2005) studied ways in which mathematicians solve problems, including the associated emotional perspectives, one of which was excitement, and developed a multidimensional framework to be used for "investigating, analyzing, and explaining mathematical behaviour" (p. 69). These mathematicians demonstrated a belief in the importance of making sense of the problem and the mathematics and "the effective management of frustration and anxiety … was an important factor in these mathematicians' persistent pursuit of solutions to complex problems" (p. 70). Misfeldt and Johansen (2015) found that Danish mathematicians' choice of mathematical problems was strategic. The problems needed to contribute to the mathematicians "identity as a mathematician" (p. 368), be interesting and potentially fruitful, fit within the mathematicians' skills and competencies, and have an audience for the work. School teachers also choose problems that are interesting and potentially fruitful and fit within their students' skills and competencies.

Leikin, Zazkis, and Meller (2018) found that mathematicians whose students included prospective teachers focussed their teaching on developing professional mathematicians rather than teachers. The four mathematicians interviewed said that teachers would use some of the mathematical content, problem-solving strategies, and techniques of proof in their teaching and would find understanding the meaning of theorems and definitions, mathematical language, distinctions between problem-solving strategies and algorithms, the beauty of mathematics, mathematical history, and abstraction useful in their work as school teachers. The focus of these mathematicians, who were also mathematics teacher educators, on teaching prospective mathematicians rather than prospective mathematics teachers, suggests they had not deeply engaged with their role as mathematics teacher educators or considered the potentially differing needs of prospective mathematics teachers.

MATHEMATICS EDUCATORS' BELIEFS ABOUT MATHEMATICS

Studies of mathematics educators' views of mathematics generally do not distinguish between whether secondary or elementary school education is the focus of their teaching and research. Mathematics educators have been more willing than mathematicians to share their views about mathematics, suggesting they are predisposed to consider "philosophical and historical questions" (Mura, 1995, p. 394) about mathematics. Mathematics educators may consider the philosophy and/or history of mathematics as part of their research or teaching where, with prospective teachers, they consider what mathematics "is." This could be because of "the demands of their profession" (Mura, 1995, p. 394) or the influence of Schoenfeld's (1992) influential definition:

> Mathematics is an inherently social activity in which a community of trained practitioners (mathematical scientists) engage in the science of patterns – systematic attempts based on observation; study, and experimentation to determine the nature of principles of regularities in systems defined axiomatically or theoretically ("pure mathematics") or models of systems abstracted from real world objects ("applied mathematics"). (p. 34)

The most common themes for the definition of mathematics offered by the 51 Canadian mathematics educators Mura (1995) surveyed were that it involves patterns, logic, and models constructing of reality (Mura, 1995). Australian mathematics educators tended to agree, that mathematics was a "beautiful and creative human endeavour" (Beswick & Callingham, 2014, p. 141) which is consistent with Ernest's (1989) problem-solving view of mathematics.

Several researchers have developed theoretical frameworks for mathematics teacher educator knowledge, and these can be useful tools for mathematics teacher educators to use to reflect on their teaching. Two frameworks that build on those that mathematics teacher educators use when working with prospective students are those of Zaslavsky and Leikin (2004) and Chick and Beswick (2018). Zaslavsky and Leikin developed the *Teaching triad for mathematics teacher educators*, to facilitate mathematics teacher educators' reflection and to enhance their "growth-through-practice" (p. 29). The framework developed by Chick and Beswick (2018) focussed on the pedagogical content knowledge that mathematics teacher educators use and includes mathematics teacher educators' beliefs about the nature of mathematics.

The views of the mathematics educators in Mura's (1995) and Callingham et al.'s (2012) studies and Schoenfeld (1992) definition align with school mathematics curriculum where students are encouraged to problem solve and reason (e.g., National Governors Association Center for Best Practices & CCSSO, 2010).

COMPARISON OF MATHEMATICIANS' AND MATHEMATICS EDUCATORS' BELIEFS ABOUT MATHEMATICS

Although research on mathematics teacher educators' beliefs about the nature of mathematics has revealed variety in those beliefs there appears to be much in common between the groups as well. For example, the key themes emerging from mathematicians' definitions of mathematics from Mura's (1993, 1995) study were that it involves formulating models of reality, logic, and formal systems (Mura, 1995) and the most commonly mentioned characteristics of mathematics mentioned by mathematics educators were patterns, logic, and constructing models of reality (Mura, 1995, p. 392). Mura identified both groups as having formalist views, that is, beliefs that mathematical symbols can be regarded as physical objects that are useful but the mathematical statement itself has no meaning. Thompson (1992) labelled the gap between this formalist view of mathematics and reform views of the discipline as, "the single greatest obstacle to achieving mathematics instruction as envisioned in many reform documents" (Mura, 1995, p. 397). Nevertheless mathematics educators considered that mathematics is "both an art and a science, both a language, [that is] a form, and a set of specific contents" (p. 394) compared with mathematicians some of whom described mathematics as an "art, a creative activity" (Mura, 1993, p. 390), while others as "a science … a tool of other sciences" (Mura, 1993, p. 390).

More than 20 years later Brandt et al. (2016) found agreement between United States and Canadian mathematicians and mathematics educators in describing doing mathematics as studying patterns and working on problems. Moving away from formalist thinking these mathematicians and mathematics educators emphasised the importance of individuals' understanding mathematical ideas. "[S]imply mimicking procedures or reciting phrases with no understanding was not doing mathematics. Instead, doing mathematics required some understanding of the underlying mathematical principles" (Brandt et al., 2016, p. 765). In interviews, Australian mathematicians and mathematics educators have said that mathematics requires time for thinking, exploring ideas, answering questions, and solving problems (Marshman, in press).

In the current study there was no statistical difference between the two groups of Australian mathematics teacher educators, with 79 (96%) agreeing or strongly agreeing with the item "mathematics is a beautiful, creative and useful human endeavour that is both a way of knowing and a way of thinking." Sixty-eight (83%) disagreed or strongly disagreed with the item "mathematics is computation." In interviews four of the five mathematicians and one of the two mathematics educators explained that the process of writing mathematics, on a whiteboard or on paper, supported thinking, and aligned with Burton's (1998) report that mathematicians discussed how problems could take years to solve.

MATHEMATICIANS' AND MATHEMATICS EDUCATORS' BELIEFS ABOUT TEACHING AND LEARNING MATHEMATICS

Mathematicians and mathematics educators have different experiences with mathematics and different foci in their teaching (teaching mathematics, or primarily teaching how to teach mathematics) and their research. Whilst mathematics educators have studied theories of teaching and learning, many mathematicians' beliefs about how mathematics is best learnt and taught come from their past experiences in mathematics classes in schools and universities. These differing experiences lead to different beliefs about how mathematics is learnt and should be taught, as well as what can be taught at different levels. At all levels, mathematicians have identified acquiring mathematical content knowledge as important whereas mathematics educators ranked it much lower.

Brandt et al. (2016) found that, according to mathematicians in the United States and Canada, the most important mathematical processes to be acquired by students differ with the level of the course. For example, when teaching lower level mathematics courses at university (e.g., College Algebra, Trigonometry, or Calculus) mathematicians identified problem-solving, acquiring content knowledge, and acquiring informal logical reasoning as priorities whereas, for higher level mathematics courses (e.g., Abstract Algebra, Number Theory, or Topology), mathematicians valued proving, acquiring content knowledge and conjecturing/generalising/exploring. Mathematics educators identified conjecturing/generalising/exploring, proving, and problem-solving as the most important activities (Brandt et al., 2016). Prospective mathematics teachers in Canada and the United States study courses at both levels.

The mathematicians in Leikin et al.'s study (2018) identified that prospective teachers' needs differed from those of other mathematics students. They believed that prospective teachers need more examples of where and how mathematics is used so that they could inspire interest in their future school students. They also believed that prospective teachers need a greater breadth of content from their mathematics courses so that there was greater coverage of school curriculum content.

When teaching, mathematicians tended not to share their struggle or pleasure of doing mathematics with their students (Burton, 1998). According to Burton (1998, p. 140) the belief that "students must learn before they can begin to think of mathematising, dominate classrooms at every level. … These personal flavours are entirely lost in the 'objective' mathematics they, as teachers, thrust towards reluctant learners." Schoenfeld (1994) expressed a similar characterisation of mathematicians:

> The implicit but widespread presumption in the mathematics community is that an extensive background is required before one can do mathematics. … Until students get to the point of doing research (typically in the third year of graduate school), learning mathematics means ingesting mathematics. (p. 65)

Mathematics educators, however, urge that all students can *do mathematics* and *think like mathematicians* believing that students of all levels are "capable of making conjectures, generalising ideas, and exploring mathematical concepts" (Brandt et al., p. 766). Dreyfus and Eisenberg (1986) recommended that teaching should include two or more solution paths for any problem so that students become comfortable with different solution methods and developed deeper conceptual understanding. Two solution methods also help students to appreciate the aesthetics of mathematics because discussions allow for the consideration of solution methods in terms of mathematical content or representation(s) used, efficiency, and elegance (Brandt et al., 2016).

In our study, mathematics educators were more likely than mathematicians to express constructivist beliefs about learning mathematics, in line with the reform agenda (Kilpatrick, Swafford, & Findell, 2001). Mathematics educators also believed that it is necessary for teachers to understand the source of students' errors rather than waiting for follow-up instruction to correct difficulties, and that it is important for students to be given interesting problems to investigate in small groups, and opportunities to reflect on and evaluate their own mathematical understanding.

Only 10% of mathematicians expressed a preference for traditional methods of teaching by agreeing with statements such as "listening carefully to the teacher explain a mathematics lesson is the most effective way to learn mathematics." However, this was how all of the prospective teachers in our study referred to many of their mathematics lectures. For example, they described them as having "the teacher up the front presenting the information." In this case a likely explanation is that these mathematicians know that lectures are not the most effective means of teaching but are constrained by university timetables, and also believe that they need to minimise time spent on teaching because they have other demands in their jobs such as research. The pressure from competing institutional demands may mean that the mathematicians cannot see a reasonable alternative to lectures. That is, their beliefs about mathematics teaching and learning are part of a system of beliefs that include beliefs about their role as mathematicians in a university, what is possible/desirable about teaching in that environment, and what else they see as possible as a teacher (other than lecturing) in that environment. One prospective teacher agreed, saying:

> I get the impression sometimes with lecturers that they might be a bit internally conflicted in that they'd like to present the information as a problem-solving approached, but that they're unable to, given the constraints of how many students there are and just the course design and what has to be covered.

Although mathematicians and mathematics educators had similar beliefs about mathematics their ideas about teaching and learning mathematics were different. Generally, the prospective teachers described traditional lectures and tutorials in which worked examples were presented whereas they were encouraged to adopt a problem-solving approach to mathematics teaching that is characterised by multiple solutions and having students justify their solutions.

PROSPECTIVE TEACHERS NEGOTIATING DIFFERING BELIEFS ABOUT TEACHING AND LEARNING MATHEMATICS

In interviews about how they were taught, the prospective teachers reported that mathematicians tended to use traditional pedagogy, for example, "you will write this down and you will understand it from having written it down and practicing it" whilst the mathematics educators advocated problem-solving and inquiry-based approaches. The prospective teachers were aware of the tension between how they were being taught mathematics and how they were taught to teach mathematics, but most claimed they coped because they were good at mathematics and they had the motivation and resources to get help. One prospective teacher described this as:

> I can gain some understanding from that but, I will often take that further myself, in my own study time, and tease that out a little more. Just try and give myself a full understanding so I'm not solely reliant on that particular style. I do have to engage in other learning practices to try and synthesise that knowledge.

Prospective teachers acknowledged that many students had difficulty with the teaching methods: "I imagine within a lot of students they'd have a lot of trouble just trying to put together that raw data they're given and actually understanding everything behind it." Other differences the prospective teachers described about the way they were taught mathematics at university and the way they were taught to teach mathematics related to methods of presentation of material, assumptions that students had the required prior knowledge, ways of working to develop students' understanding of mathematics content, catering for the diversity of the students in their classes, and whether the material was presented as facts and procedures or in a context. In addition, the prospective teachers struggled to discern the relevance to their future careers of the university mathematics they were learning from the mathematicians, because the content was at a higher level than is taught in school. One was frustrated that the mathematics he was taught at university did not reflect all the secondary school curriculum. In particular, he noted the lack of statistics in his university mathematics study. As they negotiated their different classes, prospective teachers thought about how they would teach. They wanted to engage their students by connecting the mathematics to the real world and to make mathematics relevant to their students.

The root cause of the discrpeancy that the prospective teachers experienced appears to be that mathematicians believe that students need to know the content before they can do mathematics, and prospective teachers are taught to teach their school students to do mathematics whilst learning mathematics, and as a means of learning mathematics. Curriculum documents (e.g., ACARA, 2018; CCSO, 2010; National Council for Curriculum and Assessment Ireland, n.d.; Ministry for Education Singapore, 2012) specify that students see mathematics as making sense of, and solving, problems. In addition, as Burton (1998) found, mathematicians appear not to be sharing their beliefs about mathematics as "personally- and culturally/ socially-related" (p. 139). These matters are problematic as we know that the way in which prospective teachers are taught mathematics at university influences how they teach mathematics (Leikin et al., 2018). Discussions with these mathematics teacher educators (mathematicians and mathematics educators) of beliefs and personal experiences with mathematics could help prospective mathematics teachers negotiate these differing beliefs about mathematics and its teaching and learning. The importance and the urgency of addressing them is amplified by the mounting pressures on mathematics teacher educators. These are described in the next section.

PRESSURES ON MATHEMATICS TEACHER EDUCATORS

Concerns about school student performance in mathematics, and consequent attention to teacher quality in many countries have led to increased scrutiny of initial teacher education in these countries. This scrutiny has included the publication by the National Council on Teacher Quality of a series of reports critical of the quality of teacher preparation throughout the United States (Paulson & Marchant, 2012). Paulson and Marchant (2012) reported that only 7.5% of teacher preparation programs were considered strong by the National Council on Teacher Quality, and 25% are considered deficient. In the United Kingdom responsibility for teacher education has shifted from universities to schools through the *School Direct* program (Department of Education, 2015). Most respondents in Hodgson's (2014) survey of 730 English teachers claimed that *School Direct* would lead to a decline in prospective teachers "subject knowledge, understanding of educational purpose and processes, and classroom preparedness" (p. 7). In Australia there has been widespread criticism of initial teacher education, teachers and schools resulting in increased regulation (Dinham, 2013) including "national testing and reporting of student achievement, national professional standards for teachers, a national curriculum, national accreditation of teacher education courses and a national framework for teacher development and performance" (p. 91). Tatto, Lerman, and Novotna (2010) identified that 11 of the 20 countries or regions that participated in the ICMI Study 15 had national regulation of initial teacher education programs, eight had local regulation and in one country there was a combination of national and local regulation.

With accreditation of initial teacher education programs comes the need to report against mandated requirements. In England, for example, the *Self-evaluation Document* is used by accredited initial teacher training providers to evaluate their effectiveness annually and to prepare a plan for the following year which must then be evaluated (Lerman, 2014). In many countries national policies have increased regulation and therefore restructuring of initial teacher education programs (Tatto, Lerman, & Novotna, 2010). This regulation has led to "attempts to narrow down the knowledge base on which teachers' judgements can be exercised, the increasing regulation of the content and standards in the education of teachers today in the service of policy goals set by governments" (Lerman, 2014, p. 189).

Accountability requirements on teacher education have extended to licensure examinations before graduating teachers are able to teach in England and the United States (Wang, Coleman, Coley, & Phelps, 2003). In Australia, prospective teachers must pass the Literacy and Numeracy Test for Initial Teacher Education (Australian Council of Educational Research, 2018) to graduate as a teacher. In most countries there is also a requirement for practising teachers to continue to undertake ongoing professional development, but it is rarely stipulated that this needs to be subject specific (Mullis et al., 2016).

A complicating issue in some countries is the shortage of suitably qualified mathematics teachers. The shortage of adequately qualified teachers is further complicated by the attrition of teachers in their first five years of teaching: up to 50% of Australian teachers leave the profession in this period (Australian Institute for Teaching and School Leadership, 2016). In England 22% of those teaching mathematics are out of field having not obtained Qualified Teacher status (Department of Education, 2017) and in Luxemburg and Turkey 80% of students were enrolled in schools that reported a shortage of mathematics teachers (Schleicher, 2012). In developing countries, the situation is even worse. In South Africa, for example, the shortage of secondary school teachers is critical (Adler & Davis, 2006). In the South African province of KwaZulu-Natal there were nearly 8000 unqualified (enrolled for tertiary studies) or underqualified (having a diploma or degree in the subject but no pedagogical training) teachers (Jansen, 2012).

The shortage of qualified mathematics teachers contributes to pressure on mathematics teacher educators: for example, criticism for not preparing enough mathematics teachers and the demand for professional learning for out-of-field teachers (Hoyles, 2002; O'Connor & Thomas, 2019; Schleicher, 2012; Vale & Drake, 2019). Policymakers, think tanks, stakeholders and the media examine mathematics education more than other curriculum areas in part because mathematics is believed to be "culture-neutral" (Noys, Wake, & Drake, 2013, p. 511). A further contributor to the increased scrutiny is the increasing recognition of the importance of Science, Technology, Engineering and Mathematics (STEM) because of concerns for national security and economic productivity (Noys et al., 2013). In Australia, as in other countries, initial teacher education programs are accountable "through regulation and surveillance" to continually changing policy agendas (Knipe & Fitzgerald, 2017, p. 129).

Changes in the school mathematics curriculum, pedagogical expectations, and assessment processes need to be accounted for in initial teacher preparation and by the research community (Lerman, 2014) and recognised as adding to the demand on mathematics teacher educators. The challenge for mathematics teacher educators is that along with teachers, "teacher educators have been constructed in policy documents as the 'problem'" (Fitzgerald & Knipe, 2016, p. 136) and have, therefore, also been subjected to further regulation.

In summary, mathematics teacher educators globally are under increasing accountability pressures with regulators imposing onerous accreditation, evaluation, and reporting requirements. Some mathematics teacher educators need to prepare prospective teachers for external testing either so that they can graduate or can register as a teacher. All of this is occurring in the context of a focus on STEM and a shortage of qualified mathematics teachers able to engage students with mathematics.

COLLABORATIONS TO SUPPORT MATHEMATICS TEACHER EDUCATORS

Mathematicians and mathematics educators can work together to build an understanding of the beliefs about mathematics and its teaching and learning that each group hold in relation to working with prospective mathematics teachers. Fried (2014) suggested that mathematicians and mathematics educators must work together and that although there are differences in beliefs between the two communities, "One must confront these differences and try to understand them" (Fried, 2014, p. 4). He went on to argue that the differences are not that great. Wu (2011) suggested mathematicians can contribute to collaborations aimed at progressing teacher education by engaging with school curricula to ensure the mathematical accuracy of teaching materials.

To support mathematicians and mathematics educators (as well as scientists and science educators) working together to improve the quality of prospective mathematics (and science) teachers, the Office of the Chief Scientist in Australia provided 12 million dollars funding over 2013–2017 for the *Enhancing the Training of Mathematics and Science Teachers* Program (Department of Education and Training, 2016). There were five project teams and 27 universities involving faculties, schools or departments of science, mathematics and education. Two of the projects aimed to nurture sustained collaborations between the disciplines and education.

One project led by Merrilyn Goos (mathematics educator) and Joseph Grotowski (mathematician), *Inspiring Mathematics and Science in Teacher Education* (IMSITE, 2017) aimed to integrate the discipline knowledge of mathematicians (and scientists) with teacher educators' pedagogical knowledge (Goos, Grotowski, & Bennison, 2017). The framework of Akkerman and Bakker (2011) was used to analyse learning at the boundaries between mathematics as a discipline and mathematics education. Akkerman and Bakker defined boundaries as markers of "sociocultural difference leading to discontinuity in action or interaction" (p. 133). They identified four mechanisms for learning: identification – challenging the specific ways of working of the two communities; coordination of practices – the use of dialogue to move between the two worlds; reflection on differences between the ways of working; and transformation leading to profound changes.

The IMSITE project demonstrated that these new forms of interaction between mathematicians and mathematics educators led to insights in working across disciplinary boundaries. Firstly, there were curriculum redesign and community building activities in primary and secondary school initial teacher education programs (Goos & Bennison, 2018). Mathematicians and mathematics educators co-developed and co-taught courses integrating content and pedagogy (Goos & Bennison, 2018). Secondly the project identified things that enabled or hindered dialogue between mathematicians and mathematics educators. Dialogue was enabled by personal qualities such as open-mindedness, trust, mutual respect, and shared beliefs and values, as well as working on a common or shared problem. Challenges to overcome included the physical separation of departments, workload formulas and financial models that did not recognise or reward interdisciplinary collaboration, and cultural differences between the disciplines. Thirdly Akkerman and Bruining's (2016) transformation learning mechanism was "observed at interpersonal and intrapersonal levels" (Goos & Bennison, 2018, p. 272) levels.

In another project funded by the same scheme *Its part of my life: Engaging university and community to enhance science and mathematics education* mathematicians and scientists collaborated with mathematics and science educators and prospective teachers to develop lessons connecting mathematics and science that drew on the local context and satisfied curriculum requirements (Woolcott et al., 2017). Scott, Woolcott, Keast, and Chamberlain (2018) used complexity theory to develop a framework for sustainable collaboration of the researchers and the broader community involved. My current research is further highlighting the importance of such collaborations. They can lead to changes in teaching practices, for example, mathematicians becoming more supportive of student learning.

CONCLUSIONS

Mathematics teacher educators are under increasing pressure from policymakers, stakeholders and the media to prepare, within national frameworks, more and better mathematics teachers who will improve student outcomes. Alongside this governments and their agencies are increasing regulation and surveillance. Mathematics education researchers are typically the mathematics educators so their research focusses on teaching and learning mathematics rather than on educational policy. Similarly, mathematicians tend not to engage in research related to education policy. Lerman (2014) believed that the relations between the policy makers and the mathematics education research community need to be mapped in each context so that the research can be aligned with particular policy. This would enable researchers to "play a constructive part in debates about reforming or improving mathematics teaching and learning" (Lerman, 2014, p. 189). In a similar vein Dinham (2013) calls for all "educators to find their voices in this current debate and to argue from a position of evidence to counter the misinformed and misguided views that currently predominate and influence government policy" (Dinham, 2013, p. 91). Continuing research will provide evidence to push back, when necessary, against regulation and to proactively shape policy relevant policy agendas. Research that furthers our "understanding the beliefs that underpin the practice of mathematics teacher educators" as well as their influence on the work of mathematics teachers (Beswick & Goos, 2018, p. 425) is an important part of the research that is needed. The ways in which mathematics teacher educators both influence and respond to accountability and regulatory interventions represents a new aspect of the field that warrants research.

Mathematics teacher educators' beliefs and practices about mathematics and how it is best taught and learnt influence how it is taught and the beliefs that their students develop. Although studies have revealed similarities between the beliefs about mathematics of mathematicians and mathematics educators, there are differences between the beliefs about learning and teaching mathematics of these two groups of mathematics teacher educators. Understanding these differing beliefs urgently needs attention to address institutional pressures, reduce the tensions experienced by prospective mathematics teachers and external pressures in a united and coherent fashion. Goos and Bennison (2018) have described mechanisms that can promote dialogue between the two groups of mathematics teacher educators with a view to responding jointly and more coherently to external demands.

School curricula are continually changing, and mathematics educators must prepare teachers to teach the current curriculum and to be able to adapt to future curricula. Lerman (2014) suggested that initial teacher education therefore also needs to support prospective teachers to be able to take a critical perspective on future changes. Having a critical perspective allows teachers to contribute to the future discussions of curriculum and policy and share in responding to regulatory pressures. If mathematics teacher educators are going to "find their voices" and contribute to these current policy debates there will need to be boundary conversations between the various groups that act as mathematics teacher educators, in order to find commonalities and build an understanding of the differing cultures of mathematicians and mathematics educators.

REFERENCES

Adler, J., & Davis, Z. (2006). Opening another black box: Researching mathematics for teaching in mathematics teacher education. *Journal for Research in Mathematics Education, 37*(4), 270–296.

Akkerman, S., & Bakker, A. (2011). Boundary crossing and boundary objects. *Review of Educational Research, 81*, 132–169.

Akkerman, S., & Bruining, T. (2016). Multilevel boundary crossing in a professional development school partnership. *The Journal of the Learning Sciences, 25*(2), 240–284.

Andrews, P. (2013). Finnish mathematics teaching from a reform perspective: A video-based case-study analysis. *Comparative Education Review, 57*(2), 189–211.

Australian Council of Educational Research (ACER). (2018). *Literacy and numeracy test for initial teacher education students.* Retrieved from https://teacheredtest.acer.edu.au/

Australian Curriculum, Assessment and Reporting Authority (ACARA). (2018). *Australian Curriculum: Mathematics.* Retrieved from https://www.australiancurriculum.edu.au/f-10-curriculum/mathematics/

Australian Institute for Teaching and School Leadership (AITSL). (2016). *Spotlight: What do we know about early career attrition rates?* Retrieved from https://www.aitsl.edu.au/docs/default-source/research-evidence/spotlight/spotlight---attrition.pdf?sfvrsn=40d1ed3c_0

Beswick, K. (2005). The beliefs/practice connection in broadly defined contexts. *Mathematics Education Research Journal, 17*(2), 39–68.

Beswick, K. (2007). Teachers' beliefs that matter in secondary mathematics classrooms. *Educational Studies in Mathematics, 65*(1), 95–120.

Beswick, K. (2012). Teachers' beliefs about school mathematics and mathematicians' mathematics and their relationship to practice. *Educational Studies in Mathematics, 79*, 127–147.

Beswick, K., & Callingham, R. (2014). The beliefs of pre-service primary and secondary mathematics teachers, in-service mathematics teachers, and mathematics teacher educators. In C. Nicol, P. Liljedahl, S. Oesterle, & D. Allan (Eds.), *Proceedings of the 38th Conference of the International Group for the Psychology of Mathematics Education* (Vol. 2, pp. 137–144). Vancouver, Canada: PME.

Beswick, K., & Goos, M. (2018). Mathematics teacher educator knowledge: What do we know and where to from here? *Journal of Mathematics Teacher Education, 21*, 417–427.

Brandt, J., Lunt, J., & Meilstrup, G. R. (2016). Mathematicians' and math educators' views on "doing mathematics." *Primus: Problems, Resources, and Issues in Mathematics Undergraduate Studies; Philadelphia, 26*(8), 753–769.

Brown, L., Helliwell, T., & Coles, A. (2018). Working as mathematics teacher educators at the meta-level (to the focus of the teachers on developing their teaching). *Avances de Investigación en Educación Matemática, 13*, 105–122.

Burton, L. (1998). The practices of mathematicians: What do they tell us about coming to know mathematics? *Educational Studies in Mathematics, 37*, 121–143.

Callingham, R., Beswick, K., Clark, J., Kissane, B., Serow, P., & Thornton, S. (2012). Mathematical knowledge for teaching of MERGA members. In J. Dindyal, L. P. Chen, & S. F. Ng (Eds.), *Proceedings of the 35th Annual Conference of the Mathematics Education Research Group of Australasia* (Vol. 1, pp. 162–169). Singapore: MERGA.

Carlson, M. P., & Bloom, I. (2005). The cyclic nature of problem-solving: An emergent multidimensional problem-solving framework. *Educational Studies in Mathematics, 58*(1), 45–75.

Chick, H., & Beswick, K. (2018). Teaching teachers to teach Boris: A framework for mathematics teacher educator pedagogical content knowledge. *Journal of Mathematics Teacher Education, 21*(5), 475–499.

Conference Board of the Mathematical Sciences. (2012). *The mathematical education of teachers II*. Providence, RI: American Mathematical Society.

Council of Chief State School Officers. (2010). *Common core state standards*. Retrieved from http://www.corestandards.org/Math/

Department of Education. (2015). *Improving the quality of teaching and leadership*. Retrieved from http://www.gov.uk/government/policies/improving-the-quality-of-teaching-and-leadership/supporting-pages/school-direct

Department of Education. (2017). *School workforce in England: November 2016*. Retrieved from https://assets.publishing.service.gov.uk/government/uploads/system/uploads/attachment_data/file/620825/SFR25_2017_MainText.pdf

Department of Education and Training. (2016). *Enhancing the training of mathematics and science teachers program*. Retrieved from https://www.education.gov.au/enhancing-training-mathematics-and-science-teachers-program

Dinham, S. (2013). The quality teaching movement in Australia encounters difficult terrain: A personal perspective. *Australian Journal of Education, 57*, 91–106.

Dreyfus, T., & Eisenberg, T. (1986). On the aesthetics of mathematical thought. *For the Learning of Mathematics, 6*(1), 2–10.

Ernest, P. (1989). The impact of beliefs on the teaching of mathematics. In P. Ernest (Ed.), *Mathematics teaching: The state of the art* (pp. 249–254). London: Falmer Press.

Ernest, P. (1991). *The philosophy of mathematics education*. London: The Palmer Press.

Fitzgerald, T., & Knipe, S. (2016). Policy reform: Testing times for teacher education in Australia. *Journal of Educational Administration and History, 48*(4), 1–12.

Fried M. N. (2014). Mathematics & mathematics education: Searching for common ground. In M. Fried & T. Dreyfus (Eds.), *Mathematics & mathematics education: Searching for common ground. Advances in mathematics education*. Dordrecht: Springer.

Goos, M., & Bennison, A. (2018). Boundary crossing and brokering between disciplines in pre-service mathematics teacher education. *Mathematics Education Research Journal, 30*(3), 255–275.

Goos, M., Grotowski, J., & Bennison, A. (2017). *Inspiring Mathematics and Science in Teacher Education (IMSITE). Final report*. Canberra: Australian Government Department of Education and Training.

Green, T. (1971). *The activities of teaching*. New York, NY: McGraw-Hill.

Grigutsch, S., & Törner, G. (1998). World views held by university teachers of mathematics science. *Schriftenreihe des Fachbereichs Mathematik*. Universität Duisburg.

Grootenboer, P. J. (2008). Mathematical belief change in preservice primary teachers. *Journal of Mathematics Teacher Education, 11*(6), 479–497.

Grootenboer, P., & Marshman, M. (2016). *Mathematics, affect and learning: Middle school students' beliefs and attitudes about mathematics education*. Singapore: Springer.

Guo, Y., Silver, E. A., & Yang, Z. (2018). The latest characteristics of mathematics education reform of compulsory education stage in China. *American Journal of Educational Research, 6*(9), 1312–1317.

Hersh, R. (1997). *What is mathematics, really?* Oxford: Oxford University Press.

Hodgson, J. (2014). Surveying the wreckage: The professional response to changes to initial teacher training in the UK. *English in Education, 48*(1), 7–25.

Hoyle, P. (2016). Must try harder: An evaluation of the UK government's policy directions. In Australian Council for Educational Research (ACER) (Ed.), *ACER Research Conference Proceedings (2016)*. Retrieved from https://research.acer.edu.au/research_conference/RC2016/8august/1

Hoyles, C. (2002). *Some successful strategies for promoting mathematics in England*. Retrieved from https://amsi.org.au/wp-content/uploads/2014/07/CeliaHoyles_all_slides.pdf

Inspiring Mathematics and Science in Teacher Education (IMSITE). (2017). *Inspiring mathematics and science in teacher education*. Retrieved from http://www.imsite.edu.au/

Jackson, A. (2000). Declining student numbers worry German mathematics departments. *Notices of the American Mathematical Society, 47*(3). Retrieved from https://www.ams.org/notices/200003/comm-jackson.pdf

Jansen, L. (2012). Huge teacher shortages in KZN. *The Mercury* Retrieved from https://www.iol.co.za/mercury/huge-teacher-shortages-in-kzn-1282175

Jorgensen, R., Grootenboer, P., Niesche, R., & Lerman, S. (2010). Challenges for teacher education: The mismatch between beliefs and practice in remote Indigenous contexts. *Asia Pacific Journal of Teacher Education, 32*(2), 161–175.

Kennedy, J., Lyons, T., & Quinn, F. (2014). The continuing decline of science and mathematics enrolments in Australian high schools. *Teaching Science, 60*(2), 34–46.

Kilpatrick, J., Swafford, J., & Findell, B. (Eds.). (2001). *Adding it up: Helping children learn mathematics.* Washington, DC: National Academy Press.

Knipe, S., & Fitzgerald, T. (2017). Caught between competing worlds: Teacher education in Australia. In J. Nuttall, A. Kostagriz, M. Jones, & J. Martin (Eds.), *Teacher education, policy and practice: Evidence of impact, impact of evidence.* Dordrecht: Springer.

Leikin, R., Zazkis, R., & Meller, M. (2018). Research mathematicians as teacher educators: Focusing on mathematics for secondary mathematics teachers. *Journal of Mathematics Teacher Education, 21*, 451–473. doi:10.1007/s10857-017-9388-9

Lerman, S. (1990). Alternative perspectives of the nature of mathematics and their influence on the teaching of mathematics, *British Educational Research Journal, 16*(1), 53–61.

Lerman, S. (2014). Mapping the effects of policy on mathematics teacher education. *Educational Studies in Mathematics, 87*(2), 187–201.

Leatham, K. R. (2006). Viewing mathematics teachers' beliefs as sensible systems. *Journal of Mathematics Teacher Education, 9*, 91–102.

Liljedahl, P. (2008). Teachers' insights into the relationship between beliefs and practice. In J. Maab & W. Schloglmann (Eds.), *Beliefs and attitudes in mathematics education: New research results* (pp. 33–44). Rotterdam, The Netherlands: Sense Publishers.

Marshman, M. (in press). Learning to teach mathematics: How secondary prospective teachers describe the different beliefs and practices of their mathematics teacher educators In M. Goos & K. Beswick (Eds.), *The learning and development of mathematics teacher educators: International perspectives and challenges*. Singapore: Springer.

McGregor, D. (2016). *Exploring the impact of inquiry learning on students' beliefs and attitudes towards mathematics* (PhD). The University of Queensland, St Lucia.

McKubre-Jordens, M. (n.d.). Constructive mathematics. In *Internet encyclopaedia of philosophy.* Retrieved from https://www.iep.utm.edu/con-math/

McLeod, D. B. (1992). Research on affect in mathematics education: A reconceptualization. In D. Grouws (Ed.), *Handbook of research on mathematics teaching and learning* (pp. 575–596). New York, NY: Macmillan.

Ministry of Education, Singapore. (2012). *Mathematics syllabus secondary one to four*. Retrieved from https://www.moe.gov.sg/docs/default-source/document/education/syllabuses/sciences/files/mathematics-syllabus-sec-1-to-4-express-n(a)-course.pdf

Misfeldt, M., & Johansen, M. W. (2015). Research mathematicians' practices in selecting mathematical problems. *Educational Studies in Mathematics, 89*(3), 357–373.

Mosvold, R., & Fauskanger, J. (2014). Teachers' beliefs about mathematical knowledge for teaching definitions. *International Electronic Journal of Mathematics Education, 8*(2–3), 43–61.

Mullis, I. V. S., Martin, M. O., Goh, S., & Cotter, K. (Eds.). (2016). *TIMSS 2015 encyclopedia: Education policy and curriculum in mathematics and science.* Retrieved from http://timssandpirls.bc.edu/timss2015/encyclopedia/

Mura, R. (1993). Images of mathematics held by university teachers of mathematical sciences. *Educational Studies in Mathematics, 25*(4) 375–385.

Mura, R. (1995). Images of mathematics held by university teachers of mathematics education. *Educational Studies in Mathematics, 28*(4) 385–399.

National Centre for Education Statistics. (n.d.). *Advanced mathematics and science courses.* Retrieved from https://nces.ed.gov/fastfacts/display.asp?id=97

National Council for Curriculum and Assessment Ireland. (n.d.). *Curriculum online: Mathematics.* Retrieved from https://curriculumonline.ie/Junior-cycle/Junior-Cycle-Subjects/Mathematics/Aim

National Governors Association Center for Best Practices & Council of Chief State School Officers. (2010). *Common core state standards for mathematical practice*. Washington, DC: National Governors Association Center for Best Practices, Council of Chief State School Officers.

Noyes, A., Wake, G., & Drake, P. (2013). Time for curriculum reform: The case of mathematics. *The Curriculum Journal, 24*(4), 511–528.

O'Connor, M., & Thomas, J. (2019). *Australian secondary mathematics teacher shortfalls: A deepening crisis* (AMSI Occasional Paper). Australian Mathematical Sciences Institute. Retrieved from https://amsi.org.au/wp-content/uploads/2019/05/amsi-occasional-paper-2.pdf

Office of the Chief Scientist. (2012). *Mathematics, engineering & science in the national interest*. Canberra: Australian Government, Office of the Chief Scientist. Retrieved from http://www.chiefscientist.gov.au/category/archives/mathematics-engineering-and-science-report/

Office of the Chief Scientist. (2014). *Science, technology, engineering and mathematics: Australia's future*. Canberra: Australian Government.

Organisation for Economic Cooperation and Development. (2017). *PISA 2015 assessment and analytical framework: Science, reading, mathematic, financial literacy and collaborative problem-solving*. Paris: OECD Publishing.

Pajares, M. F. (1992). Teachers' beliefs and educational research: Cleaning up a messy construct. *Review of Educational Research, 62*(3), 307–332.

Paulson, S., & Marchant, G. (2012). Voices in education: Accountability in teacher education and the national council on teacher quality. *The Teacher Educator, 47*(1), 1–8.

Philipp, R. A. (2007). Mathematics teachers' beliefs and affect. In F. K. Lester (Ed.), *Second handbook of research on mathematics teaching and learning* (Vol. 1). New York, NY: National Council of Teachers of Mathematics.

Prince, G., & O'Connor, M. (2019). *AMSI occasional paper 1 – Crunching the numbers on out-of-field teaching in maths*. Australian Mathematical Sciences Institute.

Rino, J. S. (2015). *Supporting an understanding of mathematics teacher educators: Identifying shared beliefs and ways of enacting their craft* (All Theses and Dissertations, 5576). Retrieved from https://scholarsarchive.byu.edu/etd/5576

Rokeach, M. (1968). *Beliefs, attitudes and values: A theory of organisational change*. San Francisco, CA: Jossey-Bass.

Schleicher, A. (Ed.). (2012). *Preparing teachers and developing school leaders for the 21st century: Lessons from around the world*. Paris: OECD Publishing.

Schoenfeld, A. H. (1992). Learning to think mathematically: Problem-solving, metacognition, and sense-making in mathematics. In D. Grouws (Ed.), *Handbook for research on mathematics teaching and learning* (pp. 334–370). New York, NY: MacMillan.

Schoenfeld, A. H. (1994). Reflections on doing and teaching mathematics. In *Mathematical thinking and problem-solving* (pp. 53–75). Hillsdale, NJ: Erlbaum.

Scott, A., Woolcott, G., Keast, R., & Chamberlain, D. (2018). Sustainability of collaborative networks in higher education research projects: Why complexity? Why now? *Public Management Review, 20*(7), 1068–1087.

Smith, R. C., Kim, S., & Mcintyre, L. (2016). Relationships between prospective middle grades mathematics teachers' beliefs and TPACK. *Canadian Journal of Science, Mathematics and Technology Education, 16*(4), 359–373.

Sriramen, B., Roscoe, M., & English, L. (2010). Politicizing mathematics education: Has the politics gone too far? Or not far enough? In B. Sriramen & L. English (Eds.), *Theories of mathematics education: Seeking new frontiers* (pp. 621–638). New York, NY: Springer.

Tatto, M. T., Lerman, S., & Novotna, J. (2010). The organization of the mathematics preparation and development of teachers: A report from the ICMI study 15. *Journal of Mathematics Teacher Education, 13*(4), 313–343.

Thompson, A. G. (1992). Teachers' beliefs and conceptions: A synthesis of the research. In D. A. Grouws (Ed.), *Handbook of research on mathematics teaching and learning* (pp. 127–146). New York, NY: Macmillan.

Vale, C., & Drake, P. (2019). Attending to out-of-field teaching: Implications of and for education policy. In L. Hobbs & G. Törner (Eds.), *Examining the phenomenon of "teaching out-of-field."* Singapore: Springer.

Wang, A. H., Coleman, A. B., Coley, J. J., & Phelps, R. P. (2003). *Preparing teachers around the world.* Princeton, NJ: Educational Testing Service.

Woolcott, G., Scott, A., Norton, M., Whannell, R., Galligan, L., Marshman, M., Pfeiffer, L., & Wines, C. (2017). *It's part of my life: Engaging university and community to enhance science and mathematics education. Final report for Enhancing the Training of Mathematics and Science Teachers*. Canberra, Australia: Department of Education and Training.

Wu, H. (2011). The mis-education of mathematics teachers. *Notices of the American Mathematical Society, 58*(3), 327–384.

Zaslavsky, O., & Leikin, R. (2004). Professional development of mathematics teacher educators: Growth through practice. *Journal of Mathematics Teacher Education, 7*(5), 5–32.

Margaret Marshman
School of Education
University of the Sunshine Coast

INDEX

www.ingramcontent.com/pod-product-compliance
Lightning Source LLC
Chambersburg PA
CBHW071411230926
51846CB00001B/54
9789004424203